Woodland Development

A long-term study of Lady Park Wood

Woodland Development

A long-term study of Lady Park Wood

by

George F. Peterken and Edward P. Mountford

CABI

Forestry Commission
England

CABI is a trading name of CAB International

CABI
Nosworthy Way
Wallingford
Oxfordshire OX10 8DE
UK

Tel: +44 (0)1491 832111
Fax: +44 (0)1491 833508
E-mail: info@cabi.org
Website: www.cabi.org

CABI
745 Atlantic Avenue
8th Floor
Boston, MA 02111
USA

Tel: +1 (617)682-9015
E-mail: cabi-nao@cabi.org

A catalogue record for this book is available from the British Library, London, UK.

Library of Congress Cataloging-in-Publication Data

Names: Peterken, G. F. (George Frederick) | Mountford, Edward P.
Title: Woodland development : a long term study of Lady Park Wood / by G.F. Peterken and E.P. Mountford.
Description: Wallingford, Oxfordshire, UK : CAB International, [2017] | Includes bibliographical references and index.
Identifiers: LCCN 2016057957 (print) | LCCN 2017034284 (ebook) | ISBN 9781780648668 (ePDF) | ISBN 9781780648675 (ePub) | ISBN 9781780648651 (hbk : alk. paper) | ISBN 9781786392817 (pbk : alk. paper)
Subjects: LCSH: Lady Park Wood National Nature Reserve (Wales and England) | Forests and forestry--England. | Forests and forestry--Wales.
Classification: LCC SD182.L33 (ebook) | LCC SD182.L33 P48 2017 (print) | DDC 577.30942--dc23
LC record available at https://lccn.loc.gov/2016057957

ISBN-13: 978 1 78064 865 1 (hbk)
978 1 78639 281 7 (pbk)

Commissioning editor: Ward Cooper
Editorial assistant: Emma McCann
Production editor: James Bishop

Typeset by SPi, Pondicherry, India.
Printed and bound in the UK by CPI Group (UK) Ltd, Croydon, CR0 4YY.

Contents

Acknowledgements

The Lady Park study is a collective achievement. Eustace Jones first recommended the wood as a research reserve, laid out the permanent recording transects and maintained the records for two decades. Alan Orange and Tim Barfield laid the foundations of modern recording and Vanessa Williams initiated the plots that have enabled us to track changes in the ground vegetation. Without these contributors and their helpers, we would not have had a record to analyse. Without the contributions of Richard Steele and John Workman to the debate on Lady Park's future in the early 1980s, we would not have had a research reserve to record. Without the work of the Forestry Commission, first in proposing research reserves in the 1930s and then as custodians of Lady Park and Highmeadow Woods, we would have had neither a research reserve nor a study. And, without the Nature Conservancy and its various successor bodies, the study would have been abandoned in the 1970s. A succession of site managers working for Natural England and its predecessors - Saul Herbert, Helen Lancaster, Bob Silverwood, Malcolm Whitmore, Rob Wolstenholme, Terry Claxton - have maintained the reserve and facilitated the study in various ways since 1971. In particular John Thompson and Mike Doyly recognised the need for a new recording after the 1976 drought.

We have been the principal recorders since 1983, but we have been helped in the wood by various relatives, friends and colleagues. In particular, we have been greatly helped and encouraged by Susan Peterken and Anne Mountford. Many others have helped with recording: in addition to those who contributed to the earlier recordings, Graham Berry, Barry Embling, Ann Harris, Helen Lancester, Janet Lister, Jonathan Spencer, Graham Walker, Tony Whitbread, Kate Wollen and Adrian Wood have helped us. As the wood has become more natural and the record has expanded, Jens Emborg, Kris Verheyen, Richard Bradshaw, Alistair Jump and Andrew Carr have led research projects founded on the long-term record. Liam Cavin not only carried out research in the reserve but also added to the transect records. Through Kris Verheyen and Alistair Jump, Lady Park has contributed to multi-site analyses of environmental change on a European scale.

Jonathan Spencer, having helped with the recording in the 1980s, returned in a more senior role to encourage and, through Forest Enterprise, sponsor the present analysis and write-up, as well as help us arrange for the record to be archived, probably in Oxford at the Bodleain. This book has benefited from Alan Crowden's and Peter Savill's valuable advice in the early stages, and contributions by Alan Orange (also returning in a different guise) on lichens, Sam Bosanquet on bryophytes, Keith Alexander on invertebrates and Andrew Carr on bats. Jorg Brunet and a group of European and North American ecologists organised by Kris Verheyen provided context on related records elsewhere; Ian Standing advised on local history and geology; Ian Statham advised on geomorphology; the Dean Fungus Group led by Keith and Val Davis compiled a list of fungi within the wood; Monmouth Museum gave access to S.G.Charles' records of wild plants; Sir Charles Burrell provided a different perspective on rewilding at Knepp Castle; and Rob Wolstenholme, Saul Herbert and Bob Silverwood gave access to records held by Natural England. We have also benefited from the responses of the many visitors we have shown round over the years.

We would also like to thank Ward Cooper, Emma McCann and James Bishop of CABI and Chris McEnnerney, all of whom helped to make the processes of writing and publishing this book straightforward and enjoyable. In particular, we appreciated Ward Cooper's interest and encouragement from the outset.

Apart from the early images of Lady Park, photographs were taken by GFP.

Summary

This is the story of Lady Park, an ancient wood in the lower Wye Valley that, uniquely, became an ecological reserve as the only outcome of a pre-war Forestry Commission response to criticism of afforestation in the Lake District. Since 1944, its 36 ha of mixed beech, sessile oak, ash, lime, wych elm, birch and yew woodland have been allowed to grow without direct human intervention. The record of the subsequent changes is now one of the longest and most detailed of any wood in temperate Europe.

Until 1870, the wood was an ordinary coppice fuelling the local ironworks. Parts of the wood have been allowed to grow naturally since 1870, save for light thinning in the early 20th century, but the rest was felled in 1942 and has since been allowed to regenerate naturally. By 2016, the growth and fate of over 20,000 individual trees and shrubs had been recorded in increasingly natural stands of 2–73 years growth and 75–145 years growth.

Since 1944, the principal processes have been growth and competitive thinning. Stem density has generally decreased, while basal area has generally increased. At all times, the larger individuals have generally grown faster, while smaller individuals have generally died or remained slow growing. To each of these generalisations, we have observed interesting exceptions.

This orderly process has been interrupted by various kinds of natural and semi-natural disturbance, most notably elm disease from 1971 onwards and severe drought in 1976. As the stands have aged, large trees have fallen at an increasing rate under the influence of wind and the chronic effects of unbalanced growth on steep slopes. Grey squirrels, fungal attacks, fallow deer and a heavy snowfall have also contributed to tree mortality.

Mortality from all causes has increased the volume of dead wood greatly, which is now a conspicuous feature of the wood, and has reached volumes recorded in so-called virgin forests. Natural thinning generated small pieces of rotting timber, but elm disease, drought, windthrow and slope instability generated large snags, fallen logs and rotten cores of still-standing trees.

Disturbances also created gaps in the canopy, varying in size, shape and degree of differentiation from nearby undisturbed stands. Gap creation accelerated after elm disease and the great drought, and subsided between 1990 and 2000, but is accelerating again as trees become larger and less vigorous with age.

Regeneration has been limited by shade under a closed canopy, and increasingly by severe browsing by fallow deer. However, the wood was fenced against deer in 2007 and since then vegetative regeneration from established trees, shrubs and fallen trees has been substantial, whereas seedling regeneration has developed slowly. All tree and shrub species are regenerating, especially ash and beech from seed, and hazel, small-leaved lime and large-leaved lime vegetatively.

Stand structure that must have been uniform before 1870 has differentiated into several distinct structures and dynamics under natural processes. For example, the stand on the steep ground below the cliff has developed an open structure with a denser underwood than elsewhere, due to the concentration of elms and the high frequency of tree falls. Stands on the steep, dry upper slopes were disrupted by drought-induced mortality of beech and birch.

Each tree and shrub species has responded individualistically to these developments: each has its particular character. Thus, beech has been dominant, not just because it is abundant but also because its dynamics have influenced the performance of other species. The short-lived pioneer trees, mainly birch, but also sallow, gean, aspen and alder, have declined steeply.

Until 1976, the successional trajectory in the old-growth was towards beech dominance, due to relatively fast growth, low mortality and an ability to regenerate in shade, but the drought killed so many large beech and so impaired the growth of the survivors that ash and limes have gained since 1976. In the young-growth, the drought first struck birch, then beeches were debarked by grey squirrels, leaving ash, lime and hazel to prosper. Below the cliff, the trend before 1971 towards wych elm–ash co-dominance was replaced with a trend to co-dominance by ash and large-leaved lime. If ash is struck by disease, the trends will be towards co-dominance by small-leaved lime and beech above the cliff and large-leaved lime dominance below the cliff.

The mortality and recruitment rates of all species have fluctuated substantially over the seven decades of observation. They have varied between species, being generally high for birch, aspen and other pioneers but generally low for wild service and yew. In general, recruitment rates have fluctuated more than mortality rates. While larger individuals in a population have generally survived better than smaller individuals, some small ash, lime and other canopy dominants have shown great resilience in adversity.

Radial growth rates (i.e. ring widths) have been extraordinarily variable within populations. While large trees have generally grown faster and populations have generally grown slower with age, the growth rate of some individuals has varied with time, out of step with the population as a whole, influenced by topographical position, canopy position, drought, debarking and other factors. Counting only the trees that survived, ash, beech and oak generally grew fastest. Basal area increments show that larger trees have retained vigour longer than ring width indicates, but otherwise show similar responses.

In 1984, the reserve was expanded to include similar stands that would be thinned periodically and allowed to regenerate naturally, thereby providing a managed comparison with the minimum-intervention stands in the original reserve. These have twice been thinned, but the stand has not been fenced. Regeneration has thus been negligible, but the growth rates of the retained trees have increased slightly, despite a surge in the debarking of beech by grey squirrels.

The responses of the wild flora have varied between major groups. Flowering plants have become much less diverse at both small and whole wood scales, due primarily to the retreat of light demanders. Some shade species have been lost, and light demanders have shown a marked reluctance to colonise tree-fall gaps. Ferns and spring geophytes have prospered, while bryophytes and lichens seem to have changed little. Fungal diversity is high, and may have increased as the supply of dead wood has increased. The managed stand was expected to retain its floristic diversity, but unrestricted grazing has reduced many species.

Changes in the fauna are poorly known, but butterflies have almost vanished, while bat populations have prospered. Hole-nesting birds are understandably well represented. The high diversity of molluscs discovered in the 1980s has probably remained unchanged

Considered together with observations in woodland elsewhere, experience at Lady Park points to several lessons of general significance. First, long-term ecological research is difficult to sustain in an environment of organisational change, short-term funding for research, the changing enthusiasms of researchers and the disparity between the timescales of woodland processes and the duration of careers and human lifetimes. Further, its methodology is akin more to historical research than short-term, hypothesis-driven scientific research. The initiation and survival of projects based on permanent plots depends on individuals, opportunism, a degree of altruism, proper attention to archiving and the institutional memory and willingness to accept that the long-term value of the records cannot be foreseen at the outset.

The existence of the record has facilitated other scientific research projects dealing with: carbon sequestration; the impacts of disease, drought and introduced species; and long-term changes in woodland ground vegetation. Lady Park has contributed to continent-wide analyses of vegetation responses to changing forest management, nitrogen accumulation and climate change.

The concept of natural woodland takes several forms. Only future natural woodland is attainable, but we cannot predict the form and composition that would develop under strict non-intervention. Past natural woodland, based on species that were present before the Neolithic, can only be sustained by evicting potentially invasive species. Neither is attainable in practice because the effects of people cannot be totally excluded. Lady Park is merely the nearest approach we can achieve to past natural woodland.

The composition and structure of natural woodland in Britain is unpredictable. We cannot predict the sequence and intensity of future disturbances. We know much about the responses of stands and individual species to disturbances and the ageing of stands, but woods are always liable to surprise. We can state the probability, but not the certainty, of a particular outcome over a moderate timescale. The character of natural woodland is also uncertain while debate continues about the impact of large herbivores.

Nature conservation policies in natural (i.e. minimum intervention) stands cannot be prescriptive. If we desire a particular outcome, then intervention may be necessary. Minimum intervention would be inappropriate for most lowland woods, which are generally too small and isolated to sustain species associated with open spaces without physical interventions.

The implementation of near-to-nature forestry requires some knowledge of how natural forests actually perform. Lady Park and other such reserves show that natural structures, patterns and dynamics are many and various. No one-size-fits-all approach would be appropriate.

The modern enthusiasm for rewilding is a mixed blessing when applied to woodland. While rewilded woodlands may become mature and impressive within decades, they can become untidy, dangerous and eventually less mature and impressive. Further, they may not necessarily compensate for the wild species evicted by rewilding. Partial rewilding may be optimal: wilder, not wild.

Postgraduate forestry students from Imperial College, London, under the leadership of Dr Julian Evans, enter Lady Park unconventionally in 2007.

Ecology students from Leuven (Belgium) spend a day in Lady Park, 2008. This visit led to a re-survey of the ground flora plots in 2009.

Citations of Archived Documents

PRO = a document in the Public Record Office, identified by class mark.

GG = a document in the Gage Papers, held at Gloucester Record Office under D1677, and identified by class mark.

LPW, followed by a class mark = a document in the Lady Park Wood archive, held by GFP at the time of writing.

Autumn leaves fallen from a mixed stand of Sessile oak, Small-leaved lime, Ash and Wild service.

Fig. 1.1a. The view downstream from the Seven Sisters rocks with Bloody cranesbill and Common rock-rose representing the rock-garden flora.

1 Understanding Woodland

> To know fully even one field or one land is a lifetime's experience. In the world of poetic experience it is depth that counts, not width. A gap in the hedge, a smooth rock surfacing a narrow lane, a view of a woody meadow, the stream at the junction of four fields – these are as much as a man can fully experience.
>
> Patrick Kavanagh, Irish poet, quoted by Robert Macfarlane in *Landmarks* (2015: p. 63)

> What (Nan Shepherd) learns – and what her book (*The Living Mountain*) taught me – is that the true mark of long acquaintance with a single place is a readiness to accept uncertainty: a contentment with the knowledge that you must not seek complete knowledge.
>
> Robert Macfarlane (2015: p. 71)

> Rewilding recognises that nature consists not just of a collection of species but also of their ever-shifting relationships with each other and with the physical environment. It understands that to keep an ecosystem in a state of arrested development ... is to protect something which bears little relationship to the natural world. Rewilding has no end points, no view about what a 'right' ecosystem or a 'right' assemblage of species looks like. It lets nature decide.
>
> George Monbiot (2013: pp. 8–10)

The woods we see around us reflect our past and our present. Swathes of the uplands are still covered by masses of conifers, uniform and geometrical, an arboreal analogue of the concrete, brutalist architecture that characterises many institutional buildings of the 1960s. The lowland farmlands are still studded with small, angular woods and belts planted for the fox hunters and country estate owners of the 18th and 19th centuries. Mixed with them we find numerous, less regular woods with oak, ash, hazel and other native trees that still clearly bear the marks of centuries of coppicing. And, in a few locations, park-like woodlands among heaths and pastures survive that would still, just about, be familiar to our medieval ancestors. In all these woods – their shape, pattern, composition and structure – much of the fascination lies in why people created them or allowed them to survive; how they used and misused them; what they meant to people; and how the fauna and flora managed to thrive in places that were usually not designed for their well-being.

But what happens when we take people out of the picture and allow woods to grow naturally? Do they take on different forms that might once have been familiar to our Mesolithic ancestors? Do trees grow, die and regenerate differently once they are freed from the moulding imposed on them by generations of woodmen and foresters? Will the woods degenerate? After all, foresters have long tried to improve them. Can they recover if they have been exploited? Does the wild fauna and flora prosper or languish once the habitats generated by people are no longer there? And if, as many people have found in recent years, the human element is so interesting, do they just become tedious and dull?

This book is about one attempt to answer questions such as these. Towards the end of the Second World War, a wood in the borderland between England and Wales, standing in the upper gorge of the lower Wye Valley (Fig. 1.1a and b: see pp. 1–2), was set aside by the Forestry Commission for study as a natural reserve (Fig. 1.2). While the war still raged in the East and Britain entered years of austerity and

Fig. 1.1b. View upstream from the Seven Sisters rocks, featuring the Biblins campsite and, to the right, Lady Park Wood. Horseshoe vetch and Bloody cranesbill represent the rock-garden flora.

Fig. 1.2. This is the outcome so far of nearly 150 years of leaving a freshly cut coppice with standards woodland to grow naturally. This beech-dominated, old-growth stand in Lady Park Wood has received very little management since it was last coppiced in 1870.

recovery, a small band of forestry students from Oxford University mapped and measured the trees and recorded the plants on the ground. By the end of 1945, Lady Parke – the ancient name for that part of the Highmeadow Woods – must already have been the best-recorded wood in Britain, at least as far as the trees were concerned. The records were stored at Oxford, and then 10 years later another group of forestry students sallied forth to record the wood again, and this exercise has been repeated at intervals ever since, often in the face of limited encouragement and negligible resources. Later, in 1980, adjacent woodland – also historically part of Lady Parke – was added, so that managed woodland could be compared with the untouched woodland originally set aside. Now, over 70 years later, long after those original forestry students have passed on, we have the individual histories of 20,000 or so trees and shrubs on record and can trace in detail how the wood and its inhabitants have performed when left to themselves.

Why did we bother? What have we learned? Was it worth the effort? And, is it worth continuing? One answer is entirely personal. GFP still remembers clearly the damascene moment when he first started re-recording the transects in 1980. Suddenly, with the charts and measurements from previous recordings in his hand, he could see the wood as a living, changing entity, not a statuesque, apparently unchanging collection of trees. Instead of the broad understanding of British woods that his survey and advisory work had given him, he realised that he could understand woods in depth, armed with both the detail of individual trees and an opportunity to understand their interactions across the long timescales over which they played out. EPM had much the same experience: once Lady Park had taught him how to recognise the marks of past events and the evidence of how stands were developing, this enabled him to recognise equivalent evidence in other woods.

The general answer requires some consideration of the context in which the study has taken place. After all, the assumptions, aims, concepts, knowledge and methods of research that form the foundation of ecology have changed a great deal since 1944, and so have the preoccupations and objectives of forestry. Although Alan Turing was simultaneously starting down the path that led to computers, the originators of the Lady Park recording could not possibly have envisaged the kinds of analysis that would become possible today. Equally, though the record may now have a role in, say, understanding climate change and the impact of Chalara ash disease, the originators could not have anticipated these at the outset. Back in Oxford, it must have seemed worthwhile in 1944 to initiate a study that could not have yielded results for many years, but in 2016 university ecologists have to publish quickly and in quantity to survive in their jobs.

The initiator of the study and leader of those recording expeditions in 1945 and 1955 was Dr Eustace Jones. Like Dr Alex Watt, one of the founders of British ecology, he was both a forester and an ecologist associated with Cambridge University. In the 1930s, he was recruited to the Oxford forestry department, where he taught various aspects of forestry to Masters graduates. He was also a keen bryologist, with prior knowledge of the lower Wye Valley, so there were mixed motives when he chose Highmeadow Woods for courses in field survey techniques and later selected Lady Park Wood for the study of natural woodland. Not for the first time, the chain of causation behind a scientific study was as much personal and subjective as it was objective.

Eustace Jones died in August 1992 (Obituary (1993) *Forestry* 66(2), 221–223), so what he started, others have necessarily continued. GFP became involved in the late 1970s, so he had opportunities to talk to him about the work and, late in his life, to take him back to the wood. Yet we have only a very general statement of his aims in starting the study. Today, we come up with a list of interest groups that have, or might benefit, ecologists, environmentalists, nature conservationists, foresters and the general public, not necessarily in that order of importance. In particular:

- **Ecologists** might test models, have them confirmed, or not, by long-term observation, or make observations that lead on to experimental and other kinds of fixed-term research (Figs 3.4, 14.2).
- **Nature conservationists** can see what minimum intervention achieves and assess whether this is the right management for woodland nature reserves elsewhere.
- **Environmentalists and land managers** have a reference point of near-natural woodland against which to measure human impacts on managed woodland and other land.
- **Foresters**, in particular, can measure and demonstrate the effects of their management and gain inspiration for near-to-nature forestry (Fig 16.2).
- **The general public**, and those managing land on their behalf, have a chance to witness a kind of 'rewilding' and assess whether this is desirable on a wider scale (Fig 12.10). To quote Ward Cooper of CAB International after a site visit, Lady Park 'makes you realise how managed other woods are'.

Lady Park is far from being the only place where we know how habitats have developed in the long term. The Park Grass fertiliser plots at Rothamsted were started in an ordinary meadow in 1856 and continue today without a break (Silvertown *et al.*, 2006). Plots in the Boubínský Prales, a virgin forest in Bohemia (Czech Republic), were first recorded in 1847, then relocated and repeated and thereby became a long-term study (Řehák, 1968; Peterken, 1981: p. 229). Likewise, in Britain, a plot recorded in 1921 by R. Hansford Worth at the southern end of Wistman's Wood, high on Dartmoor, was relocated and repeated (Mountford *et al.*, 2001). The record at Lady Park is shorter than these, but more detailed. In fact, Jorg Brunet, who has continued the study of Dalby Söderskog in southern Sweden, which was started in 1909 (Brunet *et al.*, 2014), tells us that it is one of the most comprehensive data sets in temperate Europe and North America, matched only by permanent plots at Bialowieza and a couple of sites in the USA.

Lady Park is now a National Nature Reserve. It occupies a spectacular position overlooking the

Wye from above a 60 m Carboniferous Limestone cliff in one of the wildest landscapes in southern Britain. It is unusual in containing a substantial population of large-leaved lime, Britain's rarest native major forest tree, and almost unique in being an ecological research reserve established by the Forestry Commission before official nature conservation came into being – and one that has actually been used for research ever since. It also offers a wide range of topography, composition and stand conditions for study.

But, attractive, invigorating, varied and well understood though it is, does Lady Park justify a whole book? After all, it is a sample of one and may be representative only of itself. Our response is that understanding a particular place can have general significance as part of a spectrum of various scales of understanding. At a large scale – regional, national or international – one can survey a large sample of woods and understand the full range of features encountered: how composition correlates with soils and geology; how relationships between species change over large distances; how climate controls composition; how management varies between regions. This is the kind of understanding achieved through the National Vegetation Classification (Rodwell, 1991). At a district scale, one can study all the woods within a limited area, including the small, apparently insignificant woods that tend to be ignored in large-scale studies, and recognise that each has its own history and individuality – analogous, perhaps, to understanding people through familiarity with a village community or small town, rather than a national survey. (An example would be our study of the flora of woods in central Lincolnshire (Peterken and Game, 1984).) At a single wood scale, one can study the detail, become familiar with all aspects over a long period and how different parts of a wood fit together and interact, much like understanding human relationships through a long marriage. One can spread such a study to several woods scattered over a district or region, but in practice one cannot match the detailed familiarity that can be achieved by concentrating on a single wood, nor familiarity over many years. This is, perhaps, analogous to the fabled 'wife in every port' lifestyle, which (we assume!) gives one a different understanding of human relationships. One can also study part of a single wood (or small samples of several woods), which is the scale at which experimentation and manipulation becomes practicable, and the kind of understanding achieved by studying pollen in deposits within small hollows within woods (e.g. Bradshaw, 1981).

Lady Park is also part of the spectrum of approaches to understanding woods. First, we can study how woods work, the processes operating and the patterns produced in stand structure, soil structure, nutrient flows, hydrology, energy exchanges, associations and interactions between species. Second, we can find out how they developed; that is, bring the time element into a study of processes. This brings in natural succession, natural disturbances, forestry, woodmanship and all other past usage, as well as the origins of the wood, and brings decidedly non-scientific historical sources into play. Third, we can enquire what woods mean to people as resources, playgrounds, symbols, freedom and wilderness, and, fourth, observe what they mean to wild species as habitat, shelter, refuge and food. Some of these forms of understanding can be advanced by scientific methods, including experimentation, manipulation, and indeed forest operations, but observation and the different methods of historical research are necessary and just as valid.

In practice, many ecologists have combined these scales and approaches to give a more complete understanding of woods. GFP first encountered this with Colin Tubbs (1986), who developed an intimate understanding of the New Forest wood pastures through historical studies combined with long familiarity with the woods and the culture that shaped them, but the exceptional exponent of recent times has been Oliver Rackham, whose work has changed our whole perception of woodlands. Initially, his work centred on the historical aspects of woods in a single district, East Anglia (Rackham, 1980), but the district was large and he combined it with knowledge of woods elsewhere (Rackham, 2006) and an element of experimentation through familiarity with nature reserve management. He also had knowledge in depth through his Hayley Wood (Rackham, 1975) and Hatfield Forest (Rackham, 1989) studies, and in time through his personal observations over 50 years in the woods near Cambridge (Rackham, 2003).

In Patrick Kavenah's terms, Lady Park is a study in depth. Detailed observations have developed over decades into an intimate historical study that tells us something about how woodlands work and how they develop. It can be placed in context by reference to other long-term studies and complements survey-type observations collected over district and larger scales, which represent studies in breadth. The problem is less the sample of one and more the limits to communicating the detail on which an in-depth understanding rests. The analysis that follows is forbiddingly quantitative, but that is the point. Instead of generalised statements that, say, regeneration was intermittent, that the ash population thinned out rapidly, that disease devastated the elms, or that the ground flora had declined, we can compare mortality and regeneration rates over different periods; show exactly how individual ash react to increasing competition as the stand grows; identify exactly what parts of the elm population suffered most; and, through the efforts of a team from Belgium, show how much the ground flora has changed, and what drives it. We know trees as individuals and how they have fared over seven decades, and appreciate how chance affects their fortunes.

Lady Park is not the only natural reserve, and there are quicker, if less direct, methods of answering the questions, so we are not greatly surprised by what we have discovered. But in this instance, the observations are more detailed in some respects than elsewhere and have been sustained for longer, so the study can act as a template and an example for other studies. It also demonstrates clearly that, however much one tries otherwise, there is still a human element. We cannot erase all the changes brought about by people in the past; we cannot seal off the reserve completely from the activities of people beyond its borders; the very act of studying it leaves some mark on the wood; and there is a real human dimension to the organisation of a study that has necessarily been handed on from one ecologist to the next.

The study is a collective enterprise. Looking back, the most prominent contributors have been Eustace Jones, who started it; Alan Orange, whose meticulous records saved it after a period of neglect; Vanessa Williams, who recorded the ground flora; GFP, who helped save it from the chainsaw and did much of the recording in the 1980s and recently; EPM, who undertook detailed surveys from 1992 to 2007 and started the analysis; and Liam Cavin, who combined research with recording. Many others have helped along the way; many officials have made crucial contributions; and a succession of visitors has shown us what others find of interest.

Our aim in this book is primarily to describe Lady Park and what has happened there since 1944, but we also put Lady Park in context and use the observations as a point of departure from which several topics of general interest will be discussed, such as the individuality of tree species, open spaces within natural woodland and the changing nature of remoteness. We also consider what Lady Park and other minimum-intervention – or rewilded – woods can tell us about the character of unfettered woodland, the nature of prehistoric natural woodland and the whole idea of natural woodland; what is likely to be achieved by rewilding woodland and whether, and where, rewilding is likely to be appropriate. We also reflect on the limits of our ability to study long-term ecological processes; consider whether natural reserves indicate how we might conduct near-to-nature forestry; and review the wider significance of remoteness and rewilding for people. In the course of this, we will discover that George Monbiot puts it well. Natural woodland is, indeed, not just a collection of species but also a setting for ever-shifting relationships with each other and with the physical environment; and that rewilding or natural succession certainly 'has no end points, no view about what a "right" ecosystem or a "right" assemblage of species looks like. It lets nature decide.'

(a)
(b)

2 Lady Park Wood and its History

Lady Parke was one of the five parts of Hadnock Wood, a large ancient wood in the parish of Dixton, Monmouthshire. The modern reserve includes most of the historic Lady Parke coppice and a fragment of the ancient Patches Wood in the parish of Staunton, Gloucestershire. Today, Hadnock Wood, Patches Wood and a large tract of secondary woodland planted on farmland in the 19th century form part of Highmeadow Woods. They are managed by Forest Enterprise (England) as part of the Forest of Dean, even though much of the ground lies in Wales.

Climatically, Lady Park lies right on the boundary between the oceanic western uplands of Britain and the somewhat drier and warmer lowlands. For the record, mean annual temperature is 9.3°C, mean annual precipitation is 719 mm and nitrogen deposition is 14.4 kg/ha/year (Verheyen *et al.*, 2012).

2.1 Physical Environment

Lady Park Wood stands on the left bank of the Wye as it passes through the upper Wye Gorge downstream of Symonds Yat (Fig. 2.2). This part of the river concludes an amazing sequence of great meanders formed in Devonian Old Red Sandstone that starts near Hereford and continues even after the river encounters the Carboniferous Limestone hills after Goodrich. The Wye enters the hills at Walford, flows out again below Symonds Yat and turns smoothly at Goodrich back into the hills below the settlements of Symonds Yat east and west, before bending round the Dowards and out into a wider valley as it approaches Monmouth. The crowds of visitors to Symonds Yat Rock thus witness a double view in which the Wye flows north on the right and back south on the left.

The gorge was formed by the downcutting of the Wye through Carboniferous Limestones. These constitute a sequence of strata (Table 2.1) from Lower Limestone shales (the oldest), through Lower Dolomite and Crease Limestone to Whitehead Limestone (the youngest). Because the rocks dip to the east and north-east, the oldest are exposed to the west and south-west. Thus, the Lower Limestone shales are exposed near the upper slopes of Lady Park and Whitehead Limestone is exposed towards the eastern end of the managed compartment and on the Doward opposite. The great cliffs and rock towers of the Seven Sisters on the right bank and the cliffs fronting Lady Park Wood are formed from Lower Dolomite and Crease Limestone. Drybrook Sandstone, which dominates the Forest of Dean, is exposed on the summit of the Doward and on the margins of the Whippington Brook. The older Tintern Sandstone and Quartz Conglomerate are exposed in the core of the Highmeadow Wood above the margins of Lady Park.

The downcutting of the Wye allowed tributary streams to incise into the sides of the gorge. Two such valleys define the northern and southern end of the reserve, the southern being the valley of the Whippington Brook. In both cases, they have cut minor cliffs in the limestones and deposited colluvium in the lower parts of their valleys.

Geologically recent alluvium and terrace gravels cover the valley floor. But at some earlier stage in the downcutting, the Wye deposited material derived from the Devonian Sandstones higher up the Wye catchment on to the sides of the gorge. These remain in Lady Park as superficial lime-free material smeared over the top of the limestone, which is nevertheless deep enough near the cliff to generate strongly acid soils.

The effect of water on the limestones has changed through their long geological history. Below the

Fig. 2.1. Charcoal burner's hut (a) and charcoal pit (b) in the Forest of Dean. (From Cooke, 1913.) This may be similar to the condition of Lady Park Wood before the 19th century.

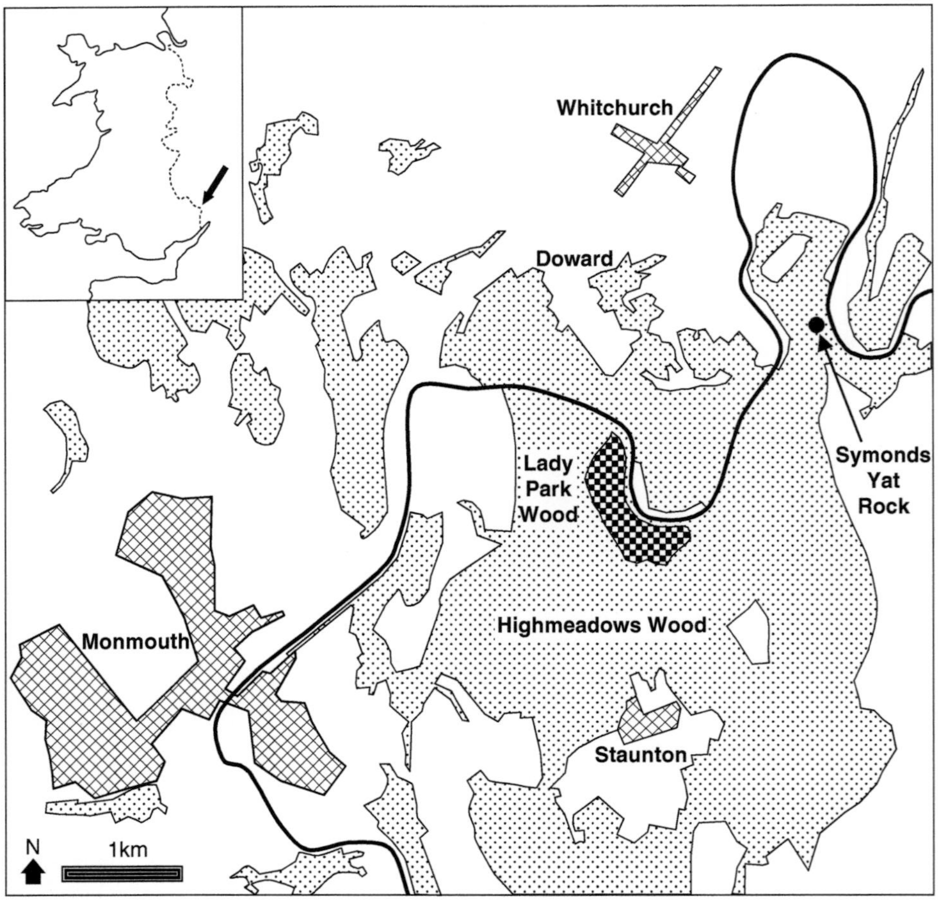

Fig. 2.2. Location map showing Lady Park Wood in its local context as part of the extensive Highmeadow and Doward woodlands that straddle the Wye below Symonds Yat. The River Wye (black line), wooded areas (stipple), open countryside (no fill), nearby towns and villages (cross-hatch) and the location on the border between England and Wales (inset) are shown.

Doward cliffs is the Madawg rock shelter, where the river once undercut the base of the cliff. Further back, Kings Arthur's Cave was excavated by water flowing in the Crease Limestone, but downcutting has left it dry. Caves and minor undercuts can be seen in the main cliff within the reserve, but these too are dry.

Today, the only water flowing within the reserve is a lime-rich seepage below the cliff, where tufa is deposited, although the amount is miniscule compared with the spectacular deposits at the Dropping Wells across the river. Water sinks freely into the ground and even after downpours does not flood over the surface; although a few soils near transect V betray subsoil seepage. The Whippington Brook flows intermittently; dry in droughts but disappearing into sinks and issuing from springs after wet weather.

The Crease Limestone on the Doward is finely fissured enough to form limestone pavement, though it is hard to make this out under the woodland. Much the same seems to have happened at the top of Lady Park. Indeed, throughout the higher ground of the managed compartment the soil is too shallow for marker stakes to be kept upright. Along the top of the reserve, limestone rubble and substantial weathered rocks outcrop under the trees. Some have been quarried, although only in small quantities, but the karst in the managed compartment seems to be intact. Immediately above the cliff, quarrying for ironstone has produced true scowles, albeit on a much smaller scale than the scowles along the exposure of the Crease and Whitehead Limestones running across the west of the Dean towards Lydney.

Table 2.1. Outcrops and superficial deposits in and near Lady Park Wood.

Period	Strata	Occurrence within Lady Park Wood
Superficial deposits	Tufa	Small seepages below the cliff on transect IX
	Colluvial deposits	Northern corner on low ground
	River-borne deposits from Old Red Sandstone	Extensive deposits in the core of the wood above the cliff
Pleistocene and recent	Alluvium	Flood plain, almost entirely outside the reserve
	First terrace gravels	Very narrow band along parts of lower fence
Carboniferous	Drybrook Sandstone	Small outcrops exposed on cliff overlooking the Whippington Brook
	Whitehead Limestone D2	Forms the main cliff overlooking the Whippington Brook and underlies the eastern part of the managed stand
	Crease Limestone D2	Forms the main cliff overlooking the Wye and underlies the middle levels of the wood above the cliff
	Lower Dolomite D1b	Mostly along the top within the reserve, forming upper steep slopes, but also underlying a small area above the cliff near transect V and the lower steep slopes below the cliff
	Lower Limestone shale D1a	Above and outside the reserve to the west and south-west
Devonian Old Red Sandstone	Tintern Sandstone group	Above and well outside the reserve to the west and south-west
	Quartz Conglomerate	Narrow outcrop west of Tintern Sandstone, including the Far Hearkening Rock

The main cliffs and the minor cliffs in the side valleys, together with the stable scree of limestone blocks below, are products of the geologically recent downcutting that produced the gorge (Fig. 2.3). In addition, the side streams have generated small cliffs at the north end of the reserve and along the Whippington Brook valley. The core of the reserve also shows signs of much earlier periods of erosion, which have long since ceased, in particular a miniature gorge that runs dry across the high ground of the managed compartment towards the Whippington Brook valley and a long, deep and more or less straight depression that runs like a hollow way across the core of the managed compartment but ends high on the cliff overlooking the Whippington Brook. The latter is so straight that it looks artificial, but Dr Ian Statham, a specialist geomorphologist, has decided that it must be natural, if only because the base level (thalweg) of the 'hollow way' is undulating and the feature ends as a notch in the cliff. He considered two possible explanations: that:

> this feature has formed by erosion along the line of a major weakness, possibly a hitherto unmapped fault, and that the closed depressions are solution features. As to its age, I would expect that it pre-dates the rapid downcutting to form the Wye Gorge because it is truncated by the steep valley slope. However there is also an outside chance that this feature may be a tension crack marking the backscar of a bedding plane slip towards the Wye/Whippington Brook incised valley intersection. If this is the case, it would be a relatively young feature and the area of rock sliding discussed below would be very large indeed.

Ian Statham also gave an opinion on the karstic outcrops, which were:

> almost certainly an area of natural slope instability. In the field we saw ample evidence of minor rock slipping, shown by joint-defined rock masses which had clearly slipped away from the outcrop by one or two, occasionally several, metres. We also saw blocks that had tilted, some showing evidence of subsidence into below-ground solution features. All this looks pretty old and unrelated to the steep gorge section of the valley (though see comment above on possibility of large scale bedding plane slip).

2.2 Soils

In broad terms, Lady Park is a mixed wood of mainly beech, sessile oak, ash, wych elm, both native limes, both birches forming a canopy over an underwood of mainly hazel, yew and holly, all growing on predominantly dry soils formed over Carboniferous Limestone. In detail, it includes a variety of soil types and a moderately correlated variety of stand types.

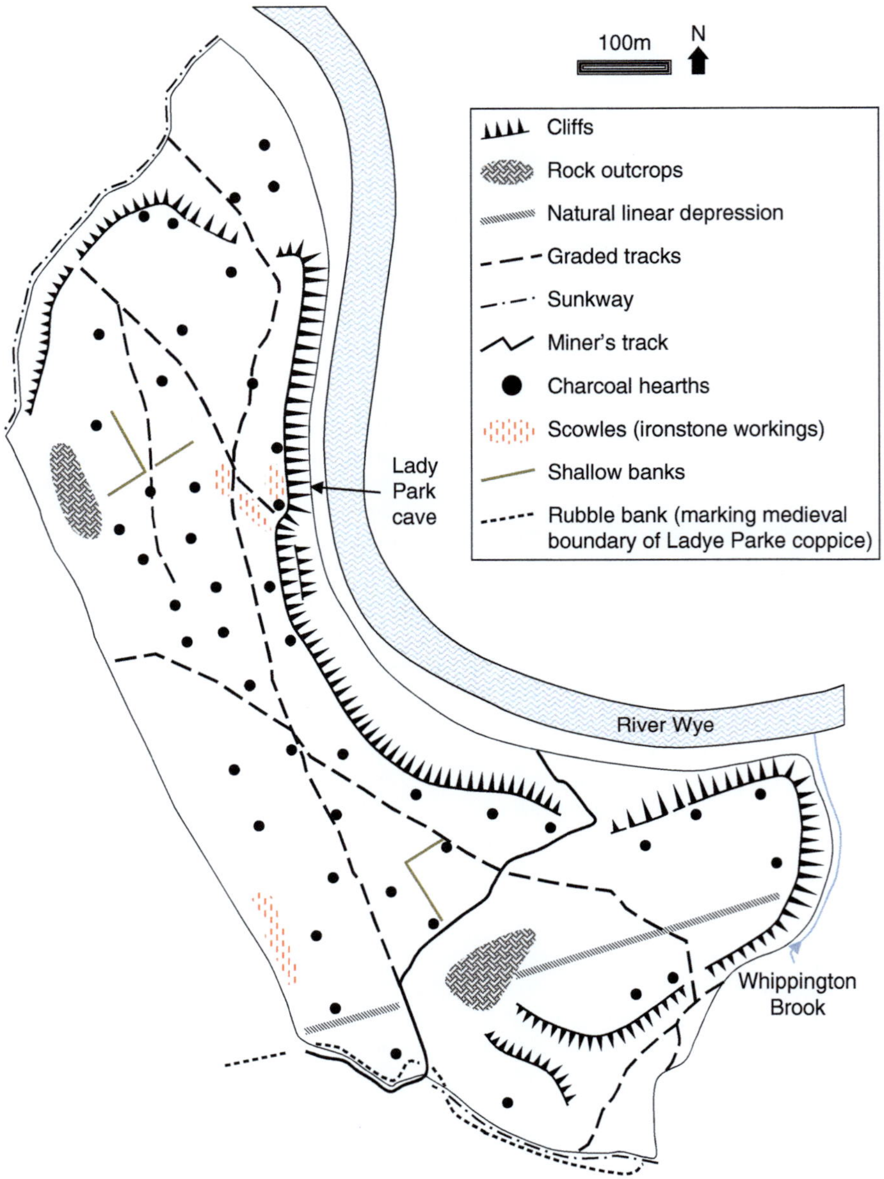

Fig. 2.3. Natural features and historical artefacts in Lady Park Wood. The locations of principal charcoal hearths have been fixed by GPS. The main concentrations of pits dug for ironstone are shown, but numerous small, shallow depressions that may be exploratory diggings are omitted.

2.2.1 Woodland soils in general

Woodland soils are many-splendoured things, but not many ecologists bother to look at them. When GFP was assessing woods throughout Britain, he routinely recorded vegetation in representative 900 m^2 samples, then dug a pit to 30 cm depth and examined the soil. Before long, he could predict the pH, texture and drainage class accurately just from the list of plants. More than anything else, this convinced GFP that soils – and, through them, geology, hydrology and landform – really did control which plant species grew where.

Using these 864 samples – which were reasonably representative of British woods as a whole – GFP was able to characterise British woodland soils in terms of texture (heavy to light), drainage and pH (acid to alkaline). The most acid sites at pH 3.2 were on sands in Tumby and Hogg woods on the northern edge of the fens in Lincolnshire, where downy birch, pedunculate oak and rowan were the main species, closely followed by parts of Staverton Park, Suffolk (pedunculate oak, holly), Great Ridge Wood, Wiltshire (pedunculate oak), and Wormley Wood, Hertfordshire (sessile oak, downy birch, hornbeam). At the opposite extreme, the most alkaline woods at pH 8.6 were hangers on the Downs in Hampshire near Selbourne (beech, ash, hazel, field maple, sycamore) and Norbury Park, Surrey (beech, ash, yew), but part of Perry Woods, Kent (sessile oak, downy birch, hornbeam), ran them close. Highland pinewoods, surprisingly, were never quite as acid – pH 3.7 at Glen Tanar was the lowest – and a few, including Ryvoan Pass and Shieldaig, were almost base rich. Woods on heavy clays and clay loams were generally poorly drained, but collectively they covered most of the pH range. Woods on light sands, loamy sands and sandy loams were mostly acid, but exceptions could be found, such as parts of Yockletts Wood, Kent, and Oxwich Wood, West Glamorgan, both of which included alkaline (pH 8.3) loamy sands. Wet woods also covered the whole pH range, from strongly acid birch scrub to fen alder carrs. Some individual woods were remarkably uniform, particularly such contrasting woods as the beech–oak wood pastures of the New Forest and the ash–hazel–pedunculate oak coppices on clays in the Midlands. Others spanned almost the full pH range within their borders, usually because they happened to straddle a geological boundary, and in these instances, soils sampled only 2 or 3 m apart might be more than 3 pH units different. For the record, the greatest spans GFP found were all in the east Midlands at Bentley Park Wood, Warwickshire (pH 3.6–7.9), Kings Wood, Bedfordshire (pH 3.4–7.7) and Bedford Purlieus, Peterborough (pH 4.0–8.2).

2.2.2 Soils in Lady Park Wood

Lady Park stands out as one of the diverse woods. Being on limestone, one might expect the soils to be alkaline and well drained throughout. The highest pH (7.9) is indeed well into the alkaline range, but most of the soils are neutral to mildly acid, and locally the pH descends to 3.8, which gives Lady Park one of the widest pH spans in British woods. The nine samples GFP recorded in 1977 and the 288 points at which Vanessa Williams measured pH demonstrated that the strongly acid, strongly alkaline and intermediate soils were equally represented within the reserve. What the wood lacks are truly wet soils, for only the lowest slopes remain moist, and permanent seepages are few and far between.

The broad pattern is straightforward, but the detail is complex. With only minor exceptions, the entire wood grows on limestone, but the middle levels are smeared with base-poor material, deposited by the Wye in an earlier geological era, that become deeper towards the edge of the cliff. The general zonation therefore runs from dry, stony calcareous soils at the top through increasingly acid, less stony soils in the mid-levels, to moist calcareous soils on the rubble-strewn slopes below the cliff, and on to deep, moist, stone-free soils in the colluvial fan and the flood plain. The complexity has arisen from the irregular outcropping of limestone and the patchiness of the acid overburden.

In more detail, much of the plateau at the top of the slope is neutral to mildly acid at the surface (pH 4.6–7.4, median 5.9), but the subsoil is in contact with limestone bedrock and the whole plateau is well or excessively drained. The bedrock is exposed locally, its fissures filled with small pockets of organic soil. The upper steep slopes also have exposed limestone outcrops and some limestone rubble, yielding well or excessively drained stony loams and clay loams of pH 4.1–7.7 (median 6.2). In effect, this is karst, patchily overlain with acid material, and dry enough to generate woodland rendzinas with steep pH gradients in the profile.

The lower boundary of the upper steep slopes coincides with the boundary of abundant Dog's mercury. Below this, the gradient diminishes and the overburden of acid sandstone material becomes deeper. The mid-slopes are strongly acid to neutral (pH 3.8–6.3, median 4.9), and the lower slopes, where the sandstone deposits are deeper, are fractionally more acid (pH 3.8–6.7, median 4.7). Limestone rubble is still occasionally seen on the surface, often where a tree has tipped over and dragged up subsoil or where the ground has been disturbed by ironstone prospectors. The orange-brown loams and clay loams are poorly structured and locally marked by faint mottles indicative of impeded drainage. Lower still, the almost-level ground approaching the cliff edge is the lower plateau, where the sandstone deposits are even deeper and the soils become still more acid (pH 3.9–7.0, median 4.6).

The ground also becomes more heterogeneous because limestone appears more often at the surface, especially where the sandstone deposits have spilled over the cliff.

The cliff itself is sheer limestone, with cracks filled with organic matter and narrow ledges with skeletal limestone soils. Where it borders the re-entrant valley to the north and grades into the Whippington Brook to the south-east, it becomes a rugged, steep, rocky slope – our lower steep slopes – where the soils are remarkably heterogeneous (pH 4.5–7.0, median 5.4). Here, we have everything from limestone outcrops and loose rubble to pockets of sandstone washed down from the core of the wood. On the very steep slopes below the cliff, the strongly alkaline (pH 6.1–7.9, median 7.4) organic loams and clay loams form the matrix between limestone boulders and stones. Almost all are well drained and actually dry where they thinly cover rock outcrops, but the slopes generally remain moist in the shade and locally they are irrigated by seepages, especially towards the foot of the slope. The colluvial fan in the north corner of the reserve has gentle slopes and deep, stone-free clay loams, moist but well drained. They must incorporate some sandstone material washed down the re-entrant valley, for they range from mildly acid to alkaline (pH 5.1–7.2, median 6.2). The flood plain, which is outside the reserve, has deep, freely drained alkaline loams.

The small-scale heterogeneity can be measured in the Williams samples, which were collected in the corners of 72 200-m^2 plots. The median span of pH within a plot was 0.9 pH units, but in one plot on the lip of the cliff it ran from 3.9 to 7.0–20 m within one wood, spanning a considerable part of the entire range in British woodlands. All of Lady Park's soils are diverse and patchy by most woodland standards, but the most diverse zone must be along the top edge of the cliff, where one can find strange juxtapositions of acknowledged calcifuge species, such as bilberry, heather and hard fern, growing in the shade of whitebeams and large-leaved limes next to dog's mercury and fingered sedge.

2.3 Stand Composition and Ground Vegetation

Woodlands were the first habitat to be brought to order by the National Vegetation Classification (NVC) (Rodwell, 1991). Most of Lady Park falls within W12b *Fagus sylvatica–Mercurialis perennis* woodland, *Sanicula europea* subcommunity with the *Taxus baccata* subcommunity (W12c) dominant on the limestone outcrops. Below the cliff, where beech is a minority component, W8g *Fraxinus excelsior–Acer campestre–Mercurialis perennis*, *Teucrium scorodonia* subcommunity prevails. On the flood plain, fragments of W7 *Alnus glutinosa–Fraxinus excelsior–Lysimachia nemorum* woodland have survived the railway.

Local examples of NVC communities invariably differ from the national averages. The most conspicuous difference in Lady Park is that wych elm is frequent in W12, and the associated oak is *Quercus petraea*, not *Quercus robur*. Among the numerous minor differences in the ground vegetation, sanicle is absent and ferns such as the male fern and hart's tongue are common. The national description for W8g corresponds well with the Lady Park stands, even though it is based mainly on samples from Derbyshire. Even the northern species, great bellflower and mountain melick, grow nearby, though not currently within the reserve.

In terms of stand types (Peterken, 1981), most of the wood is a mixture (Fig. 2.4) of both the alkaline and acid forms of 8E sessile oak–ash–beech woodland above and on the cliff, with a mosaic of this and stand type 4C sessile oak–ash–lime woodland below the cliff. In this case, the descriptions correspond closely, not least because they are based mainly on samples from the southern Welsh borderlands.

Ground vegetation is dominated by dog's mercury on limestone; great wood-rush and – very locally – bracken on the acid drift; bilberry on the dry acid soils just above the cliff; hart's tongue on the limestone rubble below the cliff; and ramsons on the colluvial fan, but zonations are far from sharp and any of these species can be found among the others. Wood anemone and bluebell are mixed into all but the bilberry patches, and the various microspecies of bramble can swamp them all if given the chance.

Both lime species are common and locally dominant in Lady Park but are rare in other remnants of ancient semi-natural stands in the former Hadnock Woods, which suggests that Lady Park was grazed less heavily during the medieval wood pasture phase. Much the same can be seen lower down the Wye, where small-leaved lime is generally common but rare in the woods derived from the former Wyeswood wood pasture. It also correlates with the condition of Lady Park in 1608 (below), which then had stronger coppice growth and fewer standards than elsewhere in Hadnock Woods.

Fig. 2.4. During bud-burst each spring, the woodland below the cliff is seen clearly to be a mixture of tree species.

2.3.1 Wider relationships

The particular characteristics of Lady Park Wood fit well into a much broader pattern. The oceanic climate of the western seaboard extends east to the Forest of Dean and lower Wye Valley: hence the abundance of ferns. The wood lies at the intersection of the native occurrence of beech, which stretches into south Wales from its centre of gravity in the south-east lowlands, and the borderland zone. As GFP has defined it (Peterken, 2008a, 2015), this lies between the lowlands to the east and south and the oceanic uplands to the north and west, where hard upland Palaeozoic rocks, including upland limestones, and the rugged landforms of the Highland zone combine with the richness of tree, shrub and herb species associated with the lowlands.

Woods similar to Lady Park occur mainly in the borderland, especially the southern Welsh borderland, where beech combines with ash, sessile oak, wych elm, field maple and both native lime species. Other examples can be found lower down the Wye in the Hudnalls, Caswell Woods and on the Blackcliff and Wyndcliff; further west at Cwm Coedycerrig, Coome Wood (Monmouthshire) and Garth Wood (Glamorgan); further south-west on Chudleigh Rocks (Devon); further north at Chanstone Wood in the Golden Valley, Halesend Wood near Malvern, Tick Wood (Shropshire), some Peak District woods on limestone, Anston Stones Wood, Kings Wood, Edlington Wood (southern Yorkshire); even further north at Gowbarrow (Cumbria), Castle Eden Dene (Durham) and in an attenuated form at Cragbank

Wood (Roxburghshire). In all these woods, ash, wych elm, sessile oak, yew and one or both limes combine on base-rich soils, but beech is regarded as native only in the southern Welsh borderland. Similar stands occur in the lowlands at Collyweston Great Wood and Bedford Purlieus (Northants).

Lady Park falls broadly within the *Fagetalia*, the diverse mixture of mixed deciduous woods dominated by beech, ash, maples, limes, elms and hornbeam that spread widely across mesic soils in temperate Europe (Ellenberg, 1988). Within this, it conforms more to the *Fagion* (beech and mixed beech woods) on the mid-slopes and to *Tilio–Acerion* (sycamore, ash and lime mixed woods) on the limestone outcrops, the steep slopes below the cliff and the colluvial fan. The Continental *Tilio–Acerion* woods often combine large-leaved lime, small-leaved lime, ash, wych elm and a limited representation of beech and oaks, which exactly describes the Lady Park stands below the cliff. These links are all the more convincing when we consider the ground flora, for several of the species that are uncommon in a British woodland context are characteristic of at least some of the associations within these alliances: narrow-leaved bittercress, fingered sedge, glaucous sedge, lily-of-the-valley, wood fescue, wood barley and mountain melick. In many ways, Lady Park Wood is an attenuated outpost of some specialised central European woodland types. On the lower fringes near the flood plain, one can see traces of *Alno–Ulmion* (mixed alder woods).

2.4 History

Lady Park has been transformed several times. From the 16th to the late 19th centuries, Lady Park was treated like so many other ancient woods as coppice with standards, but before that it was a wood pasture. Prehistorically, it must have been populated by people and large, now extinct mammals, whose bones were found in King Arthur's Cave and the Madawg rock shelter just over the valley (ApSimon *et al.*, 1992; Price, 2003). Since 1870, it has been developing as natural woodland.

2.4.1 Early history to 1608

The key historical source is:

> A Survey and Valuation of his Majesties woodes and Coppices being parcel and belonging his highness' Duchy of Lancaster in the Countie of Monmouth. Made and don [sic.] by virtue of A Commission hereunto annexed and directed to the Right Honourable the Earle of Worcester by Sir Willm. Pwell knight, Charles Somerset Chas and Matthewe Nesson gent. Substitutes and deputies of the said Earle. 31 die July Anno Dni. 1608.
>
> (PRO MPC 108)

Running to several pages, including a map, a schedule and a valuation of timber and some additional comments, it is worth considering in detail.

In 1608, Lady Parke was the smallest of five coppices in the 762 acres of Hadnock Wood. The stand below the cliff in the reserve was known as Saers Grove and the flood plain below the reserve was a meadow. Significantly, the report affirmed that the local tenants had recently been expelled from their rights in the wood:

> Hadnocke wood was always common for the tennents of Hadnocke: till of late they were expelled their rights by Mr Willm. Hall who [?holdeth it] by lease for 40 yeares yet to come (as he affirmed) the herbage, pannage, loppe, topp and underwood. Worth per annum 40s.

The state and value of the woodland is summarised in Table 2.2. Remarkably, the whole wood had been coppiced between 1599 and 1604. Lady Parke, on the steep slopes above the Wye, stood out as virtually a simple coppice with the most valuable underwood, a thin scatter of low-value standards and a complete absence of decayed trees. On the plateau, in contrast, timber trees were commoner and many were beeches, which must have cast enough shade to inhibit the growth of underwood. Moreover, the plateau woodland contained decayed trees, presumably old pollards and shred trees. If a proper succession of standards was being maintained, the 'great trees' – being younger – would have been more numerous than the timber trees, but in fact it was the reverse. Perhaps the preceding tenancy agreement had prohibited timber tree felling.

Material gathered by Paul Courtney (1983) on the history of the rural landscape of eastern and lower Gwent from the 11th to 18th centuries confirmed (p. 95) that the woods and wastes of Hadnock Wood were leased out in the 17th century, and this led to the loss of common rights and much litigation. The wood had been farmed out by the medieval owners, the Duchy of Lancaster, but Hadnock and Monmouth tenants claimed they still had common rights.

Table 2.2. A summary of Matthew Nesson's 1608 survey.

	Lady Parke	Skirred	Carvall quarter	Buckholde	Prishettes
Area (acres)	106	183	140	141	192
Underwood					
Years growth and coppicing year	4 (1603/4)	7 (1599–1600)	6 (1600/1)	6 (1600/1)	5 (1601/2)
Value/acre at 20 years (s/d)	33s 4d	20s	20s	20s	20s
Worth every fall (£/s/d)	£176 13s 4d	£183	£140	£141	£192
Timber stock					
Timber trees	2	1000	844	951	1118
Great trees	18	637	498	240	373
Decayed trees		55	12	52	18
Beeches	18	200	35	5	13
TOTAL	38	1892	1389	1248	1522
Density of all trees (n/ac)	0.36	10.34	9.92	8.85	8.03
Value per tree (s/d)					
Timber trees	5s	10s	13s 4d	10s	10s
Great trees	2s 6d	5s	6s 8d	5s	5s
Decayed trees		1s	1s	1s	1s
Beeches	2s	2s	2s	2s	2s
Stock value (£/s/d)					
Timber trees	10s	£500	£562 13s 4d	£475 10s	£559
Great trees	£2 5s	£159 5s	£166	£60	£93 5s
Decayed trees		£2 15s	12s	£2 12s	18s
Beeches	£1 16s	£20	£3 10s	£10	£1 6s
TOTAL VALUE (£/s/d)	£4 11s	£682	£732 15s 4d	£538 12s	£654 9s

It was the Monmouth tenants who went to law and thereby left a record of medieval management. The custom had been to leave underwood to grow for at least 12 years before cutting. Parts of all five coppices had been cut in the years leading up to 1584/85, 50 acres of Carvall quarter in 1580/81, 50 acres of Buckholde in 1581/82, 80 acres of Prishettes in 1582/83, and in 1582/83 and 1583/84 30 acres of Skirred and 60 acres of Lady Parke. Lady Parke was well fenced and enclosed, but Skirred was not fenced, and Buckholde and Prishettes were suffering damage by cattle grazing. Some of the underwood must have been made into charcoal, for there is a reference to 'colliers', but the timber had been taken and sold in Monmouth to build a weir, make and repair hedges and make trenchers. At about the same time, the Duchy accused William Hall (the lessee?) of spoliation; that is, cutting oak and beech timber trees, cutting out of season, blocking a track, breaking enclosures and letting in cattle (PRO DL44/376; 1604/5 PRO DL4/48/58). Hall admitted nothing, except that some of his young beasts had grazed in Lady coppice for about a month, so they could eat the browse wood that had already fallen. He claimed also that he never gave anyone leave to collect mast, except one poor woman.

The Gage Papers shed further light. By his Will of 1580, Henry Mershe of Whitecliff, a smith, left to his wife Dorothy and son Edward all his smythes tools, as well as those belonging to the ore smith and those to the hand smythe; and his part of the woods which he had in Hadnocks Wood growing during his lifetime there ... (GG495). He must have worked both a forge and a smelting furnace, for which he burned fuel from his wood. In 1592 (GG533), Christopher Humphrey of St Clement Danes, taylor, assigned to Thomas Phelippes of London ... for £150:

> a lease by the Queen under the seal of the Duchy of Lancaster ... of all that woodground and soyle called Hadnocks Woodes containing 160 acres extending from fields of the Lordship of Staunton, south-east from Staunton towards ... the river called Alsons Well ... also the herbage and pannage of the said wood. Reserving all timber trees and 2 loads of wood yearly for the expense of the receiver ... And sufficient woods for the repair of weares and water works within the manor of Monmouth.

The bounds include 'the river called Alsons Well' (= Oldstone Well, Whippington Brook) and therefore the lease must include Lady Park or woodland nearby.

On this basis, much of Lady Park had been coppiced 20 years before the coppicing recorded in the 1608 survey. Hadnock Woods had been grazed under common rights and they were still being grazed, albeit allegedly illegally. Given all these features, Oliver Rackham thought that four of the woods had probably been a wood pasture common on which there were grazing rights, whereas Lady Park, being on the site of a medieval park, was free of common rights and had been abandoned as a park long enough to have acquired a normal coppice-with-standards structure without pollards. Blast furnaces fuelled by charcoal were introduced into the district in the late 16th century (Standing, 1981): Hadnock Woods had been acquired by the iron-making Hall family. So presumably both the disputes and the astonishingly extensive cutting were consequences of the conversion from wood pasture to coppice.

Taking the story back before the 16th century becomes more difficult. The woods certainly existed in the 14th century, for the Gage Papers (GG107) include the grant of a field close to Lady Park in 1381, which was bounded partly by the 'waste land of the Lord Duke of Lancaster called Hodenakswode'.

The 13th century was important. The four Lordships of Monmouth and the Three Castles were united in 1267 as part of the Duchy of Lancaster estate (Courtney, 1983: p. 85). The Close Rolls record that King John gave permission to John of Monmouth to enclose a certain wood of his called Hodenach to make a park, which has not yet been enclosed, which wood is in [the forest of Dene] (Hart, 1966: p. 36) and permission was given to enclose it. In 1281, Roger de Clifford, then Hadnock Wood's owner, granted the right of common in his wood to the knights and free tenants at Monmouth Court (Courtney, 1983: p. 94). They were allowed to graze livestock, but not goats. Only horses used as plough beasts were to be allowed, and no pigs in the fortnight before and after Midsummer Day. Reasonable amounts of housebote and haybote (hedgebote) were also granted. In 1282, Hadnock Wood was described as a chase in the Forest of Dean regard, when Roger de Clifford claimed the right to take deer and wood. In an account of waste in woods granted by the King, 1271–1282 (Hart, 1966: Appendix IIIB), the chase of Monmouth lay in two parts, one held by Roger de Clifford. Evidently, there was 'new waste' because 'they still cause charcoal to be made there, but all the lords of Monmouth have at all times been accustomed to take vert there at their will'. Lady Park Wood must have been within this chase or park.

2.4.2 Recent history, 1608–1945

During the 17th and 18th centuries, Lady Park and other parts of the Highmeadow Woods were coppiced and the wood was made into charcoal for the local metal industries (Fig. 2.1: see p. 8). Edward Dayes' *View of the Doward rocks on the Wye*, engraved by F. Jukes, 1790s (Hereford Museum, shown on p. 18 in Peterken, 2008b), shows that the flood plain was pasture grazed by sheep, the cliff was largely bare with scrub on its summit and that the woodland on the colluvial fan was indeed coppice with standards. The colliers understandably preferred to haul logs down the slope, for the hearths were concentrated along the lower edge above the cliff, but were absent from the topmost levels. The smoke rising from the hearths was witnessed by William Gilpin as he boated down the Wye in 1770:

> Many of the furnaces, on the banks of the river, consume charcoal, which is manufactured on the spot; and the smoke, which is frequently seen issuing from the sides of the hills; and spreading its thin veil over a part of them, beautifully breaks their lines, and unites them with the sky.
>
> (Gilpin, 1782)

Charcoaling continued in the vicinity until very early 20th century (Cooke, 1913), but it presumably died out in Lady Park in or before 1870.

When Viscount Gage, a successor of the Halls, lost a great deal of money gambling in Dieppe in 1817, the Highmeadow Estate was bought by the Crown and its woods could again be managed as part of the Dean. As explained by Edward Machen, Deputy Surveyor of the Dean, in evidence to a Select Committee of the House of Lords:

> Mr Huskisson, who was at that time First High Commissioner, thought it would be an advantageous purchase for the Crown, because of the situation, and because they had a large quantity of timber 30, 40 and 50 years old that would be very useful, the timber of the forest being then either very old or very young.
>
> (Machen, 1849)

Evidently, numerous young oaks had been reserved when coppices were felled between the 1760s and the 1790s.

According to Machen, the chief produce of Highmeadow Woods was 'Timber, bark, cordwood, miners' timber, and poles' (Machen, 1849: pp. 145–146). In contrast to the well-spaced Dean woods, timber was sold after felling 'because in woods so close together, we are afraid that if we trusted the purchasers' woodmen to go out and cut timber, they would injure the other standing timber'. 'But the principal reason for felling the timber (ourselves) in High Meadow Wood is, that we may employ our own woodmen, because the coppice-wood is also cut at the same time, and the bark would be mixed if we allowed the purchasers to cut the timber.' By then, fallow deer had been released into the woods, but they had little impact. Evidently, 'Lord Dacre gave some to the Queen, and they were disposed of; 20 were sent to High Meadow Wood; I was told to turn them out there'.

This coppice with timber management continued through the 19th century. An auction of timber, bark and cordwood from the Dean and Highmeadow, held at Speech House on 15 April 1851, included: '62 oak timber trees in Lady Park, all carefully numbered, felled for the Navy in the Spring of 1850 but rejected by the Purveyor after being squared ...', '160 cords of oak miners' timber in Lady Park' and oak poles and a quantity of bark from the timber trees (Hart, 1966). The quality of the oak timber was seemingly suspect, a point reinforced by a contemporary report by a forester on 6 July 1853, who noted that 'the greater part of (Highmeadow Woods) consists of old oak coppice and not of virgin trees, and that as navy timber the trees could never come of any size' (Brown's Report, PRO F3/132). However, some grew well, for Mr Richard Hall, a land agent and surveyor of Cirencester (Machen, 1849: p. 225) had never seen better oaks in his life than the young oaks to be seen in the valley running down towards the Wye.

An invaluable description of Lady Park 120 years ago was recorded by H.C. Hill, a British forester who was home 'on furlough' from his work in India. Hill's report on the Crown woods in Dean and Tintern described each wood and recommended how they should be managed. Lady Park was:

> Coppice of beech, ash, lime, hazel and some oak 30–35 feet high. Maple, service tree, aspen, birch, willow and wych elm all represented. Good growth. Vigorous standards of oak comparatively thinly scattered of medium size. No medium aged standards. Seedlings of ash very numerous. Age 25. Density, complete.
>
> (Hill, 1897)

Lady Park was thus a mixed coppice with a closed canopy, a limited stock of oak standards, few if any conifers and an ash population that was capable of regenerating abundantly. The OS 6-inch first edition map (*c.*1880) confirms that conifers were limited to the cliff and a few at the top of what is now the managed compartment, all of which were presumably yews.

Hill's prescription for Lady Park was 'Cut out old standards with few exceptions, and store 40 oak and beech to the acre', but 'do not coppice. Merely thin out to 400 stems to the acre and underplant with beech'. Lady Park was thinned in 1901–1903 and, unlike other parts of Highmeadow, there is no note of planting (C.O. Hanson's Stock Map of 1916). (Another source, now lost, confirmed this and mentioned that the slopes below the cliff were not heavily thinned.) Ring counts of 32–35 years made by Dr Eustace Jones (EWJ) in 1941 of oak, ash, beech and cherry stumps (presumably in what is now the young-growth) confirmed the thinning date, and a larger oak of 60 rings indicated some recruitment after the 1870 coppicing. Sutton Palmer's *c.*1910 watercolour (Fig. 2.5), which was topographically accurate, suggested that the woodland at the bottom of the colluvial fan, which contained few beech, was felled about 1870 and again in the early 20th century, leaving a few narrow-crowned standards.

His working holiday completed, Hill returned to India and took over as Inspector General of Forests on 25 October 1900, but he died 2 years later at the age of 48 (Ledzian, 1991). His plans for the Dean did not last either. From 1900 until the 1940s, forestry in the Highmeadow Woods was characterised by constant changes in policy. Almost every new Deputy Surveyor abandoned the plans of his predecessor, an almost comical sequence that ended seriously with extensive emergency fellings in the Second World War. Throughout all these changes of policy, Lady Park remained unaffected, for its inaccessibility and dense stock of relatively young timber always put it late in the prescribed order of treatment.

Much of Lady Park Wood was probably singled. In 1941, EWJ described it as:

> ... fundamentally oak standards with hazel coppice. Ash standards are frequent. Hazel coppice is very vigorous, and there are considerable amounts of vigorous coppice of Tilia, Ulmus and Acer campestre, making a dense stand. Beech is rare in some areas (e.g. C15 and 16), but abundant as coppice in others (C19–20). In some areas within the past 20–30 years

Fig. 2.5. Sutton Palmer's watercolour, which illustrated *The Wye* by A.G. Bradley (1910) between pages 128 and 129. Topographically accurate, it confirms the known history of management in Lady Park Wood and shows that the flood plain was not yet wholly wooded. The same view is now obscured by riverbank trees, but in the century since, the Seven Sisters rocks have been largely obscured by the growth of trees from below.

> there has been some thinning out of stems to leave one or two stems per stool of all the larger species, with cutting out of much birch and hazel. In Lady Park there seems to have been such a cutting in 1938 or 1939.

The stock of beech and oak was 81 trees/acre, much the same as elsewhere in Highmeadow at the time. The beech cast considerable shade, yet small birch were present everywhere, and holly and hazel were also present throughout. The oaks were 53–65 feet tall, about the average for Highmeadow. In 1940, the girths of oaks on what became transect I were up to 272 cm, but mostly below 160 cm; beech up to 128 cm and ash, lime and wych elm up to 80 cm. The ground vegetation comprised mainly bramble, with some greater woodrush, wavy hair-grass and ivy.

In 1942–1944, a considerable part of the reserve was felled, leaving scattered standards of oak and beech for the sake of amenity (and perhaps because large timber of these species was not greatly needed at the time). EWJ considered that this felling must have been similar to the felling of 1901–1903, except for the absence of any deliberate reservation of tellers. In addition, large lime timber was in demand for 'special secret purposes' – evidently the frames of fighter aircraft – and this led to much felling of lime below the cliff, and even from inhospitable places on the cliff. Further, the oak coppice and some of the beech coppice in that part of the stand above the cliff crossed by transect V were singled.

The Ross–Monmouth railway was built in 1872 through the meadow along the flood plain below

Fig. 2.6. This image from the Forestry Commission's local archives shows the Doward end of the cableway with which timber was slung across to the railway on the other bank. The woodland in the background is Lady Park's stand below the cliff, looking much like it does today.

Lady Park Wood. This meadow was already marshy and scrub-ridden by 1842 (John Atkinson's 1842 map of Highmeadow and Great Doward Woods, PRO F17/26), and by the 1880s, the Ordnance Survey showed the flood plain as woodland. Timber was loaded on to waggons parked in sidings below Lady Park: the original Biblins bridge was a sling to carry timber to the railway from the Doward (Fig. 2.6).

2.5 Prehistory

The medieval woodland was clearly used by a wide variety of people as a source of both pasture and wood, and one can only suppose that this is much as it had been used in prehistory. Summarising:

- The Little Hadnock Roman villa stood near the Wye to the west of the wood (opposite Wyastone) during the 2nd to 4th centuries, and there is evidence of iron smelting at this time, which implies the charcoal was made in nearby woods.
- Encampments were built in the Iron Age on the Little Doward and at Symonds Yat.
- The caves and rock shelters in the limestone overlooking the Wye were frequented by a spectacular array of large fauna (Peterken, 2008a, where original sources are detailed).
- People occupied the gorge throughout the post-glacial millennia.
- Nevertheless, small mammal remains indicate that the area remained densely wooded once the trees returned after the last glaciation (Price, 2003).

2.6 Archaeology

The banks, ditches, pits, mounds and other relicts of past use often fill the gaps in a wood's written

Fig. 2.7. Shallow workings for ironstone in old-growth just above the cliff.

history, and they may be the only source of information on prehistoric activities. The most obvious artefacts in Lady Park are the charcoal hearths (see Fig. 2.3), which, having been used for 250 years, survive as flat, level, usually treeless platforms notched into the slope, with a scarp on the upside and a steep bank below, and usually still showing dark, almost black soil where, say, a badger has recently disturbed the ground. Scattered through the wood, they are evenly spaced, implying an understandable reluctance on the part of the colliers to carry wood further than they had to, but absent from the topmost levels and concentrated towards the lower edge above the cliff – they preferred to carry wood downhill rather than uphill.

Another obvious feature is the scatter of ironstone quarries and mines. Those within the wood are mostly small and shallow (Fig. 2.7). Indeed, several are no larger than the pits and mounds left by big trees thrown over in high winds. The miners prodded and probed throughout the wood, but their richest finds were made at the top of the wood and in and above Lady Park cave. Today, one finds mounds, pits and small rock exposures (i.e. scowles) filled with leaf litter. Ian Standing, geologist, historian and one of the Verderers of the Dean, thinks most, perhaps all, could be the work of one family, living in the wood while eking out a living.

The tracks through the wood have a mixed origin. The main track, which runs the length of the wood and continues beyond, is part of the wider network of unmetalled roads through the Highmeadow Woods and surrounding lands. It has a constant width and has been levelled carefully by cutting into the slope above, building up the lower margin and reinforced by small stone walls where

Fig. 2.8. The medieval boundary between Patches Wood (left) and Hadnock Wood (right), part of which now runs within the Lady Park reserve fence. This was once a wall, but it has collapsed into a line of rubble.

necessary. At one point, it has also been cut through a rock outcrop. The 'blind' track and its fork must be woodmen's tracks, but the mystery is why one fork ends above the main cliff. Perhaps they threw material down to the Wye for transportation. The Miners' Track, now the main public route through the reserve, also appears to be recent, made as a direct route to the railway siding – in contrast to the old roads, which climb diagonally up the slope. The sunkway at the southern end of the reserve must be contemporary with the boundary remains, and these seem most likely to be medieval.

Ancient woods usually had sharply defined boundaries, marked by banks and ditches and reinforced by walls or hedges, or some combination of these. Lady Park, however, has only fragments (Fig. 2.8; see also Fig. 2.3). The lower boundary with the meadow coincided with the break of slope at the edge of the flood plain, and if it ever had a boundary bank, it has been destroyed, presumably when the railway was built. A line of rubble near Biblins Bridge may be all that remains. The northern boundary is a sunkway beside a cliff, which was never the historic boundary. The southern boundary is the Whippington Brook, a deep valley along which a forest road now runs and a tramway once ran. The upper boundary is now a forest road, straightened in 1960. It was never the historic margin of the wood.

The only old boundary is a convoluted, collapsed wall that once defined the boundary between Patches Wood and Hadnock Woods. The associated sunkway ran within Lady Park from near where the Bellman's Oak once stood, and runs on down to the Whippington Brook as the south-western bank of the sunkway. This must once have been the boundary between Gloucestershire and Monmouthshire, but at some point that boundary was replaced by a straight, arbitrary line that ran unmarked within the managed part of the reserve down the narrow, blind valley (see geomorphology) to a point below the Oldstone Well marked in the 19th century by a boundary stone. Thus, Gloucestershire somehow 'stole' a small part of the original Lady Park Wood, which was once wholly in Monmouthshire, and the present fenced reserve incorporates a small corner of what was once Patches Wood.

This leaves three features (Fig. 2.9) that all imply human activity within the reserve:

1. Two instances of wide, shallow banks, straight and turning at a right angle, which would form two sides of a 'field' if the other two sides existed. First noticed on the LiDAR images, they are difficult to see on the ground. The southern example has a charcoal hearth at both ends. Were they used by medieval cowherds when Hadnock Wood was a wood pasture?

2. A short length of rubble bank in the core of the reserve between transects IV and V, which runs up towards the corner of the northern example of 1 above. Roughly 2 m wide and no more than 30 cm high, it is associated with a seemingly even-aged group of yews, the largest of which (234 cm GBH (girth at breast height)) must have originated before 1850. Beech stubs and stools on and near the bank must be over 200 years old.

3. Irregular enclosures and compounds, just west of the reserve, all clearly visible on LiDAR but now obscured below bramble and scattered saplings. This abuts the largest tract of ironstone mining within the reserve.

One suspects that the fringes of the reserve may once have been cleared and that woodland on the gentle slopes in the core of the wood may have been more disturbed than the rest.

Fig. 2.9. LiDAR image, taken for Forest Research. The charcoal hearths stand out as a regular scatter of pockmarks. The features above the wood (to the left) are now obscured by saplings and bramble.

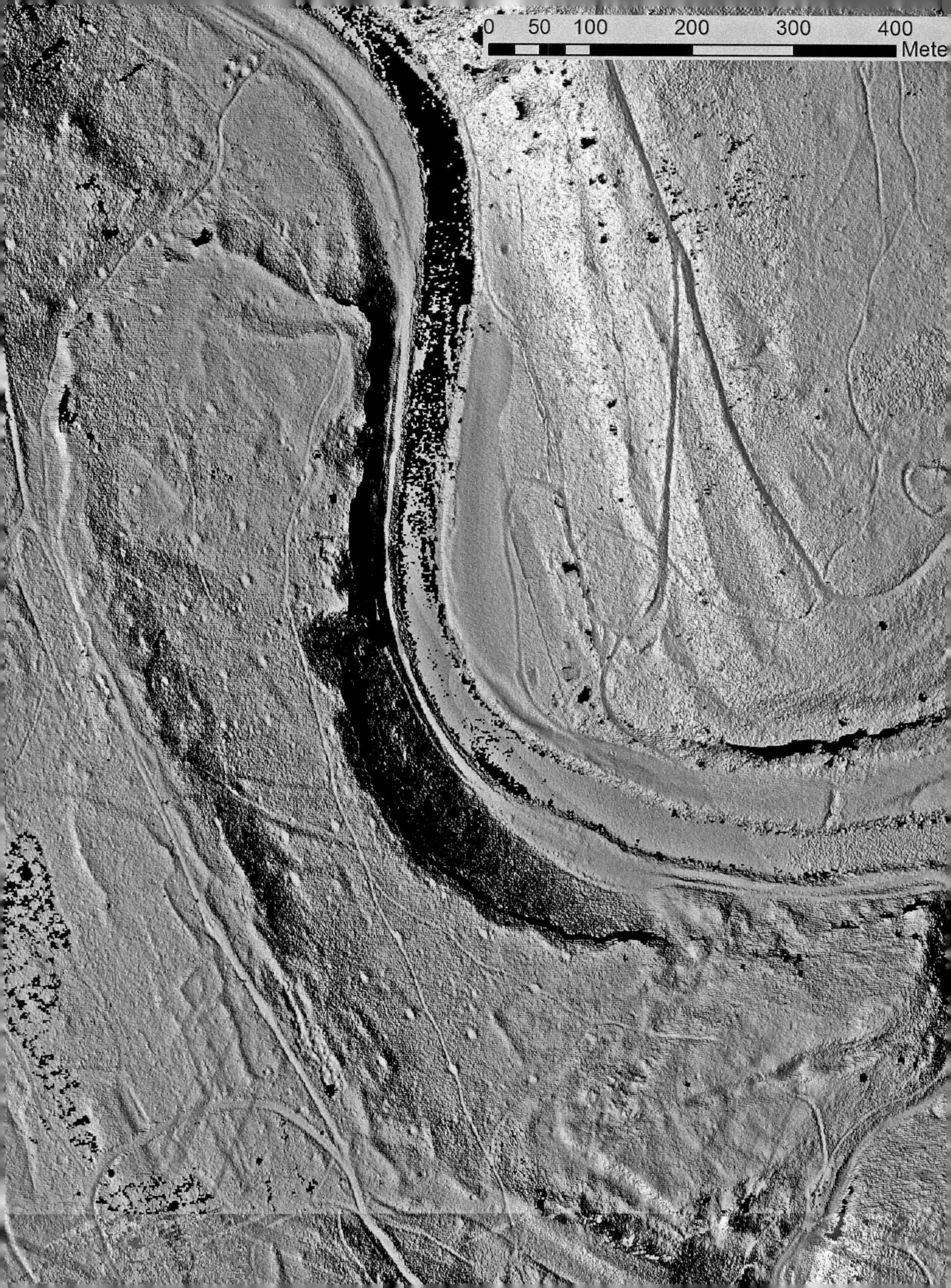
0 50 100 200 300 400
Mete

Welcome to / Croeso i'r
Gwarchodfa Natur Genedlaethol
Lady Park Wood
National Nature Reserve
NATURAL ENGLAND
The slopes of the Upper Wye Gorge are important for their long continuity of woodland cover; much of the area has been wooded since the last ice age.
Lady Park Wood is one of the few woods in Britain where long term research has been carried out to study the changes which take place when people do not interfere with natural processes.
Please help look after the site by following the Countryside Code and in particular, keep dogs under control. We hope that you enjoy your visit.
This Reserve is owned by the Forestry Commission and managed by agreement with Natural England
Please see the back of this sign for further important information, or call Natural England on 0300 060 6000.
Forestry Commission England
www.gov.uk/natural-england

3 The Ecological Reserve

Lady Park is an attractive place and an ecologically interesting wood, but its main significance for ecologists lies in its early establishment as a research reserve and the long and detailed record of its trees. From 1945 onwards, we have the curriculum vitae of some 21,000 individuals, from huge, spreading trees of 18th century origin to saplings that only briefly grew large enough to be recorded before they perished. The creation of the reserve and the circumstances in which the record has been compiled and preserved is a story of shifting institutional relationships, scientific endeavour and human and organisational limitations spanning the entire post-war decades.

3.1 Origins

Lady Park Wood is almost unique in becoming a research reserve before nature conservation was officially established in Britain, and it remained an anomaly for 40 years as an 'ecological reserve' on Ordnance Survey maps, but neither a Site of Special Scientific Interest (SSSI) nor a National Nature Reserve (NNR), even though these designations were ostensibly for scientific purposes.

The idea of a reserve in or near Lady Park emerged a century ago. Roy Robinson, then assistant inspector for forestry at the Board of Agriculture, and Sir William Schlich, under whom he had studied forestry at Oxford University, visited Highmeadow on 12 April 1915 (PRO F3/1395) and reported that 'The woods are to be worked so as to produce the best financial results: on the steep slopes overlooking the Wye only are aesthetic and similar considerations to influence the scheme of operations.' By that date, the notion of ecological reserves in forests was already established in Scandinavia and Germany, and had in a very broad sense been linked to Highmeadow Woods and other forests by a German forester writing about the care of natural communities (Conventz, 1909). The importance of the Upper Wye Gorge was also emphasised by the British Ecological Society, which, through the initiative of Eleanor Armitage, held its first excursion there in July 1914 (Meeting report in *Journal of Ecology* 2, 202–205).

Apart from the inclusion of 'Symonds Yat' in the British Correlating Committee's 1929 list of 66 areas 'where nature reserves are most required' (Sheail, 1976), the sequence of events that led to the reserve started when the expanding Forestry Commission ran into a storm of criticism for afforesting Upper Eskdale in the Lake District with conifers. Faced with both criticism in the House of Lords and a critical book (Symonds, 1936), Sir Roy Robinson, by then chairman of the Forestry Commission (FC), rose at the British Association meeting in Cambridge in 1938 to invite ecologists to join the FC in studying the feasibility of reserves for monitoring ecological changes and for experimental work, seeking thereby cooperation from ecologists but also favourable publicity (Sheail, 1976: pp. 116–117). A preliminary committee met on 7 February 1939 and formed an official committee, which circulated a statement to heads of university botanical departments – an interesting reflection, incidentally, on the dominance of botany in the early years of ecology in Britain.

Further work was abandoned with the outbreak of war in September 1939, but the idea remained alive for, in an internal memorandum dated 24 October 1942 (PRO F18/291), the FC mentioned that they maintained no formal ecological reserves but that informal reserves had been made at Wretham, Thetford (Norfolk Naturalists' Trust 33.5 acres); New Forest (arrangement made *c.*1930 not to drain one or two bogs to preserve their flora); Mendip; Bramshill; and Kings Forest. In the last three cases,

Fig. 3.1. The boundary fence of the non-intervention part of the Lady Park Wood National Nature Reserve, erected early in 2007.

'arrangements have been made with the Ecological Reserves Committee that certain areas should be left un-planted so that changes in the vegetation could be kept under observation by the committee'.

The Forestry Commission's willingness to establish ecological reserves was confirmed by Robinson when he summed up a discussion on 'Ecological Principles Involved in the Practice of Forestry' held at the Linnean Society, Burlington House, Piccadilly, London, on 2 July 1943.

> I agree as to the necessity for laying down ecological reserves. The Forestry Commission in association with ecologists have already made a beginning and are prepared to go further. To get the full value out of reserves it will be necessary to undertake and periodically repeat a great deal of descriptive work. The Commission can and will afford the facilities but foresters have not the time to spare from routine duties nor the specialised training. I trust therefore that the Society will formulate its ideas in some detail so that the venture may go forward smoothly. I hope, too, that the problems will be regarded, as our Chairman [Sir Arthur Tansley] has pointed out, as not only botanical but also as biological in the widest sense.
>
> (Meeting report, *Journal of Ecology* (1944) 32, 83–115)

In a report of their own Nature Conservation and Nature Reserves Committee, the British Ecological Society recorded (*Journal of Ecology* (1944) 32, 55) 'with pleasure' that the Forestry Commission had 'already set aside two areas of old woodland, one in the High Meadow Woods by the banks of the Wye, the other in the New Forest [which would] be left completely undisturbed as permanent forest reserves', and this appears to be the earliest mention in print of both Lady Park and Denny Inclosure as reserves.

Lady Park was suggested by E.W. Jones, who for several years had taught forest survey techniques to Oxford students in Highmeadow Woods and who already knew the district well as a bryologist. He evidently regarded Robinson's invitation as both an opportunity and a damage-limitation exercise, for he told GFP that he had been shocked to hear that the FC intended to replant the wartime fellings there with sycamore and red oak 'for amenity'. Sir Arthur Tansley, the Chairman, wrote to EWJ from Dairy House, Elsted, near Midhurst, Sussex, on 17 January 1943 [GFP has the letter], mentioning that he had received a memorandum from Jones about a reserve in the woods overlooking the Wye and that he had been in touch with Taylor of FC about this. Tansley evidently 'asked for Skirrets and Lady Park in accordance with your memoranda though I did not expect to get the whole area!' Tansley also urged Jones to 'follow up this opportunity as energetically as possible' with the FC, adding that he had 'told Taylor that a fair block is necessary to include representatives of the varied tree flora, ?......? as well as steep slopes, and to avoid the marginal effects in a small area'. He pushed Jones to 'get as much as you can, and emphasise the importance of the area and the tree flora, which you can do with perfect truth'. Jones noted that '100 acres have been felled but not replanted; two small strips [had been] reserved, partly out of amenity and partly because of [the] precipitous slope'. We now call the felled areas and the reserved stands the young-growth and the old-growth, respectively. Jones was supported by the Deputy Surveyor, W.H. Guillebaud, who suggested that part of this area would make an excellent ecological reserve and that he and EWJ should visit High Meadow with Hope (Acting Deputy Surveyor) in the spring to select a suitable block 'which need not be extensive'. It was this restraint that until 1984 excluded what is now the managed compartment.

Lady Park was not the only place to be nominated in response to Robinson's 1938 invitation. Some 25 botanists from 20 departments proposed 22 sites (Salisbury in PRO HLG/93/24), but many had only vague ideas of what they might want, and several proposed unwooded ground where they wanted to study natural succession to woodland (Papers of Huxley Committee, undated, but about 1946 (WLC/Inf/23 in PRO HLG 93/48)). One of the specific suggestions was Denny Inclosure in the New Forest, put forward by Imperial College London, which was eventually established as a research reserve for Southampton University, complete with transects that, like Lady Park, have been recorded ever since. On 17 January 1944, M. Tetley from Bristol University put forward some broad suggestions for native woodland conservation in the whole of the lower Wye Valley, which foreshadowed the suggestions GFP made in 1969 on behalf of the Nature Conservancy for woodland conservation in the proposed Area of Outstanding Natural Beauty (AONB).

By early 1944, the British Ecological Society had reported on National Habitat Reserves and Scheduled Areas. The 19th Scheduled Area proposed was: 'Wye Valley Woods. These include woodlands of great ecological interest and scenic beauty. Large areas are controlled by the Forestry Commission, which has already reserved one area for ecological purposes' (PRO F18/291).

Two reserves were evidently proposed, for, on 5 June 1944, WLT (FC) comments to GFHS that Bristol University 'mis-state the case regarding the Highmeadow Woods. My Dept have not set aside this area as suggested, but only an area facing the Gorge, which has been demarcated in consultation with the School of Botany at Oxford'.

The offer of ecological reserves in state forests was repeated in 1946 (PRO HLG/93/24) as part of the deliberations of the Huxley Committee, which led eventually to Cmd 7122 (1947), the Nature Conservancy and National Nature Reserves. By then, Lady Park was already a reserve and the basic recording was already in progress, so the recommendations for recording as set out by the Huxley Committee in *Ecological Reserves in State Forests* (PRO HLG 93/48, Papers of Huxley Committee. WLC/Inf/23) had been anticipated. At the Linnean Society on 2 July 1943, Eustace Jones suggested that the two main kinds of ecological work that were desirable in these reserves were: (i) that devoted to the elaboration and study of ecological theories and concepts, about which very little was known about the nature of the climax in the North Temperate Zone; and (ii) the close study of environmental conditions and their effects on the individual, including here, autecological work. In detail, the Ecological Reserves Committee suggested:

> the basic data which should be obtained in all such areas. Whilst it is evident that the best use cannot be made of this opportunity without considerable extension beyond the minimum indicated, it is felt that the direction of elaboration must depend in large measure upon the particular interests of individuals and on the facilities at their command.
>
> The precise area chosen for a reserve should be one typical of the locality in respect to soil and preferably the maximum of change in a reasonable period of time. In general, such areas will be between 10 and 30 acres in extent and should where possible be chosen within existing deer- or rabbit-proof fencing. It may be emphasised that the only liability that FC undertake is the provision of the enclosure with any necessary fencing.

In order to incorporate a measure of uniformity, the standard record would be:

1. Ground map of vegetation; topographical description
2. Vascular plant list
3. Permanent enumeration quadrats in each community
4. Transects across vegetation zones
5. Soil profiles
6. Photographs from permanently marked positions
7. List of animals related to plant communities
8. Periodic re-survey to observe changes
9. Deposit of records in some Institution 'is essential to the permanent value of such reserves'.

The attitude of the FC was stated by Robinson at a meeting with the National Parks Committee and Wild Life Conservation Special Committee on 21 February 1946:

> There is no difficulty about setting aside certain areas of woodland where wild life conservation was the main aim, rather than timber production. The FC was becoming more and more interested not just in trees but in the whole ecology of woodland.
>
> (PRO HLG/93/24)

This must have sounded reassuring, but it was vague, and FC staff evidently saw this as just a limited concession. Thus, in an internal FC letter of 5 December 1947 to GFHS, WLT noted that:

> Our National Forest Parks [Dean had become a Forest Park before the war] are a distinct species. The raison d'etre of the FC is to grow timber ... but we are developing public access. We do not regard nature reserves in our NFPs [National Forest Plans] necessarily as exceptional. We give facilities for extensive reserves in the FD [Forest District] (Highmeadow) ...
>
> (PRO F18/291)

Very few of the ecological reserves proposed to the FC actually materialised. Lady Park and one of the native pinewoods in Scotland, the Black Wood of Rannoch, became 'ecological reserves'. The latter was nominated as a national nature reserve by the Ritchie Committee (Cmd 7814, 1949), and Lady Park was merely included within the proposed Forest of Dean and Wye Valley Conservation Area by the Huxley Committee (Cmd 7122, 1947). Lady Park was marked as an ecological reserve on OS maps and used for long-term research, but was not designated as an SSSI, possibly because the core area was divided between three Nature Conservancy regions. Nor did it become a Forest Nature Reserve like Denny in the New Forest and the Blackcliff and Wyndcliff lower down the Wye, and we can think of no reason for that.

With the establishment of the Nature Conservancy in 1949, nature conservation became the responsibility of a separate bureaucratic entity. Competition between bureaucracies ensured that the FC increasingly came to regard nature conservation as a constraint on timber production, and this generated a degree of distrust between the two interests. The Nature Conservancy quickly developed a series of NNRs whose purposes prominently included ecological research, but Lady Park remained a glaring anomaly, an ancient, semi-natural wood actually used for ecological research that attracted no designations. Eventually, on 24 May 1984, Lady Park Wood did indeed become an NNR, and by that time, the FC and the Nature Conservancy Council (NCC) were negotiating the accommodation that became the Broadleaves Policy of October 1985, backed by the Ancient Woodland Inventory.

From 1945 onwards, the reserve was limited to what is now the fenced (non-intervention) reserve, but in 1984 it was extended to the south-east to take in what is now the managed compartment, bringing the total area to 44ha within a Nature Reserve Agreement that runs to 2083 (Fig 3.1).

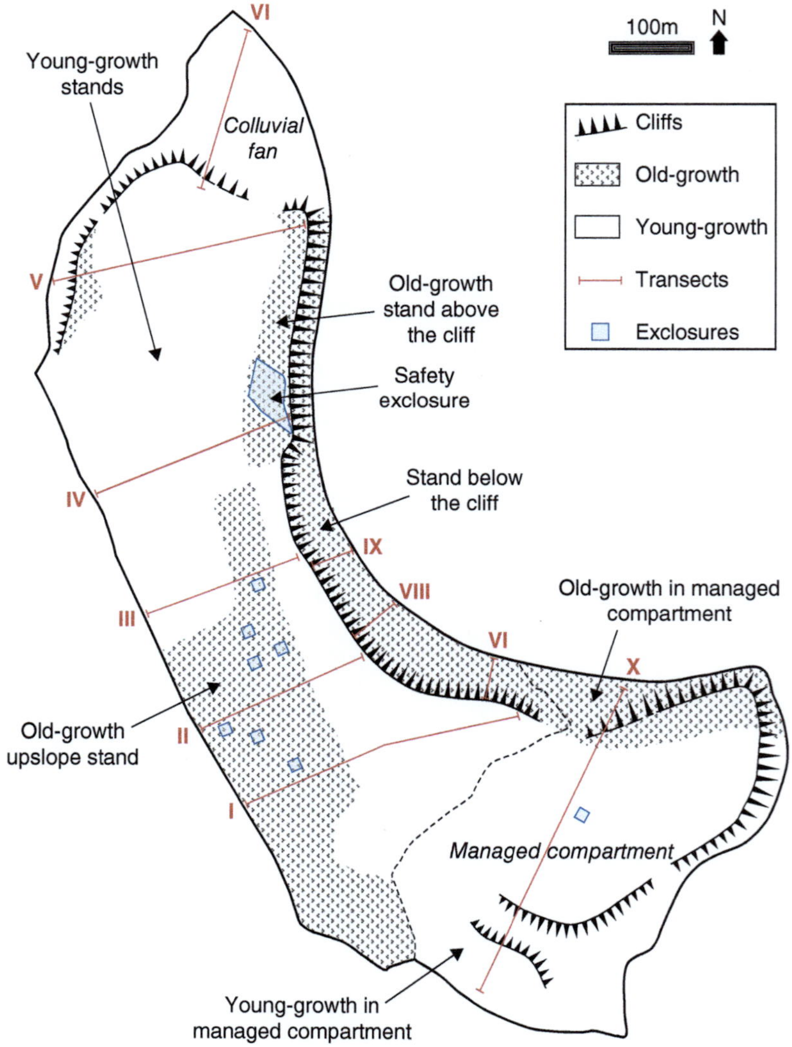

Fig. 3.2. Chart of Lady Park Wood Reserve and managed compartment. Woodland on unshaded ground was felled from 1942 to 1944, leaving a shelterwood of large trees; this now forms the 'young-growth' stands. Woodland on shaded ground was not felled (and now forms the 'old-growth' stands), but the 'stand below the cliff' was thinned. The sinuous Miners' Path defines the boundary between the managed compartment and the fenced minimum-intervention stands. Roman numerals label the ten transect lines used to record stand change.

3.2 The Managed Compartment

The expansion of the reserve to include a managed compartment was the outcome of an interesting and sometimes tense debate between NCC and FC. As the impact of the 1976 drought became ever more obvious, the FC became increasingly concerned that valuable beech and oak timber was going to waste, so they had it valued (£79,000) and then asked NCC whether the sacrifice of this timber was justified by the scientific output. NCC's response was to 'confess' that nothing had yet been analysed and published in the scientific literature and to point out that this was a long-term project, from which the research benefits would be apparent sometime in the future. This must have seemed lame to the FC, who promptly asked NCC to buy the timber or watch it being felled and extracted. NCC said in reply that this would destroy the purpose of the reserve and break the terms that FC had proposed and agreed 30 years earlier. Exchanges between directors followed, which prompted FC to ask their Regional Advisory Committee for an opinion. By the time we convened in the wood in 1983 and both FC and NCC put their case, GFP had re-recorded samples of the transects (Chapter 4) and could demonstrate how the wood was changing and this was evidently enough to convince John Workman and one or two other committee members that the project was worthwhile. They spoke strongly in favour of the original concept of the reserve and the rest of the committee agreed.

The background to this debate was that several FC officers were hostile to the existence of the reserve. According to documents on file, its condition had been a source of anxiety for years, due to loss of production and danger to the public. Being 'under negative management' it would 'take on the appearance of a devastated wood for many decades'. Nevertheless, FC accepted the RAC's advice and agreed to retain the reserve as an unmanaged stand for study, whilst NCC agreed to expedite initial analysis and publication. In addition, FC proposed that the adjacent compartment, which had also remained largely untouched silviculturally since 1945, should be added to the reserve as a managed comparison with the unmanaged original reserve. In essence, the FC would treat it as a mixed deciduous stand, retain the existing mixture of species, but carry out such thinning and coppicing operations as they saw fit, and extract the utilisable timber. NCC agreed to add a tenth transect to the nine in the original reserve and record the changes by the same methods (see chapter 4). Tony Joslin, Deputy-Surveyor of the Dean added spice to the agreement by offering GFP a wager that he would make 'his' reserve more diverse than NCC's, a kind offer that GFP declined, saying that NCC knew he would win in the floristic diversity sense he meant it, and in any case encouraging high species diversity was not the point of the unmanaged reserve. [In the event, the 'result' three decades later was that both sides lost.]

This debate and the resulting agreement came as the FC was reviewing its policy towards native woodland. Critical public response to the emphasis on timber production and return on investment, expressed most forcefully at the Broadleaves in Britain Conference at Loughborough in 1982, led eventually to a new Broadleaves Policy in October 1985 and a remarkable transition to multi-purpose forestry. Lady Park did not contribute to this broad and fundamental change, but the broader change contributed to the beneficial outcome in Lady Park.

3.3 Fencing

The reserve is a small part of the Highmeadow Woods and, like the rest of the woods, was unfenced when it was set aside. Fallow deer and people had unrestricted access to the reserve – indeed, neither would have known that the reserve was in any way exceptional – and, until the 1980s, unfettered access by humans and beast was not seen as a problem, even though deer were obviously influential in the regenerating stands. Few people walked through the reserve, partly because the cliff separated it from the main concentration of walkers beside the Wye.

In the 1980s, however, it became increasingly clear that deer were preventing natural regeneration, even in the considerable openings that developed after the 1976 drought. In response, in 1988, several small exclosures were erected in an effort to measure and demonstrate their impacts, and a much larger exclosure was added in 1990 round old ironstone workings, which were deemed to be a liability under the health and safety legislation. Further small exclosures were added in 1999.

Until 1993, the brambles had been abundant enough to oblige us to mimic John Cleese's 'ministry of silly walks' when walking off-path, but then they suddenly vanished. Actually, they were still widespread as bitten-down stubs. Fallow deer, it seemed had learned that the reserve was the only part of Highmeadow Woods where they were not 'controlled', i.e. shot. Shortly after, repeat surveys of some of the Williams plots confirmed that the ground vegetation had become markedly less diverse, and by then the small exclosures confirmed that the main culprits were the deer.

Accordingly, the no-cull policy was abandoned and deer were occasionally shot within the reserve, but there was little discernible effect on the vegetation, and certainly no recovery of bramble or growth of saplings. So, when the Wye Valley AONB office joined with others to develop an EU-funded project to protect *Tilio–Acerion* woodlands, a bid was included for funds to fence the whole reserve, and this was successful. The entire perimeter of the reserve, but not the managed compartment, was deer-fenced in 2007 (Fig. 3.1: see p. 26).

The rationale for the fencing had been considered carefully, for the complete absence of large herbivores would be as unnatural as the presence of excessive and uncontrolled deer populations. Briefly, red and roe deer were originally native in the Wye Valley, and their numbers must have been limited by wolves, whereas fallow deer were long-established introductions. Fallow were, however, a reasonable substitute for the original natives, for they were present in previous interglacials.

Against this background, three options were considered: (i) we could witness the progressive conversion of Lady Park to wood pasture under the influence of uncontrolled deer populations; (ii) foresters could control the populations at much lower levels than were currently present; or (iii) the wood could be fenced. The first was the easy option and would no doubt have been most acceptable to the general public. Moreover, there was a research case for recording the conversion of the reserve from closed woodland to wood pasture. The second might have maintained populations at natural levels (whatever they are) and would have yielded a product for sale, but it was limited by the resources available to the foresters and others, and was compromised by the diversity of ownerships and recreational activities within the range of the local deer herds.

The third option was taken because it was more likely in practice to keep grazing and browsing levels closer to natural than the other options. Deer are past masters at wriggling under flaws in the fence, and they would get in anyway when, as was likely, the fence was crushed by falling trees Although three deer leaps were installed just inside the fence to allow any deer inside to get out, we concluded that total deer exclusion was improbable. In any case, the risks from insufficient browsing for a couple of decades seemed less than those from indefinite excessive browsing.

In practice, very few deer had entered the reserve before 2014 and there was no sign that the deer leaps had been used. In summer 2014, the leaders of groups of small ash saplings were browsed in one of the new canopy gaps, and in January 2015, wild boar finally broke in. The fence has been repaired, but muntjac can wriggle under tiny gaps and feral pigs are strong and persistent. The reserve has thus been largely free of deer for a decade, but at the time of writing (2016), muntjac droppings have recently been found above and below the cliff, and a few wild pigs still get in.

The managed compartment has never been fenced, save for a small exclosure erected in 1999 to demonstrate the effect of deer. Ever since, fallow deer have been seen almost every time one walks off the main track, and in 2014 wild boar seemed to become especially numerous. By then, it was obvious that deer would make tree regeneration impossible at the population levels prevailing from the early 1990s onwards.

3.4 Recording and Research

Recording in Lady Park has been reasonably faithful to the recommendations of the Huxley Committee – hardly surprising, since EWJ presumably helped to develop them. Details of the baseline recording of trees and shrubs are given in Chapter 4, this volume, but this is a convenient point to discuss the links between recording and research.

Two general interest articles describe the wood, how it has changed and the successes and failures of attempting to study natural developments in a 'natural' reserve (Peterken and Mountford, 1995, 2005). In addition, sections in two books describe the wood and the findings (Peterken, 1996, 2008a). More formally, two early papers describe how the wood has developed (Peterken and Jones, 1987, 1989).

While a simple account of changes in the stand would probably not be published today, this baseline has formed the foundation for papers by the original recorders on more specific aspects of long-term change. The impact of the 1976 drought

Fig. 3.3. Liam Cavin taking a core from an ash.

on beech (Peterken and Mountford, 1996) and the response of wych elms to the arrival of disease in 1971 (Peterken and Mountford, 1998) seemed the most interesting of many potential accounts of the performance of individual tree species. The impact of grey squirrels on stand development through their habit of stripping bark from beech also seemed of wider interest (Mountford, 1997, 2006; Mountford and Peterken, 1999) (see Chapters 7 and 8, this volume). Mountford (2004) included experience from Lady Park in his assessment of the major processes and structural changes associated with natural stand development across north-west European temperate woodland reserves.

The increasingly 'natural' condition of the wood, combined with the existence of the long baseline record, has attracted ecologists from elsewhere. The first of these was Vanessa Williams from Cardiff University. In 1979, she marked and recorded plots throughout the wood that have become the basis for studying change in the ground vegetation. These eventually attracted ecologists from Belgium, who drew Lady Park into a network of long-term studies that spanned temperate Europe and eastern North America. In addition to a summary of changes in Lady Park (Vanhuyse *et al.*, 2012), these repeat records contributed to methodological innovations (Baeten *et al.*, 2012) and to Europe- and north temperate-wide meta-analyses of long-term change in woodland ground vegetation (Verheyen *et al.*, 2012; Bernhardt-Römermann *et al.*, 2015; see Chapter 14, this volume).

Increasing interest in 'coarse woody debris' attracted Patrick Green to measure dead-wood volumes and distribution in the managed woods of the lower Wye Valley and compare these with the

Fig. 3.4. Karen Hale and Richard Bradshaw studied carbon sequestration in Lady Park.

'natural' baseline in Lady Park (Green and Peterken, 1997). The estimates for Lady Park later contributed to a Europe-wide assessment of dead-wood volumes in beechwood forest reserves (Christensen *et al.*, 2005; see Chapter 12, this volume).

The existence of the record of growth attracted Liam Cavin from Sterling University to include Lady Park in a dendrochronological study of the influence of climate on the growth of beech. This highlighted the relationship between beech and oak and how this had been changed by the 1976 drought (Cavin *et al.*, 2013) (see Chapter 7, this volume). The cores taken in Lady Park also contributed to an analysis of drought impacts on 46 populations throughout the European range of beech (Cavin and Jump, 2017; Fig. 3.3). This revealed, surprisingly, that over the past 40 years or so, the growth of beech has been more sensitive to droughts in its core temperate range (and especially southern Britain) than elsewhere. Further north, drought had relatively little impact. Further east, in continental temperate populations, drought did restrict growth, but beeches were better at recovering their pre-drought growth rates. Further south, at the southern edges of its range, beech showed a degree of drought resistance, though beech failed to recover if drought suppressed growth severely. There was a general trend to slower growth in the south, but increased growth in the core and northern populations, which is consistent with climate change predictions.

The long-term record has also formed a foundation for studies of carbon sequestration by Richard Bradshaw and Karen Hale (Fig. 3.4) of Liverpool

University (Hale, 2015; Allen *et al.*, 2016). Between 1945 and 2010, the carbon store in the old-growth increased from 8.92 kgC/m^2 to 17.50, of which 60% was in the tree mass, 38% in the soil and only 2% in dead wood. This and equivalent evaluations in the young-growth were used to check a model of carbon sequestration in forests, which indicated that unmanaged semi-natural woodland in Europe could be a major potential carbon store.

The existence of near-natural, managed semi-natural and plantation stands together is currently allowing Andrew Carr of Bristol University to study how structure influences the feeding habits of bats (see Chapter 13, this volume).

4 Recording Trees and Expressing Change

Since 1944, recording has developed as a relay. Foresters from Oxford University initiated recording and kept it going until 1960. The baton was passed to the Nature Conservancy between 1971 and 1987, by which time others had been involved intermittently. Since 1992, recording has been maintained by ecologists working independently, but with the cooperation of the successor bodies of the Nature Conservancy and the Coleford office of the Forestry Commission (FC), now Forest Enterprise (FE).

4.1 Oxford University

The project was started by Dr Eustace Jones (Fig. 4.1: see p. 36), a lecturer in the forestry department of Oxford University. A keen bryologist, he had searched out mosses and liverworts in the Wye Valley and had organised forestry field courses in Highmeadow Woods during the 1930s. There, and in the New Forest, Hampshire, he had taught survey techniques to forestry students based on long transects through representative stands. Several crossed Highmeadow (the records survive with GFP), including one running from the Bellman's Oak to the Whippington Brook, which was close to the later transect I in the new reserve. EWJ referred to these transects as 'lines' in the early records, but he quickly dropped this for 'transects', symbolically changing from forestry to ecological nomenclature.

In 1944, with student help, EWJ laid out six transects in Lady Park running down the slope at approximately even distances from transect I in the south to transect V towards the northern end (see Fig. 3.2). Transect VI could not fit into this pattern, but instead ran across the triangle of colluvium from the subdued extension of the cliff to the Wye flood plain. Transects VII–IX were added in 1950 below the cliff, running perpendicularly up the slope from the bottom edge to the cliff at regular intervals, from VII opposite the Biblins Bridge to IX close to the main cliff; that is, they effectively extended transects I, II and III above the cliff. Thus, I–V are 'above the cliff' and VI–IX are 'below the cliff', but VI has always been recorded with I–V.

These transects were marked on the ground by oak stakes at 100-foot intervals. They were numbered by sections (e.g. '100–200') starting from the top (I–VI) or the bottom (VII–IX). Their width was taken to be one chain, but the margins were never defined. The stakes proved to be durable – indeed, some survived in 2017, 73 years since they were first driven in – but some fell in the first decade and had to be re-erected exactly where they were before. The stakes, incidentally, were all carried into the reserve from the nearest public road on the backs of Dr Jones and his students. They recorded transects I–VI in 1945 and 1955 (but could not finish one until 1958) and VII–IX in 1950 and 1959.

These initial records (Fig. 4.2) were made by EWJ or recorded by the students under his supervision, and no doubt under his direct gaze. The results were recorded on foolscap sheets in pencil, and some were duplicated in ink and even typed in Oxford. They show, as one would expect, signs of tiredness, short cuts and quixotic layouts brought about by finding more individuals to record than could be accommodated in the space allowed on the page. There is unmistakable evidence of post-war paper shortage – not quite 'turning the page upside down and writing on the underside of the lines', but as notes squeezed into odd corners and 1930s lecture notes preserved on the reverse sides.

The actual record takes the form of:

1. Systematic records of individual trees, seedlings and ground vegetation in the transects.
2. A few supplementary investigations, such as ring counts of felled trees.

Fig. 4.1. Dr Eustace Jones examining the oak stake that he used to mark the transects in 1944 during his last visit to the wood 40 years later. The stake was still standing in 2016.

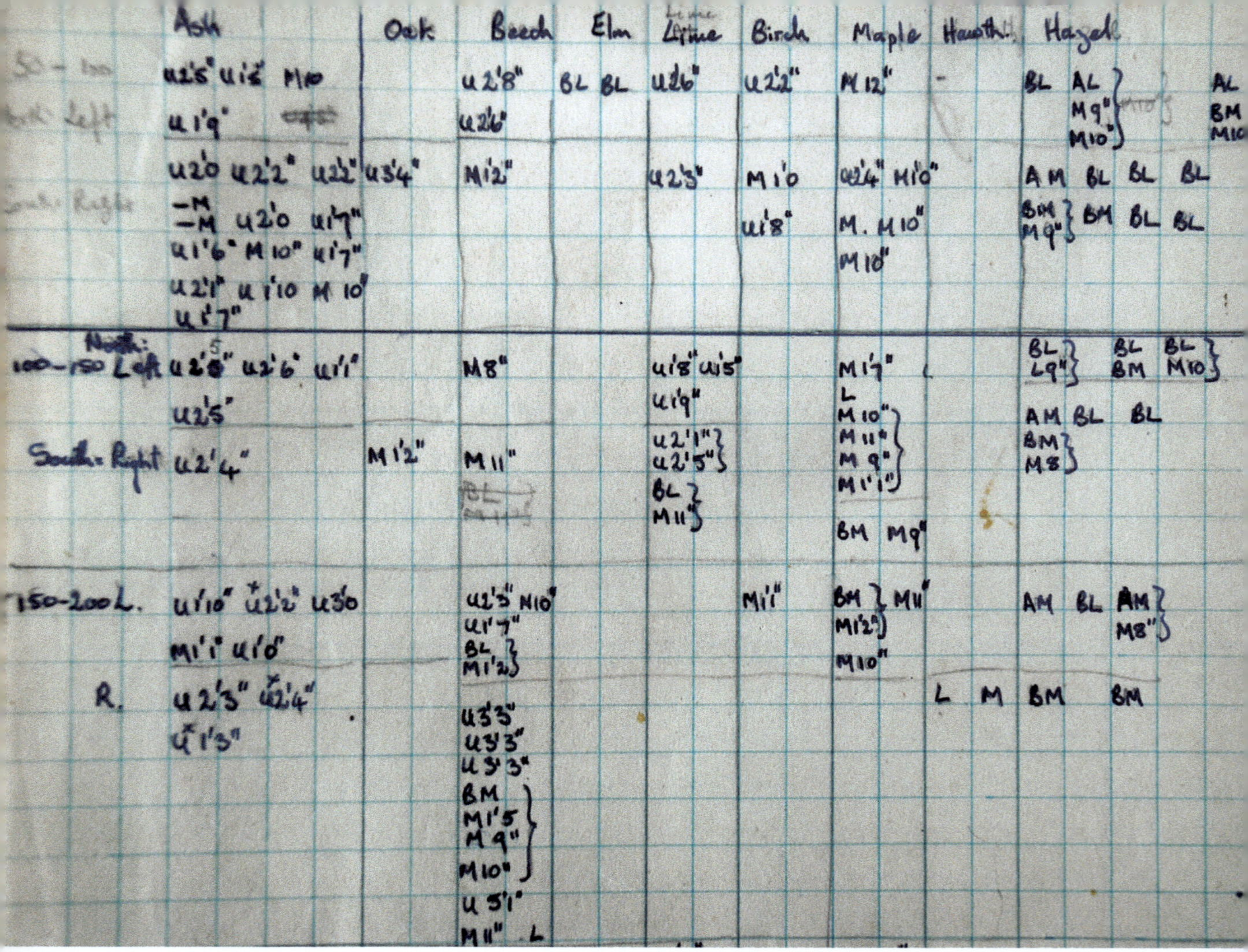

Fig. 4.2. Part of the 1945 record of the old-growth stands. No maps were made. Rather, for each quarter section, we have a list of the trees according to their canopy position and girth at breast height in feet and inches. Multi-stemmed trees are indicated with brackets. Small trees are indicated merely by canopy position. Multi-stemmed small trees are indicated by 'A' (more than 10 stems) or 'B' (2–10 stems).

3. Field notes, mostly relating to the transects, but some general material about the wood.
4. Summaries and preliminary analyses.

In the old-growth and the stand below the cliff, the systematic tree record consists of measurements of trees within quarter sections. Thus, for example, individual trees in II: 300–350 north side are listed and measured, but no chart shows their exact location within this quarter section. For each tree, we know the species, GBH in feet and inches and its position in the 'upper', 'middle' or 'lower' stratum of the stand. EWJ also recorded which trees grew from one rootstock, i.e. were of coppice origin (though he did not record which single trees originated as coppice). The smallest recorded was 8 inches girth (6.5 cm diameter at breast height (DBH)). When individuals (often hazels) possessed several small stems, they recorded whether there were 10 small stems, or more.

The young-growth was recorded quite differently. The large retained trees were sometimes measured, but the main record comprised a systematic survey of tree seedlings and saplings and ground flora species, recorded on a DAFOR (Dominant, Abundant, Frequent, Occasional, Rare) scale for each transect section, half section or sometimes quarter section. Their height range was also recorded in feet. In 1948, EWJ also mapped two quarter sections in detail (I: 600–650N, V: 800–850N, see Section 6.1 in Chapter 6, this volume).

Writing in 1960, EWJ left a clear picture of the realities of recording on such a large scale (LPW C11/1–8):

- Small differences in judgement were inevitable between recorders in deciding the edge of the transects and what constituted a single coppice stool of hazel or lime.
- Recorder error. EWJ evidently considered some students to be less careful than others. He found differences between 1945 and 1955/58 (e.g. in the ash and lime in III) that could only be explained by mistaken identity. The apparent loss and gain of substantial individual trees also seemed to be due to omissions on one occasion.
- It was just not possible to record all the birch and ash saplings, so the recorders economised by recording only those in the canopy. However, decisions on which individuals to count were somewhat inconsistent between recorders and between dates.
- In the earliest records, made in September 1944, particular attention was paid to seedlings of forest trees, but other vascular plants may have been recorded hastily.

Reviewing the records with the advantage of hindsight, we are impressed with the amount of work completed against the odds. For its time, the record was substantial, but it is never clear what Dr Jones hoped it would reveal. He was particularly careful to record each tree's position in the stratification, which led GFP to believe that he was not just interested in how natural woodland worked but also wanted to explore the parallels between natural woodland and silvicultural systems. Certainly, that was a theme of his classic 1945 paper.

4.2 Nature Conservancy and Successor Bodies

EWJ's last records were made in 1960. He started an analysis, but was overtaken by an increasing workload of teaching and administration, combined with what proved to be an unfinished attempt to write a forest ecology textbook (GFP has the surviving manuscript). In 1971, after a decade of inactivity, staff of the south-west region of the Nature Conservancy measured all the trees in the wood – literally! The scribe marks are still visible and have been used as the heights at which girths have since been measured.

In 1976, Lady Park was subjected to the great drought. Early in 1977, John Thompson, then regional officer for the Midlands region of the Nature Conservancy Council (NCC), realised that big changes were afoot, but that EWJ would not be recording them. In fact, by then, EWJ had given most of the records to Dick Steele of the NCC. Accordingly, he (John Thompson) used some regional funds to commission Alan Orange to resurvey the transects (Fig. 4.3).

Alan knew Mike D'Oyly, the NCC's local Assistant Regional Officer, but his reliability as a surveyor was not known when he disappeared into the wood. However, at the end of the summer he emerged with outstandingly clear and accurate charts of the trees and shrubs on transects I–VI (Fig. 4.4), as well as other observations, including lists of plants. Moreover, he marked the 100-foot points on each transect with sturdy aluminium angle irons, leaving the residue of EWJ's oaken posts in place. Alan's charts have been the basis for field recording ever since, and his stakes have proved to be durable, visible and vandal proof. In effect, Alan saved the entire long-term study.

By the early 1980s, the NCC came under pressure to allow the FC to harvest the drought-damaged trees. The NCC resisted this on the grounds that in 1944 the FC had undertaken to leave the wood to grow naturally in the interests of ecological research, and the issue was referred to the FC's Regional Advisory Committee (RAC). In the event, the RAC backed the original concept of the reserve, but with strings attached. As a comparison with the original reserve, the adjacent compartment would be managed by the FC as productive native woodland based on natural regeneration, while the NCC would record the changes.

The NCC's case was prepared by GFP, then the senior woodland ecologist on the Chief Scientist's team. In preparation, he resurveyed sample sections using Alan Orange's charts, and this proved to be transformative. Having surveyed so many woods at a superficial level, it was a revelation to stand in the wood, hold up the record and see how individual trees were changing over decades. Subsequently, with the help of Susan Peterken and Jonathan Spencer (Fig. 4.5), he re-recorded transects I–VI when opportunity allowed from 1983 to 1987; and, in 1984, Tim Barfield set up and charted transect X through what was now the managed compartment. Tim's accurate 'copperplate' charts were drawn so carefully that they would remain unsurpassed as a

Fig. 4.3. Alan Orange girthing an ash in 2013.

baseline for recording change. In 1985, GFP fully recorded six representative sections of the young-growth transects – there was just no time to do more – and in 1991, he recorded the state of trees in a few old-growth sections of I and III.

Until late in this burst of activity, no attempt had been made to resurvey transects VII–IX below the cliff. Indeed, we thought they had been lost. In 1984, GFP managed to persuade EWJ to return to the wood, partly in the hope of finding the 'lost' transects, but, after climbing carefully along under the cliff, EWJ was still unable to find the marker he had placed at the top of VII. However, on the way to a half-term Youth Hostelling week with his family in the Black Mountains in early 1985, GFP stopped off at the wood, passed the last known directions to the transect lines to his sons and left them to search. With Andrew directing from below, Michael climbed the wood and eventually tripped over a stake marking VII: 200, and at that point we realised that Alan Orange had found and marked the 'missing' transects, but had not had time to make a new record. We even found on a cliff ledge the small stone with which EWJ had marked the top of VII (still there in 2013). Later that summer, GFP returned with several members of the team compiling the Ancient Woodland Inventory to chart and measure transects VII–IX.

The record from this period consists of: a complete record of the old- and young-growth transects in 1983, but only at the level of 'alive' or 'dead'; remeasurements of trees in selected sections in 1986; a complete survey of transects VII–IX; and various associated field notes, including plots of canopy gaps within each transect. The immediate objective was to prepare two papers (Peterken and Jones, 1987,

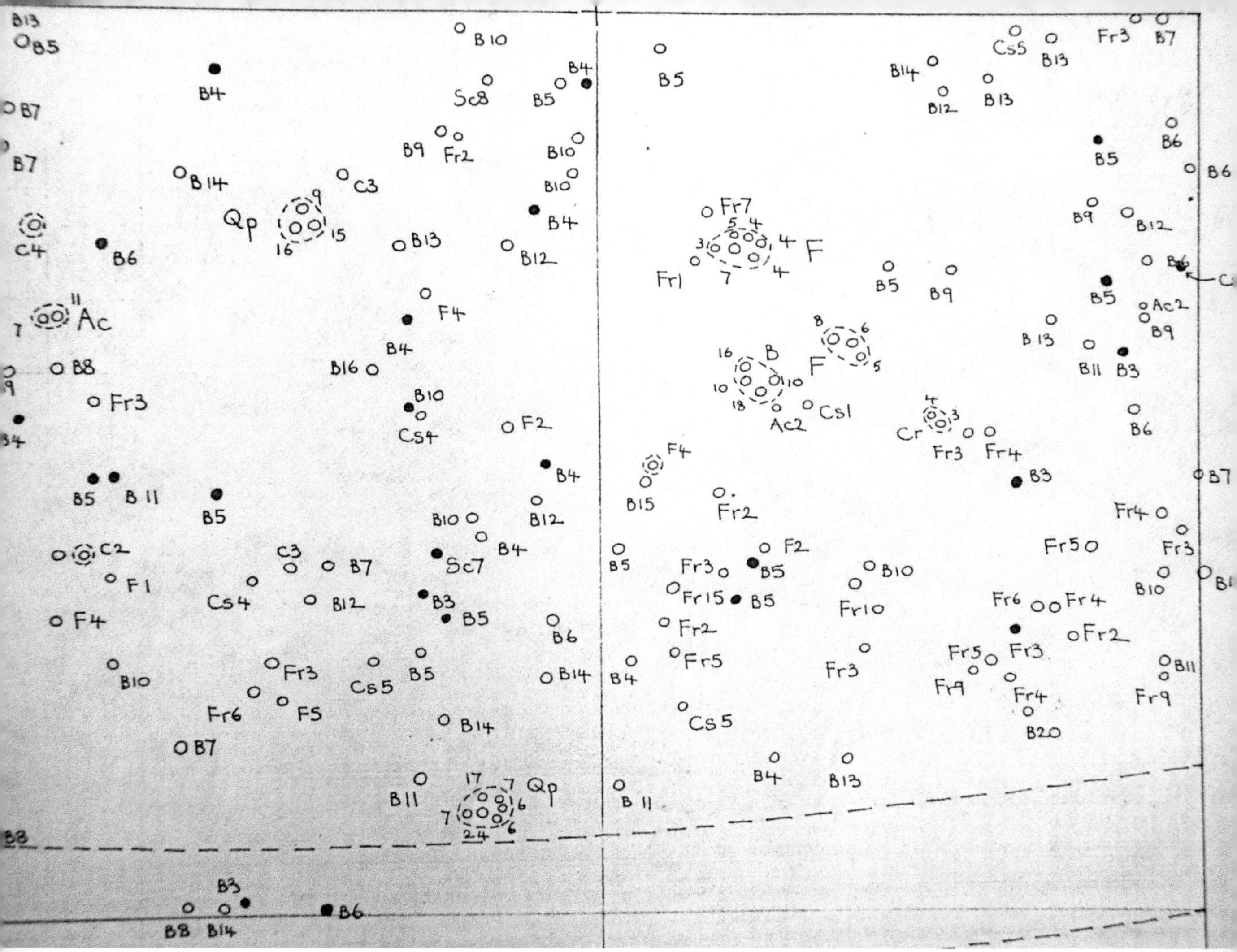

Fig. 4.4. A 50-foot-long (*c.*15 m) sample of the 1977 record for the young-growth stands. The centre line runs from top to bottom. Each tree is plotted precisely and labelled according to its species (e.g. F = beech) and diameter at breast height in centimetres. Coppice stools are delimited by pecked lines. Trees that were dead in 1977 have a filled circle. This half section happened to be crossed by an old forest track.

1989) that tracked the changes since records began, but the longer aim was simply to continue the record through what was obviously a period of change.

4.3 Recording since 1992

Almost all the transect recording from 1992 to 2005 was undertaken by Ed Mountford (EPM). Initially, this formed part of his diploma and degree studies in field biology and conservation management, but later he returned while on attachment to the Forestry Institute at Oxford University as part of an EU-wide project examining 'nature-based management of beech forests'. The final leg of his recording was undertaken on a voluntary basis while between jobs. Again, the fieldwork was undertaken with the help of others. Key among these were Anne Mountford and Graham Berry; other helpers were Kate Wollen and Barry Embling from the Forestry Commission, and Helen Lancaster from English Nature.

The records and field notes made in this period represent the most detailed and accurate yet. All transects (I–IX) in the minimum-intervention area were recorded twice (between 1992/93 and 2000–2005), whereas transect X in the managed compartment was recorded once (in 1997). This involved measuring the girth and assessing the status (alive, dead, fallen) and crown position and crown size of stems on all trees and shrubs. More attention

Fig. 4.5. The recording team of Jonathan Spencer and Susan Peterken measuring a small-leaved lime on 6 March 1984 (a). By 15 March 2016 (b), 32 years later, the lime had grown from 97 cm GBH to 140 cm, developed substantial low branches from the trunk and fractionally increased its angle of lean. The ivy died between 1986 and 1992. The recorders, of course, have hardly changed, though Jonathan has bought a new anorak and boots.

was given to the recording of smaller stems than previously, and detailed diagrams were prepared of complex multi-stemmed stools to aid relocation (Fig. 4.6). Time was also spent on assessing the condition of mature trees, scoring debarking damage caused by grey squirrels, identifying why individual trees appeared to have died or declined, and mapping canopy gaps within the transect bounds. Detailed spreadsheets were created of all the transect records, within which 'errors' in the previous records were corrected.

In 2009, Liam Cavin started research at Stirling University on the effects of climate on the growth of beech. One of his research sites was Lady Park, to which he was attracted by the long-term record containing, inter alia, details of the impact of the 1976 drought on beech. His research involved sampling growth by means of cores taken from a large number of beech and, for comparison, oak, but in 2010 he also completed a resurvey of the old-growth stands of transects I–V.

Shortly afterwards, ash disease spread to eastern England, and we realised that various people and organisations had introduced the disease to other parts of Britain in ash transplants imported from infested parts of mainland Europe. The threat was clear: a major tree in Lady Park could be killed off in much the same way as DED had killed most

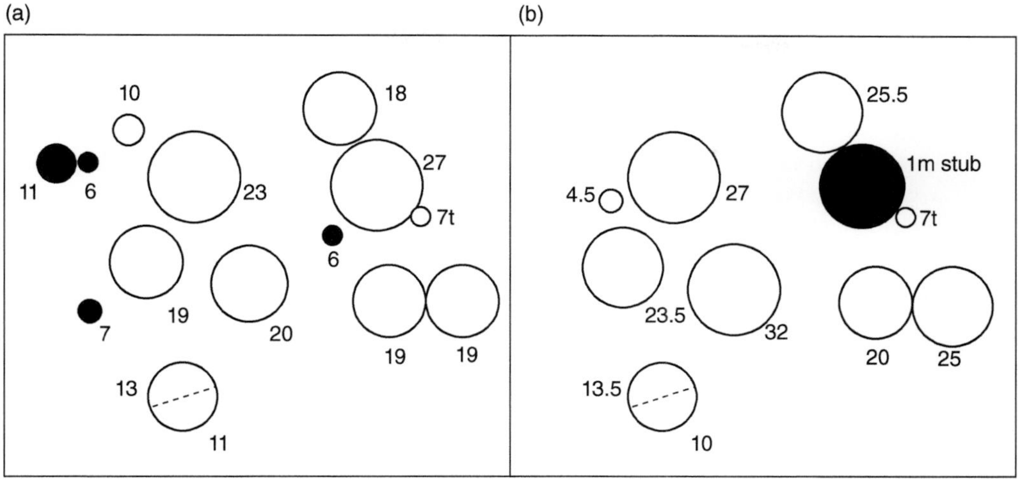

Fig. 4.6. An example of a pair of diagrams drawn to track changes in complex multi-stemmed individuals. This shows the position of stems (at ground level) on a hazel stool on transect VII. (a) = state in May 1992; (b) = state in March 2005; numbers = stem GBH cm; white circles = live stems; black circles = dead stems; dashed line = forked stem; t = trunk shoot. Redrawn from the original records made by EPM.

wych elms after 1971. Having long since regretted the failure to record the transects shortly before the arrival of DED, and remembering that a decade had elapsed since the last recording of the young-growth transects, GFP started a new recording in February 2013, this time as an unpaid volunteer. The young-growth parts of IV and VI were completed ahead of the growing season, together with the six sections in other transects recorded in 1987 and the whole of the old-growth transects. Later, in November, GFP completed transects VII–VIII with Ann Harris of Natural England and IX with Andrew Peterken. In the winter of 2014/15, GFP, with help from Adrian Wood, recorded transect X and the trees in several of the Williams plots in the managed compartment. Finally, the remaining portions of young-growth on transects I–III and the top of V were recorded by GFP in the autumn of 2015.

All these resurveys recorded the girth of living trees and snags, combined with field notes of condition. In the interests of completing the record, only the largest hazel stem was measured on each individual. (It is a daunting experience for a surveyor in a hurry to come to a large hazel stool with 20 or more living stems!) EPM had previously sketched the stem configuration on all multi-stemmed stools, recorded snag heights and plotted canopy gaps. GFP recorded the crown size and condition of all ash in 2013 and recorded snag heights.

4.4 Other Records of Trees

Several other records have supplemented the main records from the transects:

1. Photographs and field notes of tree condition and events affecting trees in the reserve as a whole. In particular, GFP has kept a diary of observations from each visit.
2. Complete enumeration of trees and shrubs (1971). This exercise was the brainchild of James Archibald, an Assistant Regional Officer with a forestry background, who believed that a total enumeration would help with reserve management. The wood was divided into 'compartments', but, crucially, the measurements were not related to the transects and could not be used as a contribution to the main record. The 1971 measurements do, however, record the state of the reserve just before elm disease took hold. In particular, they show the size distribution of each species in each portion of the reserve.
3. Alan Orange photographed the transects in 1977 from fixed points looking up and down the transects from each 100-foot marker. The prints show the general structure within and beyond the transects.
4. In 1979, Vanessa Williams, a PhD student at Cardiff recorded 72 200-m^2 plots distributed more or less evenly throughout the wood, mapping the location of all trees and shrubs and recording their

species and girth. A few overlapped the transects. In 1997, Alistair Monument recorded 24, again as part of an MSc degree in forestry at Oxford University. In 2009, many were recorded by Pieter Vangansbek of Ghent University (Belgium) for his Masters project. The 12 plots in the managed compartment, which were recorded again in 1984, 1994 and 2014/15 by GFP, provided the only estimates available of the trees removed by the first thinning.

5. In 1986, a group of four students from the silviculture department of Wageningen Agricultural University (the Netherlands) studied stand development by constructing carefully measured diagrams of their vertical and horizontal structure of representative samples of the old-growth the young-growth and the stand below the cliff. Their profile diagrams and plans of canopy extent remain the only pictorial representations of the stand profiles available.

6. In 1997, Patrick Green used the random short transect method (Green and Peterken, 1997) to estimate dead wood volumes in and around Lady Park and compared them with other stands in the lower Wye Valley.

7. In 2001, EPM undertook a study of canopy gaps across the minimum-intervention area, which included a detailed examination of tree regeneration and ground vegetation cover in four gaps in the old-growth stands. The results clearly illustrated that deer had eaten back much of the bramble and browsed most tree and shrub seedlings.

4.5 Review of Recording

Under the management plan, the NCC and successor bodies took responsibility for recording, but in the event, the role of individuals has been paramount. True, NC/NCC officers used small research budgets to underwrite recording in 1977 and establish transect X in 1984, but otherwise recording has depended on volunteer labour and short-term research and training opportunities. EPM and Liam Cavin both chose to work in Lady Park as part of a wider research project. Recording by GFP from 1983 to 1989 formed part of his job as a woodland specialist, but only as a relief from the responsibilities listed in his job description.

Recording has not only been opportunistic, but it has been achieved on a shoestring. Costs have been limited to pens, tapes, markers, photographic materials and the necessary travel and subsistence. The time of those involved has been paid mostly by employers and grants. Recording by EWJ, Alan Orange and GFP and associates was always constrained by lack of time. Much of the analysis by EPM and GFP has been completed in their 'spare' time. No institution has regarded recording as an obligation, though many individuals in institutions have seen it as desirable.

EWJ originally favoured re-recording at 10-year intervals (letter to 'Williamson', 22 June 1953, on FE files), but in practice recording has been irregular. Looking back, 10-year intervals starting in 1945 would have been adequate, and it would have provided a record just before the 1976 drought, but the last record before the arrival of elm disease would still have been 6 years before the event and there would be only one recording (1985) in the period of rapid change from 1977 to the early 1990s. Ideally, we would have made a record just before each major event. Recognising this, we anticipated the arrival of ash disease by recording in 2013–2015.

4.6 Measuring and Recording Trees

In good forestry tradition, individual trees have been measured at 'breast height' (1.3 m) by standing above the tree (when it grows on a slope) and stretching one's arms out horizontally. None of the recorders has been a dwarf or a giant, as far as we know, so this is a reasonably standard height. The 1971 scribe marks are still clear on the larger trees and could be used as a guide, except those which are obviously too high. EPM and Tim Barfield in fact checked the height of each measurement.

The actual measurement recorded was either a girth (with a tape) or a diameter (with a calliper). The former seems more reliable, for many trees are not circular in cross section, and it averages radial growth. These measurements are interconvertible if one assumes that trees do have a circular cross section, so tree size can be expressed as diameter, radius, girth and/or basal area. The early measurements were in feet and inches; measurements from 1977 onwards have been in centimetres. Small stems are not measured below a certain girth/diameter, though the minimum size threshold for measurement has changed, ranging from 6.5 cm diameter initially to 5 cm in 1977 and 3–15 cm girth from 1992.

Girthing with a tape is reasonably accurate and reproducible. Indeed, when we have remeasured a tree inadvertently, we have recorded much the same value each time. Even so, some variation in the positioning of the tape is inevitable and there are difficulties that affect a small proportion of measurements:

1. If breast height coincided with uneven bark, a gall, scar or other distortion, the measurement was taken as close as possible.
2. Trees growing out of reach on cliffs have sometimes been estimated. Such estimates are more accurate than one might imagine. In fact, when recording becomes tedious, we have often challenged each other to guess the girth before applying the tape, and it is remarkable how accurate one becomes after several hours of girthing. In any case, the estimated records have not been included in the analysis.
3. Fallen trees can be impossible to tape when they are pressed to the soil or hidden under dense brash and brambles.
4. Large ivy lianas usually allow one to feed a tape behind them, but a few cling tight. In these instances, we measured the diameter of the ivy and added it to the accessible girth of the tree.
5. Forked trees: forks well below 1.3 m result in two trees being recorded. Forks well above are treated as one tree. When a tree forks at 1.3 m, we have measured below the fork.
6. Personal bias: when a tree measures half way between two sizes, we approximate upwards. However, there are faint signs of bias. For example, Alan Orange must have had a mild predilection for even numbers in 1977!

We have also made supplementary measurements and observations on:

- condition or status (dead; alive; fallen but alive; alive, but only below 1.3 m)
- snags (height; girth if not decayed; decay state; per cent of bark cover)
- origin, especially which stems originate from the same coppice stool
- general descriptions of tree form, crown condition, lean, dieback, damage, causes of decline/death, etc.
- climbers, if they reach into the canopy
- conspicuous features, such as scars, dead branches, etc.
- degree of debarking by squirrels on a 5-point scale
- canopy position (canopy; subcanopy; underwood)
- crown size (very large, large, medium, small, very small)
- canopy spread of largest trees
- fallen trees, direction of fall.

The most detailed descriptions date only from 1983 onwards.

Most of the measurements relate to trees within the transects, but a few trees just outside have been recorded, notably uncommon species, large trees whose crowns influence trees within the transect and trees that are very close indeed.

A few trees were omitted at particular recordings, due to lapses in concentration, and in these cases we have estimated the girths. Such trees were omitted from growth calculations, but included in calculations of population change.

We have also made several attempts to chart canopy gaps; that is, places where there is an opening in the overstorey that is not filled by lower trees and shrubs. This was easy enough in the old-growth in 1983 after the drought had opened sharply defined gaps in what had been a closed canopy with little underwood. In this case, gap boundaries were sketched on to the transect charts by estimating crown projection from charted trees. Attempts have been made to repeat this since, but with increasing difficulty. As gaps are filled by crown expansion, their boundaries become more diffuse and the question of what is a gap becomes harder to resolve, particularly in the stand below the cliff, where trees lean. Regrowth in the underwood increasingly occupies and shades the ground below the gap.

4.7 Quantifying and Expressing Growth of Individual Trees

The growth of individual trees can be expressed as an increase in any of the measures and their direct derivatives:

- Girth: this is what has been measured at several recordings. Increases in girth involve no assumptions about tree shape.
- Diameter and radius: diameter has occasionally been measured, but latterly has been derived from girth on the assumption that the cross section is circular.
- Basal area: that is, the area of the cross section of the trees at 1.3 m, which gives a rough index of the bulk of the tree.

Tree rings are an easily appreciated measure of change in a tree's circumstances (Fig. 4.7). Narrow in small saplings, they usually increase quickly, then decrease slowly as the tree ages. Coppice shoots, being rocket propelled by established root systems, often have wide rings from the outset.

Fig. 4.7. Annual growth rings on this small-leaved lime, cut close to the reserve margin, vary from year to year and from radius to radius as the tree reacts to climate fluctuations and changes in physical circumstances. By calculating radial growth rates from girth measurements, this variation is integrated.

Within this general relationship, ring width varies annually (hence dendrochronology) and often reveals growth spurts and periods of poor growth lasting several years. Average radial growth rates can be calculated for each recording period, changes in which may reveal changes in the vigour of the tree, stand or population. A vigorous Douglas fir growing near Lady Park had grown to a radius of 23 cm in 42 years (radial growth rate of 5.48 mm/year), showing what was possible, but the vast majority of trees in Lady Park grew very much slower.

The other measure of growth rate, basal area increment, shows different features. If annual ring widths are constant, growth of the basal area will be accelerating. Conversely, if crown size, and thus wood production, remains constant, ring widths will decrease slowly because the new wood is distributed over a larger circumference. In practice, young trees with expanding crowns generally have wide rings and an accelerating rate of basal area growth. Later, as they sit in a mature canopy, ring width slowly decreases, while basal area growth is maintained until the tree loses vigour and both ring width and basal area growth decline. As a measure, radial growth rate has the advantage that it is a surrogate for ring width, which can be observed easily in cut trees and cores, but it gives a somewhat premature indication of declining vigour. A downturn in basal area increment indicates real decline. Both measures have been given in the chapters on the forest dominants.

4.8 Characterising Populations and Stands

Measurement is straightforward enough, but the problems really start when we consider the stand. How can one characterise a population of trees and express differences between stands and changes over time?

Stand characteristics can be measured as tree density (N/ha) and basal area (m^2/ha, the ground area occupied by the population), and differences between stands and changes in a stand can be expressed in these terms (Chapter 5, this volume). Populations can be characterised by the average (arithmetical mean), median (size of the individual in the middle of the size range) and range (smallest and largest) of girth and/or diameter of constituent stems.

Each measure tells us something, but each by itself can be misleading. Range, for example, shows potential, but depends on extreme individuals that may be very much larger (or smaller) than the bulk of the population. The average is dominated by the usually greater number of small individuals and tells us nothing about the larger individuals. Median may be unduly influenced by a gap in the middle of the size distribution, which can distort summaries of small populations. Basal area is a surrogate for volume, but not a direct measure. Density is dependent on how one records the smallest size classes (see below). Size distribution gives some indication of the generation structure. Thus, a continuously regenerating population will show most individuals in the smallest size classes and a negatively exponential decline in density as size increases, whereas an even-aged population will initially display a bell-shaped curve that becomes more skewed with time as a few individuals dominate and the smallest individuals die. However, it does not give a single figure for later comparisons. Using them all together is cumbersome and tedious.

4.9 Expressing Changes within Populations and Differences between Stands

In Lady Park, trees have been measured since 1945; changes in the woodland have been quantifiable since 1955; and changes in the rates of change have been detectable since 1977. Expressing change should be straightforward, but in fact gives rise to several difficulties and judgements.

Many species coppice or otherwise generate two or more stems from one genetic individual, so populations can be analysed at two levels, stems and individuals. However, defining individuals is not always straightforward, especially when old stools become fragmented and clones have been generated by underground root systems (aspen) and by layering (limes). We have generally quantified in terms of stems, supplemented by some analyses in terms of individuals.

The Achilles heel of the Lady Park records is inconsistent treatment of small sizes. Generally, stems that reach, or have reached, 1.3 m have been recorded. From 1945 to 1960, stems below 20.3 cm GBH (8 inches girth) were not measured but simply counted in two classes, 2–10 stems and more than 10 stems. At later dates, the smallest sizes recorded consistently were 5 cm DBH (1977), 3–5 cm GBH (1992–2005 old-growth and below the cliff), 15 cm GBH (1993, 2002 young-growth) and 5 cm GBH (2010, 2013/15). In 1983 and 1986, all stems were recorded as dead or alive and described, but not all were measured. Many smaller stems were measured in 1977, all thinner stems were counted in 2013/15 and there were few very small stems from 1992 to 2005.

Given this, all possible expressions of change have strengths and weaknesses:

1. Analysis can be confined to individuals. This allows the 1945 and 1955 populations to be included on the same basis as the later records, but is subject to uncertainties in delimiting individuals and will disguise substantial thinning of stems on stools.
2. Confined to stems of 6.5 cm DBH or greater, the minimum size invariably recorded. This, the 'convoy approach', is unambiguous but discards information about the possible substantial mortality and recruitment of small stems.

3. Accept all stems recorded, counting the multi-stemmed individuals of 1945 and 1955 as single stems. This makes use of all available information, but seriously underestimates the populations in the early years.
4. Accept all stems recorded, but adding an estimate of unrecorded stems reaching 1.3 m; for example, counting all multi-stemmed individuals of 1945 and 1955 as, say, 6 stems (if recorded as 2–10 stems) or 15 stems (if recorded as more than 10 stems).

In practice, we have mainly used the third option, counting the multi-stemmed individuals of 1945–1955 as single stems. The result in comparison with other options is illustrated in Table 4.1, which compares annual mortality rates calculated by the second and third options and estimates how much the value under the third option differs from the result that would have been obtained if all stems reaching 1.3 m had been enumerated on all recording dates. Note, incidentally, that competitive thinning normally ensured that the mortality rate for small stems was higher than for large stems, but this was reversed in 1977–1983.

Both mortality and recruitment have been quantified by taking the 1.3 m height as the qualifying threshold. Any stem that grows enough to be recorded above this height counts as a recruit, and any that remains alive only below this height is deemed to be dead. Over any period, the number at the start, less the mortality (losses), plus the recruitment (gains) gives the number at the end. While this gives a standard expression to population change, account must be taken of further ambiguities and anomalies:

- If a tree or stem is recruited and dies between recordings, this will not register in the statistics. The calculated rates are therefore minima. This problem increases with recording interval.
- Losses are generally steady and can take place at any season. It would thus be marginally better to quantify mortality in terms of the years between recordings, not growing seasons, but for clarity we have kept to seasons.
- Trees that fall or die back to below 1.3 m are counted as mortality, but they are not dead and in some cases actually regenerate new stems. They are 'Draculas', the undead that may resurrect themselves. If they do, they count as recruits, yet they are the same genetic individuals that were there at the start. In practice, the numbers are too small to affect the general picture, but they are mentioned in some species accounts. The same general considerations apply also to the few new individuals from layers and suckers, until they appear to be functionally independent.

Change can be expressed in both absolute numbers and an annual rate of change. Rates can be determined simply by dividing the number of losses or gains by the number of years, or relatively as a per cent of the start population per year. Relative rates – based on compound interest rules (Sheil *et al.*, 1995) – make more sense when quantifying mortality, since the loss of say all five individuals in a small population will be more significant than the loss of five individuals from a large one. Recruitment is better expressed against area and years, partly because quantifying it as a per cent of the start population can yield stratospherically large numbers. The number of years is itself difficult. All annual change has been expressed against the number of growing seasons between recordings (Table 4.2), which makes sense for growth and recruitment but less

Table 4.1. Annual mortality rates (%/year) in the old-growth stands, 1945–2010, for all tree and shrub species combined.

Year	Option 1: individuals	Option 2: larger stems only	Option 3: large + small	Option 4: estimate of how the value calculated by Option 3 differs from the rate that would have been calculated if all stems reaching 1.3 m had been recorded
1945–1955	1.69	1.17	1.74	Underestimate
1955–1977	2.59	1.68	2.54	Substantial underestimate
1977–1983	3.41	3.37	3.08	Little difference
1983–1986	3.23	1.47	2.10	Little difference
1986–1992	1.23		1.71	Small underestimate, due to small 1983 and 1986 stems not detectable in 1992
1992–2000	1.65	1.29	3.09	Little difference
2000–2010	1.57	1.17	2.86	Little difference

Table 4.2. Survey intervals (years) in the Lady Park stands.

Old-growth stands		Young-growth stands		Stand below the cliff		Managed compartment	
1945[a]–1955[a]	10	1950–1977[b]	27	1950[b]–1959[b]	9		
1955[a]–1977[b]	23						
1977[b]–1983[b]	6	1977[b]–1983[b]	6	1959[b]–1985[a]	25		
1983[b]–1986[c]	2.5	1983[b]–1986[c]	2.5				
1986[c]–1992[a]	5.5	1986[c]–1993[c]	7	1985[a]–1992[a]	7		
1992[a]–2000[b]	9	1993[c]–2002[c]	9			1984[c,b]–1997[a]	12, 12.5
2000[b]–2010[a]	9	2002[c]–2013[a,b]	10.5, 11.5	1992[a]–2005[a]	13		
2000[b]–2013[a]	12	2002[c]–2015[b]	13.5	2005[a]–2013[b]	9	1997[a]–2014[b]	18
1945[a]–2010[a]	65	1977[b]–2015[b]	38	1950[b]–2013[b]	63	1984[c,b]–2014[b]	30, 30.5

Notes: [a]Recorded before summer; [b]recorded after summer; [c]recorded during summer; in 2013, some of the young-growth transects were surveyed in March, others in October.

Box 4.1 Measures of trees, stands and change

Stand

Basal area of all trees (m^2/ha)
Tree density (N/ha)
Size distribution

Individual trees

Girth (cm); formerly diameter (cm)

Individual tree growth

Radial growth rate (mm/year)
Growth in basal area (cm^2/year)

Population (of a tree species; or all trees in a stand)

Density (N/ha)
Range of girths
Size distribution of girths
Median girth
Regression of log (N) versus size classes

Measures of change

Relative rate of mortality (% of initial population/year)
Rate of recruitment (N/ha/year)
Radial growth rate (mm/year), range and median
Basal area growth rate (cm^2/year), range and median
Years determined as number of growing seasons in the period

Mortality from all causes can be quantified by determining survival over a period. Trees were classified as alive or dead, but some of the living trees were either fallen or alive only below 1.3 m. For simplicity, we have quantified mortality as 'trees alive at 1.3 m', i.e. the fallen and 'live below 1.3 m' were counted as dead.

sense for mortality, which can be higher in winter. When transects have been recorded in midsummer, these have counted as half seasons.

We have examined all measures to understand how a population has changed, but have been selective in presenting and explaining the results. Thus, we have examined the records from several angles, identified the important features and presented the most informative measures. In practice, we have tended to combine several of the measures listed in Box 4.1.

Finally, we omitted a few sections from the calculations of population change. Sections I: 0–24 and II: 0–24 were damaged by the top-ride re-alignment. V: 0–100 was not recorded in 1945 and 1955. III: 400–450 was disturbed by the adjacent felling in 1942–1944. And, two sections close to the cliff, IV: 950–985, V: 1110–cliff, were recorded inconsistently.

Summarising, recording methods have varied over the decades, due to the changing character of the stands, the time available to recorders, their circumstances and their personal interests. They will, perforce, change again if the current flush of regeneration results in massive recruitment. Even with perfect and uniform recording throughout, various methods of analysis and presentation are available, each with pros and cons. This messy methodology is not ideal, but it *is* realistic and seems to be inevitable in under-resourced long-term studies.

(a)
(b)

5 The Changing Woodland

Woods are constantly changing. However fixed and statuesque they may seem to a casual, short-term observer, both the balance between individual species and the character of the stand as a whole change constantly. Thus, for example, as trees grow larger, so their numbers must decrease, for there is a fixed amount of space for them to occupy. Here, we deal with trees as a collective, the 'stand', leaving the behaviour of individual species to later chapters.

5.1 How Stands Grow

Several schemes have been devised to describe how stands grow (Table 5.1). At their simplest, stands start with 'bare ground' (which is how foresters see land devoid of trees) and a pioneer phase of saplings with plenty of space between them. As saplings grow, branches interlock and a thicket develops. Further growth results in stratification as fast-growing trees overtop slow growers and weaker trees die, i.e. they are 'excluded'. Eventually, trees grow to full height, at which point the stand is mature, with a canopy of individuals whose crowns are open to full sunlight, above a subcanopy of trees whose crowns spread below the canopy and an underwood of shrubs, saplings and slow- or poorly-grown trees which, necessarily, suffer the shade cast by canopy and subcanopy trees. If the trees have been grown for timber, they would be felled at this stage and a new stand would be established; or some of them would be felled, leaving space for subcanopy trees, underwood trees and even saplings to grow into the canopy. If, however, the stand is allowed to grow on, it develops a distinctive condition with aged canopy trees and a complex vertical and horizontal structure known as 'old-growth'.

Once a thicket has developed, recruitment ceases, a few trees get ahead of the rest and the range of tree sizes increases. Small individuals greatly outnumber the large individuals, generating a skewed size distribution whose mode is below the median and well below the average. As growth progresses:

- All trees grow, but the larger trees grow faster, so that mode, range of sizes and maximum size all increase. If there is no further recruitment, even the minimum size increases, for the smallest trees usually die.
- The skewed distribution remains, taking the graphic form of a wave moving to the right, which becomes less pronounced as stands age.
- The largest trees increase slowly, accelerate, but then decelerate as the stand reaches maturity, i.e. they execute a sigmoid curve.
- The number of individuals declines, so, while the mode increases, its height decreases.

During each period of growth:

- The large trees generally grow fastest, and none grow very slowly. These are the trees in the canopy growing with full light.
- Small trees mostly grow slowly, and none grow fast. These are the trees in the subcanopy and underwood, which are shaded by their taller neighbours.
- Individual trees maintain their growth rate if they are large, but smaller trees tend to decelerate with time, especially those that formerly competed with the fastest growing trees but lost the race for crown space, i.e. they drop back from the canopy to the subcanopy.

Eventually, the stand reaches maturity as the largest and fastest growing trees reach their top height and slow down. By this stage, the well-lit trees in the canopy are fruiting and enough light penetrates to

Fig. 5.1. (a) Wartime felling, leaving a shelterwood of oak and beech, photographed in 1942. (b) The same view 73 years later, the retained trees now almost indistinguishable among the subsequent regrowth.

Table 5.1. Stages of stand development following Oliver and Larsen (1996).

Stage	Defining processes
Regeneration or stand initiation stage	New individuals become established following felling or stand-destroying disturbance
Thicket or stem exclusion stage	Regeneration ceases; individuals compete
Understorey reinitiation stage	Regeneration resumes in the understorey
Old-growth stage	Gaps form in the canopy into which some understorey individuals grow

the ground to enable seedlings to develop into saplings, which grow slowly at first, but, as their number increases, some grow faster than the rest. As time passes, mature trees in the canopy lose vigour and some will die, leaving spaces that at first are filled by crown expansion of other canopy trees. Eventually, larger gaps appear, while surviving canopy trees thin out and lose enough vigour to allow the best placed of the new generation to compete for canopy space with the surviving small trees in the old generation. The new generation trees have youth on their side and will probably be more successful.

This orderly, inexorable, 'endogenous' competitive thinning process may be disrupted by various kinds of disturbance, i.e. 'exogenous' factors that change the structure of the stand. They range from extreme disturbances, which destroy a stand, leaving the ground vacant for a new stand to regenerate and follow the path described above, to lesser disturbances, which do not destroy stands but remove some trees and leave others. Each kind of disturbance has its characteristic impact. Some, such as the admission of herbivores or the arrival of a disease afflicting a small-growing species, take out only saplings and shrubs, thereby increasing the modal tree size but leaving the canopy unaffected. Others, such as storms, mostly remove the largest trees and create spaces in the canopy. The impact of a particular disturbance will depend partly on stand composition and the size distributions of each species. All disturbances break up the generation structure (i.e. disrupt the smooth flow of the 'wave' from left to right) to some degree.

Responses depend partly on the character, pattern and sequence of disturbances. If, for example, herbivore populations increase substantially, there will be little disruption during the growth of the present stand, though they could eventually transform the wood into parkland if populations remain high indefinitely. If, in contrast, storms blow down large trees and crush some underwood, the canopy space vacated will be distributed in patches, or 'gaps', which, if small, fill by crown expansion of neighbouring canopy trees or, if large enough, admit new regeneration. Eventually, if there is still no stand-destroying disturbance, the whole stand becomes a patchwork of small stands, each following the course described above. A succession of disturbances over decades or longer would generate patches of different ages, and thus a mosaic of gap-sized stands at different stages of development. The consequences for size distributions are exemplified by the analysis of mountain beech stands in New Zealand by Coomes and Allen (2007).

5.2 Lady Park Wood as a Chronosequence

That is the general pattern, but what about the reality? Lady Park Wood is particularly well placed to illustrate real-life stand development, for the young-growth stands show the first 73 years of stand development (Fig. 5.1: see p. 50), while the old-growth stands continue the story on the same range of sites with the same range of species from 75 to 145 years growth. (The stand below the cliff is a special case, which is considered in Section 5.6.)

Neat though this seems, Lady Park is not a true chronosequence, for the stands differ in more than just time; that is, they are not following the same succession from the same starting point. Old and young-growth are similar insofar as they cover much the same range of site conditions; both started with standards and long-established stools; and both are products of both vegetative and seedling regeneration, but they differ in that: (i) the old-growth developed from in-rotation coppice with standards with strong stools, whereas the young-growth developed from 75 years outgrowth from weakened stools and the stumps of 75-year-old maidens; (ii) the old-growth was thinned twice at 32 and *c.*60 years growth, whereas the young-growth was not thinned; (iii) oak coppice was singled in a small patch of old-growth just above the cliff; and (iv) the major influence of the 1976 drought happened at 33 years in the young-growth and 106 years in the old-growth (Fig. 5.2). Nevertheless, the similarities are close enough to show some features of a 145-year near-natural succession to contrast with the managed stand nearby.

Fig. 5.2. The great drought of 1976 was the most profound disturbance to shape the development of Lady Park Wood in the seven decades of observation. This photo of autumn colours in beech and other species was taken near the Wyndcliff in August 1976.

5.3 Early Stand Development and the Impact of Deer

Replacement stands in managed woodland inevitably inherit various features from the preceding stand. Even if every tree has been felled, some poles and saplings will survive and many trees will spring again from the stumps. This, of course, was the principal response in coppice-with-standards woodland. Other legacies will be a woodland soil, a seedbank waiting to break from dormancy and the full range of woodland herbs, fungi, etc. Further, in classical shelterwood systems, mature trees were retained as seed sources, then either allowed to grow with the new stand as two-storied high forest or felled once a good crop of saplings had been established. Today, mature trees are often retained as habitat and for amenity.

Natural stands also start with a legacy. Even where storms have levelled the previous stand, a scatter of poles and saplings is left standing and some of the fallen trees will sprout from their prostrate trunks. In the natural conifer forests of the boreal and Mediterranean zones, large trees often survived forest fires and functioned as seed trees for the next generation – a natural shelterwood.

In both 1870 and 1944, Lady Park inherited a scatter of mature trees, a stock of coppice stools and a scatter of stumps from felled standards and maiden trees. In 1870, we assume that densely packed stools regenerated vigorously, but in 1944 the stools were further apart than they had been in 1870 and were weakened by decades under a developing high forest canopy. Nevertheless, regrowth from lime and hazel stools was vigorous, some of the stumps of standard oaks sprouted and seedlings appeared in huge numbers, especially birch and ash.

In the early years, these shoots and seedlings had to compete with brambles, festoons of clematis and

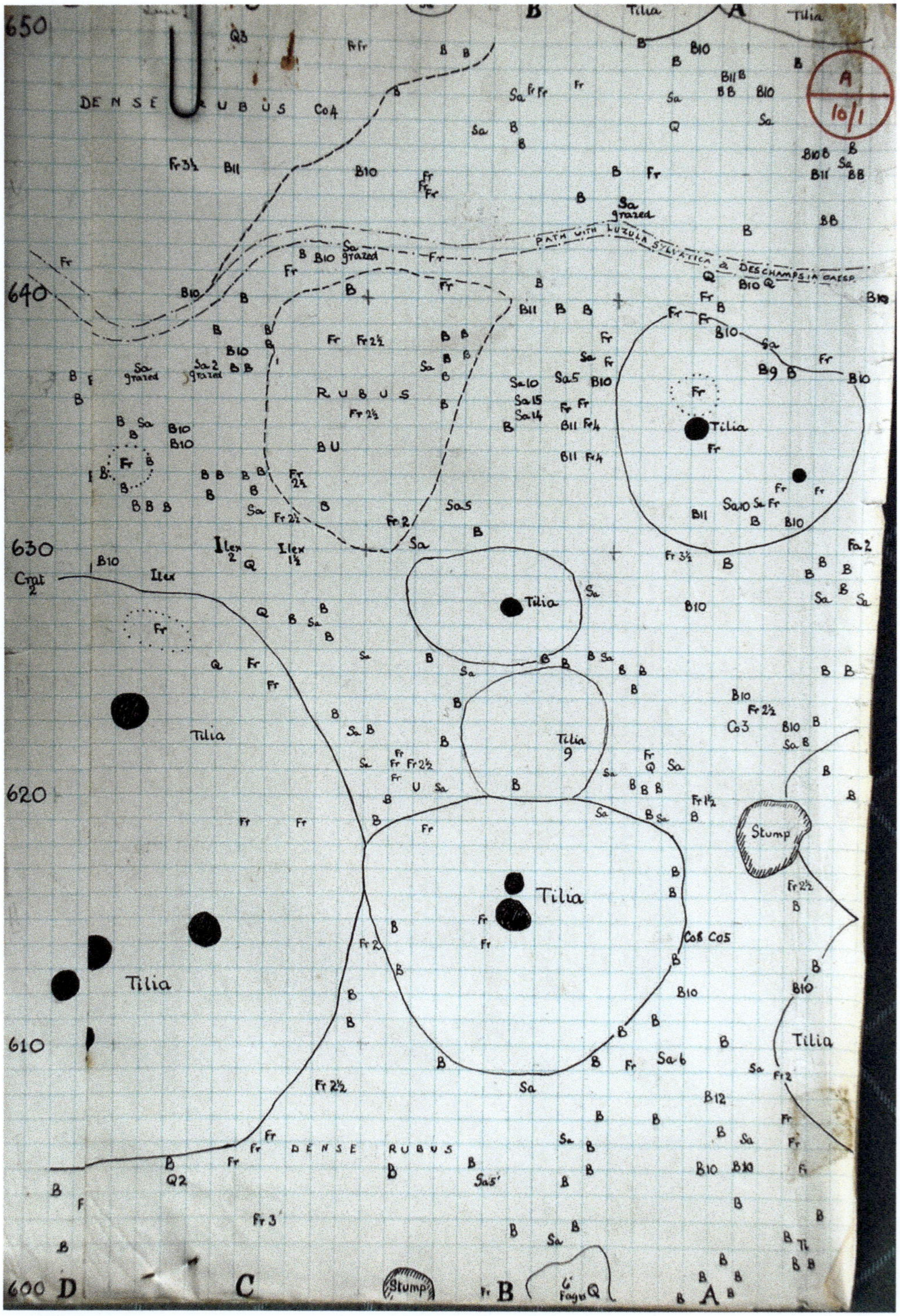

Fig. 5.3. Eustace Jones's chart of quarter section I: 600–650N in 1948. Vigorous lime coppice stools dominate in a matrix of bramble and saplings, most of which are birch, sallow and ash. A path used by deer crosses near the top and some adjacent saplings have been browsed. Some sallows were already 15 feet tall.

vigorous ground vegetation. As a result, the new stand was patchy (Fig. 5.3). Beyond the crowns of retained trees, shoots from stumps easily outpaced most of the ground vegetation. Saplings were clustered, perhaps concentrated on patches that had been bare of ground vegetation before the felling, but some forced their way up through bramble. Within a few years, the main source of patchiness was evidently the resident population of fallow deer.

Deer have become increasingly influential in temperate woodlands. Fallow deer were abundant in Lady Park before the reserve was established, yet they did not prevent a new stand closing the canopy. Eustace Jones recorded their impacts in some detail from 1945 to 1960: his unpublished analysis (LPW C10-12) is the basis for this summary.

Deer had not been reduced in numbers during the latter part of the war, and in the years immediately following they were so excessively abundant in Lady Park that it looked as though they would greatly interfere with, if not prevent, the regrowth of woody vegetation. However, from 1947 onwards attempts were made to reduce their numbers: 150 head were shot in 1947/8 and by 1949 they were no longer a major influence. Estimates of their numbers in Highmeadow thereafter varied from 50 to 100 head.

Deer impacts before canopy closure were far from uniform. Zones immediately above and along the foot of the cliffs were heavily used, probably because deer used them to find passages down to the river, their only drinking water apart from the small ponds and spring near Redding's Lodge. Along the base of transects II and III, clear passage along unfelled woodland just above the cliff apparently reduced grazing and browsing within the new growth a little higher up the slope. Deer tracks nevertheless permeated the developing young-growth stands.

Deer congregated where stand development lagged behind its immediate surroundings, notably paths, old dragging tracks, fire sites and immediately around old standards. Under standards, where they also gained some shelter, they created cores of bare earth and 'feeble' Tufted hair-grass. One such 'shaded clearing', 'much loved by deer', was on steep slopes at the top of transect V, where a large beech had been retained on very shallow, stony soil. Bounded by a low cliff, this was a site made safe from surprises from the north-west, and it is still open, even though the retained beech was much damaged in 1976 (Fig 7.14).

The consequences of this patchy grazing were clear in the distribution of bramble. By 1948, bramble had formed a tangle almost everywhere, except beneath standards on transect I. It was already widespread in 1944, but increased in density and vigour, though it did not spread. It failed to dominate beneath retained large trees, where Great wood-rush and/or Tufted hair-grass became dominant; where deer grazed heavily; on one or two very steep banks with degraded soil; and on fire sites, where Tufted hair-grass often became dominant.

Deer impacts also changed as the stand developed. By 1958, Jones noted that 'The amount of grazing steadily diminishes as crops close. Fresh bites are mainly along the margins of tracks and open grassy places.' Much of the browsing was 'about mouth-height' – 2–3 feet from the ground – leaving most saplings taller than 5 feet (1.5 m) unmolested, but as the stand grew, many tall saplings were bitten and broken, especially ash and dogwood. A small amount of bark stripping was seen, but it was trivial in its effect. On transect VI, the small saplings were browsed increasingly as the larger saplings grew out of reach, but the lateral branches of the tall ash were also browsed.

Deer browsing was not only selective in respect of size and location, but also for particular tree species. Oak was top of the menu, followed by ash, beech and birch in that order, and consistently so throughout the wood. The relative immunity of birch and beech saplings was clear by 1949, but coppice regrowth from large stools of beech, oak and ash was so closely cropped that 'they made dense rounded cushions of twigs' that eventually died. Other species were also browsed in descending order of preference: dogwood, spindle, holly, yew, maple, hazel, broom and sallow, leaving only elder untouched. Many holly and yew were closely cropped, and dogwood was so favoured that deer stretched for it. Jones concluded that oak was 'the most seriously handicapped' species, whereas the distribution of ash in the developing young-growth was unaffected because there were always individuals distant from tracks that grew rapidly beyond the reach of deer.

5.4 Growing through the Thicket: The Young-growth

Well before the second recording in 1955, the canopy closed (i.e. trees and shrubs grew together to form an upper layer of foliage that completely shaded the ground vegetation) and the thicket of brambles and close-spaced saplings and poles must have been hard to penetrate. New tree seedlings could no longer establish themselves and those that germinated just before canopy closure failed. Indeed, the chance of each year's new entrants surviving must have declined steeply, ensuring that most survivors in 1955 must have germinated in or before 1947. Once the canopy closed, those saplings that had not kept pace with the canopy suffered increasing shade, grew slowly and died, together with the weaker shoots on coppice stools. Thus, after almost a decade of recruitment, a nearly even-aged stand had been created and the phase of intense natural thinning had set in.

Throughout the late 1950s and 1960s, the stands were understandably described as 'birch woodland', for birch saplings were the most abundant and fastest growing of the new growth (Fig. 5.4). Birch, however, casts only light shade, so other species persisted in the thicket. Sallows and aspens grew almost as fast, but were far less abundant. Likewise, the new growth on lime and hazel stools kept pace. Meanwhile, the surviving standard oak and beech were free to expand their crowns and enlarge the 'shaded clearings' beneath them. Throughout this thicket phase, the stand as a whole seems to have been undisturbed, save for a severe outbreak of bark stripping by grey squirrels in 1958, which was evidently concentrated on the birches.

From 1977, it became possible to track how the stand grew and changed in composition (Table 5.2). Of the 8678 stems recorded, 18 originated as standards, 4 as layers, 4385 as seedlings and 4271 as coppice sprouts. Some 1563 of these were standing dead ('snags'), mostly birch of seedling origin that died in the 1976 drought. Thus, 1976 was a transitional year: before then, most individuals were of seedling origin, but thereafter stems of coppice origin predominated, reaching 62% by the millennium, though patches wholly of seedling origin remained. The clusters of ash and birch noted in the earliest records were still obvious in 2015.

The 1976 drought was the first major disturbance. Just 32 years old, the birches died in droves – some 30% of the basal area of birch was killed – and the canopy opened for the first time. The impact was patchy at small and medium scales: some birch groves merely thinned; others broke up to form gaps; and yet others, particularly those in the northern part of the wood where the soil was deeper, remained unscathed. When the stand was recorded in 1977, several wych elm stood dead on the upper plateau and steep slopes, and some parts of the young-growth were thickets of dead birch trees. For the first time, much of the 'birch wood' of the 1960s had been revealed to be a mixed stand of hazel, ash, beech, oak, lime, etc.

With the canopy partly opened, one might have thought that underwood stems would be released to stronger growth and seedlings would once again have had a chance to survive and develop into saplings. Underwood stems did indeed grow faster and canopy trees did expand their crowns (Fig. 5.5), but, perversely, the release of previously suppressed beech stems acted as a trigger for a prolonged bout of debarking by grey squirrels of stems that might otherwise have become canopy dominants (Mountford, 2006). Further, beech did generate a widespread scatter of saplings that looked set fair to prosper, even though they were still no more than 40 cm tall in 1985, but in that summer, bank voles were so numerous the wood rustled – one could see them swinging on small saplings – and almost all the small beech saplings were ringbarked close to the ground and killed.

By 2015, 71 years since it was initiated, the large birches, sallows and aspens that had started life in the open ground created by the 1942/44 felling were reaching their limits, losing vigour, developing rot and were easily broken or uprooted in winds. The former 'birch woodland' had unambiguously become a mixture of beech, ash, limes and oak: the long-lived had taken over from the short-lived pioneers. Further, the stand was poised for significant new recruitment. In response to the ring-fencing of the minimum-intervention reserve in 2007, hazels had started to produce clusters of vigorous sprouts; limes had produced new shoots from the base of stools; ash seedlings, which were very numerous, and beech seedlings, which were frequent, were persisting and starting to accelerate; and even a few oaks, limes and wych elm saplings had developed. The larger gaps were filling slowly with hazel sprouts, lime regrowth and new groups of birch and sallow, which were bidding fair to repeat on the scale of a gap the sequence of events that had taken place on a stand scale after 1942/44. Subject to the continued exclusion of deer, stem density would again increase.

5.5 Mature Stand Development: The Old-growth above the Cliff

The story continues with the old-growth stands (Fig. 5.6), despite the differences between the young-growth of 2015 and the old-growth of

Fig. 5.4. Young-growth in V: 850–900 in 1958. The location was identified by the small-leaved lime regeneration from a stool in the centre background.

1945, summarised earlier. In 1945, the old-growth was still recovering from an unrecorded thinning, only detectable from its low basal area (Table 5.3). By design, the canopy was dominated by beech and oak, with numerous smaller ash, elm, lime and birch in the canopy and subcanopy. The underwood was reasonably well stocked with small hazel, holly and tree saplings. Despite the lapse of 75 years, any forester would have noticed immediately that the stand had grown from coppice with (oak) standards towards a beech-dominated high forest.

Until 1976, the canopy remained closed, trees grew larger, but stem density fell as weaker trees were shaded out by stronger neighbours. Even the tall oak standards 'raised their crowns', a euphemism for death by shade of the lower crown branches. Heavily shaded by beech, the underwood thinned out substantially: the wood became 'hollow'. Saplings were unable to grow, except on the margins, where they benefited from side light. In the core of the wood, beech was becoming increasingly dominant: canopy beech were growing fastest; their mortality rate was lowest; established poles were present in the underwood that could develop into trees if need be; and there was even a sprinkling of further regeneration.

Then in 1975/76 came the great drought. Some canopy beech died outright, singly and in groups, others lost most of their crown branches, and all the survivors almost stopped growing. Other species lost most of their leaves by late summer, though in general they survived. For the first time, substantial canopy gaps appeared, principally in the upslope stand, where the soils are shallow, and they

Table 5.2. Changes in the basal area (m²/ha) of species recorded in the young-growth stands from 1944 to 2013/15.

	1944	1976	1977	1993	2002	2013/15
Stand age (years)	0	32	33	49	58	69
Standards	*2.1*	*2.97*	*2.97*	*3.19*	*3.32*	*3.57*
Beech	1.3	1.85	1.85	1.88	2.09	2.25
Oak	0.8	1.12	1.12	1.31	1.23	1.32
New growth						
Principal canopy trees	*0*	*5.97*	*5.97*	*10.41*	*12.74*	*14.15*
Ash		1.09	1.09	2.83	4.11	5.51
Beech		1.36	1.36	2.20	2.35	1.97
Limes		2.71	2.71	4.20	4.97	5.65
Oak		0.61	0.61	1.04	1.14	0.99
Wych elm		0.20	0.20	0.14	0.17	0.03
Short-lived canopy trees	*0*	*11.32*	*8.19*	*7.15*	*6.06*	*2.58*
Alder		0.08	0.08			
Aspen		0.11	0.11	0.08	0.06	0.05
Birch		10.75	7.62	6.75	5.83	2.51
Sallow		0.43	0.43	0.32	0.17	0.02
Small trees	*0*	*0.89*	*0.89*	*1.08*	*1.08*	*1.12*
Crab apple		0.01	0.01	0.01	0.01	0.02
Field maple		0.57	0.57	0.63	0.57	0.58
Hawthorn		0.10	0.10	0.16	0.17	0.14
Holly		0.02	0.02	0.03	0.03	0.03
Service		0.08	0.08	0.11	0.12	0.11
Whitebeam		0.02	0.02	0.05	0.06	0.06
Yew		0.05	0.05	0.05	0.06	0.09
Underwood Hazel	*0*	*2.36*	*2.36*	*3.13*	*2.90*	*2.62*
Shrubs	*0*	*0.09*	*0.09*	*0.05*	*0.03*	*0.03*
Blackthorn		<0.01	<0.01	0.01	0.01	0.01
Dogwood		0.08	0.08	0.08	0.01	0.01
Spindle		0.01	0.01	0.01	0.01	0.01
Others						
Larch		0.01	0.01	0.02	0.04	0.04
TOTAL BASAL AREA m²/ha	**2.10**	**23.68**	**20.55**	**25.15**	**26.29**	**24.22**
DENSITY Live stems/ha	**9**	**3899**	**3370**	**2370**	**1784**	**1108**[a]

Notes: To obtain an estimate of the stand in 1976, the dead standing birch recorded in 1977 were added to the 1977 values, with no other adjustments. [a]Only hazel stems of 5 cm GBH and larger were included. If all smaller live stems were added, the stem density climbs to 1607 trees/ha.

continued to enlarge as several tall, lollipop-shaped trees, which were exposed on their margins, were blown down and disease killed elms. By 1983, gaps occupied some 26% of the upslope stand.

The beech's loss was other species' gain. Ash and oak grew faster, and survival rates improved. Oaks in particular recovered quickly, aided in many cases by the growth of a thicket of epicormic shoots on their trunks. Meanwhile, the damaged beech grew new leaders from their crowns, bark expanded over the scars left on the trunks by heat-induced necrosis, and radial growth rates picked up by the mid-1980s, but they never regained their pre-drought vigour. Stem density continued to fall, but basal area, which had dipped after the drought, picked up again. Beech started to reassert dominance, but regeneration was limited by deer and voles.

After the millennium, tree falls became more frequent in the core of the wood. The gap creation rate thus started to increase, but measuring it was difficult, because the stand had become multi-stratal and generally complex. Most falls involved several trees. In one (between transects II and III), a beech that had been mortally damaged in the drought finally split and fell against a large oak, which itself knocked over other beech and limes in the canopy. Another on transect II started when a group of tall, slender limes toppled against

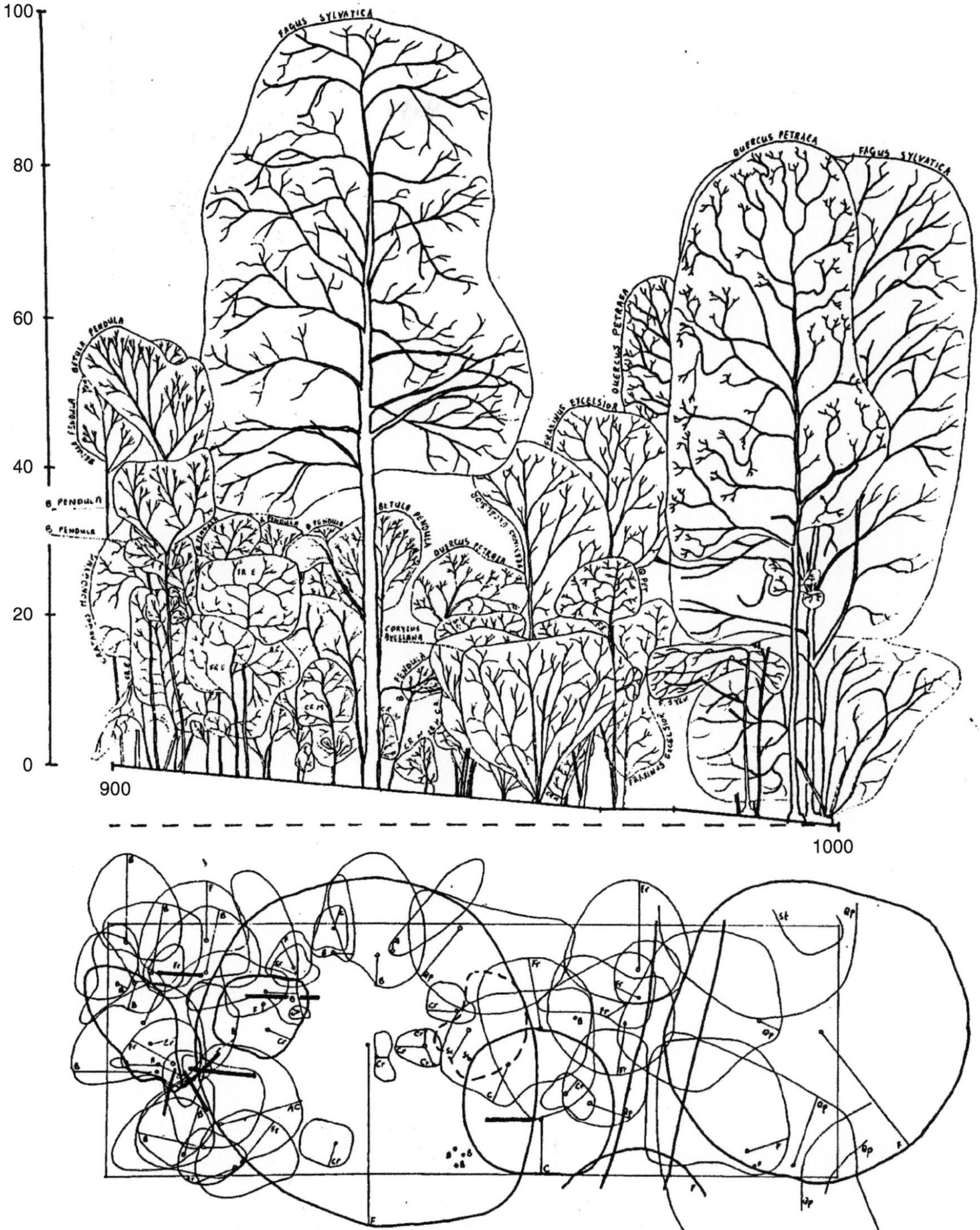

Fig. 5.5. Structure of a young-growth stand, V: 900–1000, after 44 years growth, drawn by foresters from Wageningen. The oak and beech retained in 1942 suppress the growth of nearby underwood. (From Bartelink *et al.*, 1987.)

Fig. 5.6. Upslope old-growth near transect I in 2012, dominated at this point by beech.

a tall ash and went on to knock over another large oak and a beech. Falls did not always require wind: the beech that fell in calm sunny weather in October 2013 (Chapter 7, this volume, Case study B4) was a fine demonstration of the unpredictability of natural woodland.

The 1976 drought, therefore, accelerated the ongoing thinning. Thereafter, the canopy grew higher; large trees tended to grow faster than small trees; and small trees continued to grow slowly and die in their shade (Fig. 5.7). Stand density decreased, but average tree size increased enough to restore the upward trend in basal area. As the years passed, gap-creating disturbances became more frequent. A few old standards fell or shed large branches. Groups of tall birch died en masse, apparently after wind had felled a few and exposed their neighbours. A large patch dominated by lime was blown down.

Until the reserve was fenced in 2007, saplings rarely developed and basal sprouts were browsed to stubs, but thereafter both had an opportunity to prosper. Lime in particular generated a good deal of new, healthy, if weak, growth. Tree seedlings survived longer and in larger numbers, but again they did not shoot up rapidly. By 2014, however, a new generation was poised to enter the stand, comprised mainly of beech and ash, with scattered lime, oak and wych elm. As stem density was bottoming out, basal area seemed to have reached a plateau.

In the stand just above the main cliff, where the soils were deeper, the 1976 drought stopped the beech growing and caused major dieback, but it did not kill them. Rather, the canopy remained closed and the beech resumed their progress towards dominance. The remaining canopy oak, ash and limes were increasingly compressed and several died, though the limes attempted to grow back. Initially, there was no regeneration, but, when much of the stand was enclosed within the safety exclosure, a thicket of beech saplings developed, together with a few ash, holly, maple, service and yew (Fig 7.7). This was facilitated by side light from the irregular canopy above the nearby cliff, where a large oak fell.

Table 5.3. Changes in the basal area (m^2/ha) of species recorded in the old-growth stands from 1945 to 2010.

	1945[a]	1955[a]	1977	1992	2000	2010
Stand age (years)	75	85	107	122	130	140
Principal canopy trees	*20.14*	*24.67*	*31.79*	*31.84*	*33.90*	*35.47*
Ash	4.32	5.01	6.33	6.57	7.00	7.85
Beech	9.56	12.31	16.54	16.08	17.23	17.97
Limes	1.71	2.35	2.95	3.18	3.53	3.62
Oak	4.15	4.48	5.25	5.92	5.97	5.84
Wych elm	0.40	0.52	0.72	0.09	0.17	0.19
Short-lived canopy trees	*0.50*	*0.62*	*0.28*	*0.20*	*0.13*	*0.15*
Birch	0.50	0.62	0.26	0.20	0.13	0.15
Sallow		<0.01	0.02			
Small trees	*0.82*	*0.80*	*0.72*	*0.86*	*0.96*	*1.03*
Field maple	0.30	0.27	0.12	0.11	0.14	0.15
Hawthorn	0.09	0.06	0.03	0.03	0.03	0.03
Holly	0.08	0.09	0.07	0.08	0.08	0.08
Service	0.15	0.16	0.24	0.31	0.33	0.32
Whitebeam	0.10	0.11	0.12	0.15	0.17	0.18
Yew	0.10	0.11	0.14	0.18	0.20	0.24
Underwood Hazel	*0.36*	*0.26*	*0.10*	*0.23*	*0.25*	*0.25*
Shrubs	*<0.01*	*<0.01*		*<0.01*	*<0.01*	*<0.01*
Dogwood	<0.01	<0.01				
Spindle	<0.01	<0.01		<0.01	<0.01	<0.01
Others						
Sycamore	0.01	0.01	0.02			
TOTAL BASAL AREA m²/ha	**21.83**	**26.36**	**32.91**	**33.13**	**35.24**	**36.90**
DENSITY Live stems per ha	**2135[b] (1284)**	**1781[b] (1229)**	**790**	**864**	**729**	**670**

Notes: [a]Basal area values take no account of recorded counts of small stems (<8 inches DBH). [b]Counting five stems for each individual recorded as multi-stemmed. The figure in brackets gives the density counting just one stem for multi-stemmed individuals.

5.6 Woodland below the Cliff

Like the rest of Lady Park, the woodland below the cliff had been treated as coppice with standards until 1870, but there is little sign that it had been thinned in 1902 – there were few beech to promote – so by 1940 it may well have enjoyed 70 years of unfettered growth. However, in the early years of the Second World War, it was logged selectively for lime, which was used to make frames for Mosquito fighter aircraft, and the impact of this was still clear when it was first recorded in 1950 (Table 5.4). The basal area was much lower than the old-growth above the cliff and stem density was much higher, but there were still many large trees. It also differed in composition: ash, large-leaved lime, wych elm and gean were far more prominent, beech, oak and birch much less so, and the underwood included numerous yews and a high density of hazel. Physically, it was also different: numerous small cliffs and outcrops were exposed among stable, moss-covered screes of loose limestone within a matrix of generally moist, highly alkaline loams.

Growth from 1950 onwards was very fast, but it was accompanied by numerous tree falls. Indeed, a fortnight after transect VIII had been laid out and fully recorded in early September 1950, a large beech at the top of the transect lost its roothold and fell through the ash grove below, leaving an avalanche of smashed trees and displaced rubble. Similar, if less spectacular, falls have been frequent, so surviving trees have had space to grow fast and the underwood, being well illuminated, has remained well stocked.

By 1970, the stand was dominated by tall, vigorous ash, wych elm, large-leaved-lime, small-leaved lime and a scatter of beech. Combined with a dense, but patchy underwood, stand structure was complex, both horizontally and vertically. Then, in 1971, *Ceratocystis ulmi* arrived and the so-called Dutch Elm Disease (DED) killed most of the elms, leaving a much more open stand, dominated by ash, limes and – for a few years – the gaunt remains of the elms.

Unfortunately, the transects were not recorded again until 1985, so the undoubtedly sharp changes

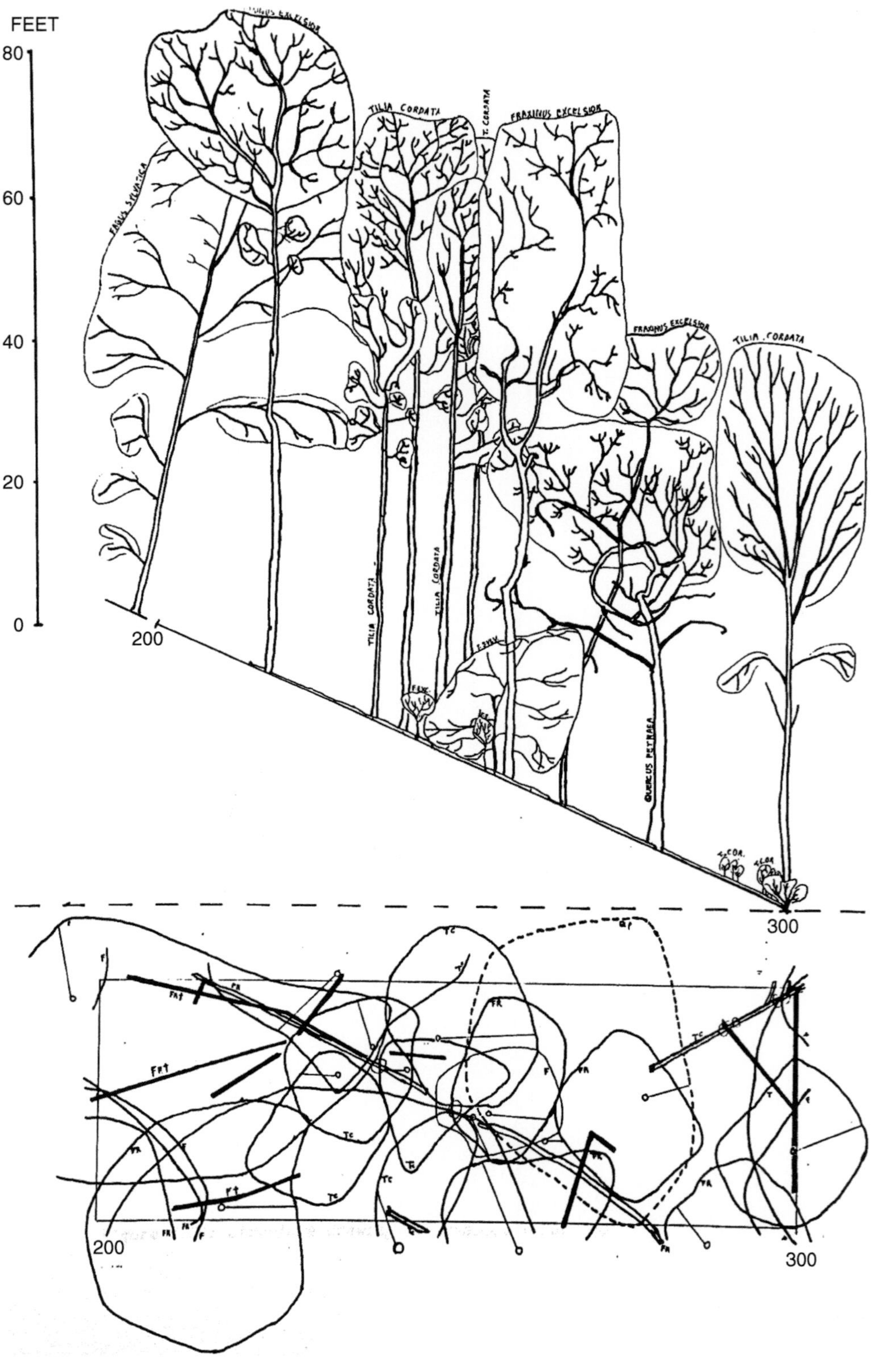

Fig. 5.7. Structure of an old-growth stand, I: 200–300, in 1986, drawn by foresters from Wageningen. The former oak standard has been overtaken by younger lime and beech. (From Bartelink *et al*., 1987.)

Table 5.4. Changes in the basal area (m²/ha) of species recorded in the stand below the cliff from 1950 to 2013.

	1950[b]	1959[b]	1985	1992	2005	2013
Stand age (years)[a]	80	89	115	122	135	143
Principal canopy trees	*12.06*	*14.74*	*14.77*	*15.24*	*17.45*	*19.32*
Ash	4.81	5.63	6.20	6.00	6.52	7.07
Beech	3.35	3.93	3.60	3.13	2.71	2.93
Limes[c]	2.24	2.83	3.40	4.52	6.51	7.51
Oak	0.13	0.13	x	x		
Wych elm	1.53	2.22	1.57	1.59	1.71	1.81
Short-lived canopy trees	*0.73*	*0.78*	*0.34*	*0.42*	*0.16*	
Alder	0.09	0.13	0.24	0.28		
Birch	0.35	0.36	0.08	0.09	0.09	
Gean	0.13	0.24	x	x		
Sallow	0.16	0.05	0.02	0.05	0.07	
Small trees	*2.04*	*2.43*	*2.58*	*2.82*	*3.50*	*3.30*
Crab apple	x					
Field maple	0.21	0.23	0.06	0.11	0.13	0.14
Hawthorn	0.01	0.04	0.04	0.06	0.06	0.04
Holly	0.01	0.02	0.01	0.03	0.07	0.03
Whitebeam	0.39	0.48	0.43	0.22	0.18	0.19
Yew	1.42	1.66	2.05	2.40	3.06	2.90
Underwood Hazel	*0.70*	*0.78*	*1.24*	*2.29*	*2.26*	*1.61*
Shrubs	*x*	*0.01*	*x*	*0.01*	*x*	*x*
Dogwood	x	x		x		
Elder	x	x	x	0.01		x
Guelder rose	x	x		x	x	
Privet				x	x	
Spindle	x	x				
Wayfaring tree				x		
TOTAL BASAL AREA m²/ha	**15.58**	**18.76**	**18.93**	**20.77**	**23.37**	**24.23**
DENSITY Live trees per ha	**3067[d] (1453)**	**3634[d] (1623)**	**3224**	**5012**	**2839**	**1937**

Notes: [a]The value for stand age is almost, but not quite, meaningless. The age given is the number of years since coppicing ceased, but the stand was high graded in about 1942. [b]Basal area values take no account of recorded counts of small stems (<8 inches DBH). [c]Since 1985, large-leaved lime has comprised about 90% of the total lime basal area. [d]Counting five stems for each individual recorded as multi-stemmed. The figure in brackets gives the density counting just one stem for multi-stemmed individuals. x = less than 0.005 m²/ha.

around 1970 can only be estimated. Basal area must have peaked at about 23–24 m²/ha in 1970, but then it slumped to perhaps 16–17 m²/ha as many large elms died outright. Combining the dead elms standing as snags in 1985 (1.35 m²/ha) with the other large elms lost between 1959 and 1985 (0.55 m²/ha), about two-thirds of the elm basal area in 1970 (3.3 m²/ha) was killed by DED in the 1970s. No doubt other trees, now exposed, fell. With the canopy now so open, the frequency with which ash fell or split increased after 1985.

Canopy trees have continued to fall, but growth and recruitment has outpaced mortality enough for the basal area to increase, albeit to much lower levels than achieved by the old-growth above the cliff. In 1985 and 1992, regeneration seemed prolific – elms, limes and hazels generated new growth, many groups of saplings became well established and minor shrub species recruited – but enduring recruitment has been limited, due probably to increasing shade and browsing by fallow deer, supplemented by damage caused by falling trees and debarking, mainly by grey squirrels. Notes made in 2005 record that 'all low growth was browsed, including all bramble, rose and sprouts on hazel stools ... deer have certainly destroyed some potential recruits and prevented further establishment'. In 2000, 'safety fellings' damaged growth along the lower margin, and elms have continued to catch DED when they grow towards the subcanopy. Since 2007, fallow deer have been excluded, but recently muntjac have colonised the wood, including the stand below the cliff. By 2015, the stand structure had become even more irregular

(a)
(b)

and again semi-open (Fig. 5.8), while many underwood limes and elms had assumed weird and wonderful forms, due to regrowth after repeated falls and crushings.

5.7 General Features of Change over 145 Years

Combining young and old-growth allows us to piece together the course of near-natural stand development above the cliff over one and a half centuries. The graphs do not 'join' because of the contrasting histories of the old and the young-growth, but they still demonstrate the astonishingly steep fall in stem density from around 30,000/ha, as seedlings and new shoots from stumps establish themselves before the canopy closes to around 4000/ha at about 30 years and 700/ha within a century, stabilising around 650/ha at 145 years (see Tables 5.2–5.3). Imperfect though this is as a chronosequence, this is a real-life example of the 'self-thinning rule' (Westoby, 1984), which, when expressed as log (density) against age, forms a straight line with a gradient of –3/2. The irregularities disappear if we confine the expression to the original cohorts by excluding later recruits.

Complementary to this is an increase in basal area as the trees get taller and larger (see Tables 5.2–5.4). The basal area started very low, but it rose quickly to over 20 m^2/ha after 30 years and eventually oscillated around about 35 m^2/ha, much the same as most other minimum-intervention stands in Britain (Mountford, 2004: pp. 407–409). The steep rise in basal area corresponds with the fiercest competition between individual trees. However, the changes are not as tidy as one might expect. First, the theoretically steady accumulation of basal area can be disrupted by disturbances, in this case principally the advent of elm disease and the impact of drought, which set the basal area back to a lower level. Second, it can also be limited by the particular trees that happen to dominate, in this case the birch. These not only died disproportionately in the drought but also the survivors die out rapidly 60–70 years after stand initiation. As a result, the basal area plateaus at a low level until regeneration fills the gaps or the long-lived trees expand their crowns. Third, the value at which the basal area plateaus varies according to conditions. This shows in Lady Park through the contrast between the old-growth and the stand below the cliff: the latter seems to be limited to 25 m^2/ha because the rate of tree falls on the steepest slopes is higher.

Changes in stand structure deviate from the neat trend of events in the single-species stands of light demanders studied by Coomes and Allen (2007). For example, the typical trends in even-aged stands undergoing competitive thinning have been complicated by copious vegetative regeneration in stands of 70 and 140 years following the exclusion of deer. This is illustrated by some of the young-growth stands, whose canopies were not disrupted during the process (Fig. 5.9).

5.8 Disturbances and Gaps

In natural stands, stability is a mirage. The basal area is repeatedly reduced by various kinds of tree- and stand-destroying disturbances. The greatest disturbances in Lady Park have been the 1976 drought and the 1971 outbreak of elm disease, but tree falls associated with wind and the instability of large, unbalanced trees on steep slopes have also been important (Table 5.5; Fig. 5.10). Each disturbance impacts unevenly, affecting some species more than others and operating in some parts of the wood but not others. It is this that largely determines the variation in the time and place of the plateau value of basal area. If we exclude the exceptional openings associated with disease and the drought, the frequency of disturbances has increased as stands have aged, which is much to be expected, for – to take one kind of disturbance – older, taller trees are more vulnerable to blowdown than shorter, younger trees.

Disturbances that kill tall trees create gaps in the canopy. These are notoriously difficult to measure, especially in natural stands where the stratification is complex. In young stands, where individual trees occupy very little canopy space, gaps are occupied quickly by the expanding crowns of their neighbours. In mature stands, large trees occupy much more canopy space and are liable to smash other trees if they fall, so they leave correspondingly larger gaps.

In Lady Park, gaps formed by the loss of single trees are rarely circular (Fig. 5.11). Falls on steep slopes generate elongated gaps, because falling trees knock over others below them on the slope. Elsewhere, when groups of trees die, gaps are usually irregular and often leave any survivors within them in

Fig. 5.8. Varied structure in the stand below the cliff. While one part (a) is still recognisably coppice with standards, with beech over large-leaved lime, another (b) is a rapidly thinning hazel underwood with scattered lime.

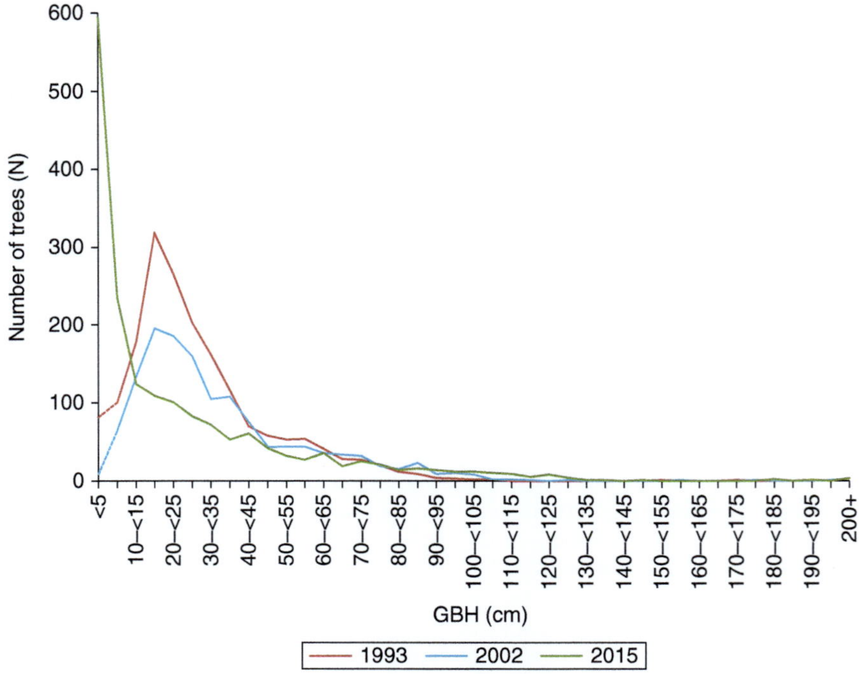

Fig. 5.9. Changes in the girth distribution of trees in the young-growth transects I–III, 1993–2015. Stems <5 cm GBH were only partially recorded in 1993 and 2002, as shown by the dashed lines. Competitive thinning between 1993 and 2002 continued in 2002–2015, but a new cohort developed by sprouting from the base of established trees. The pre-1993 population gradually lost most of its small stems, but most large individuals continued to grow rapidly.

grand isolation. Gaps have been created in pulses, notably after the 1976 drought, which came at 32 years in the young-growth and 106 years in the old-growth, and in 1971–1975 as elms died below the cliff. These pulses have somewhat obscured the anticipated long-term trend to increased rates of gap creation as trees age in the old-growth, but this now seems to be gathering pace. Total gap area was still substantial in 1983, but had been reduced by 1992, only to increase again after the millennium.

5.9 Population Changes

The fall of great trees is spectacular, but most mortality takes place in the subcanopy and underwood, where competition for light and perhaps moisture and nutrients weakens and kills the smaller stems. In the subcanopy, this exclusion process usually starts when neighbours shade part of the crown of the lower trees, thereby reducing vigour and eventually killing the tree, leaving a snag. Before death, some trees fight back by sprouting from the trunk, often close to the roots, but without a change of circumstances they are still doomed.

The countervailing process is regeneration and recruitment. The first is the initiation of new individuals; the second represents their growth to a threshold size, in our case 1.3 m high. Regeneration is obviously the key process at the start of stand development, but it finishes remarkably quickly. Seedlings established immediately after the gap forms or, better still, before it forms, face little competition from ground vegetation, but those germinating in later years face more competition, not only from burgeoning ground vegetation but also from older and taller saplings and the expanding crowns of trees on the edge of the gap. 'Last in, first out' is the rule, so by the time the canopy closes, the new stand may be almost even-aged.

Regeneration is not just a matter of seedlings. Many of the Lady Park trees started as stump sprouts from coppice stools, felled standards and fallen but living branches. (The distinction between coppice and felled standards is less clear than it might seem, for many standards are derived from coppice.) Regeneration is usually patchy, not only because some develops from coppice stools but also because seedlings tend to pick out what was bare

Table 5.5. A summary of disturbances in Lady Park Wood since 1945.

	Natural or influenced by people	Episodic or chronic	Frequency	Impact
Wind	Felling on the margin increases windthrow	Episodic	Very frequent, with rare large blowdowns	Small patches usually
Drought	Natural	Largely episodic, since ordinary dry spells are rarely lethal	Frequent mild droughts, but rare major droughts	Large selective effects on particular species (beech, birch)
Disease	Major outbreaks due mostly to introduced fungi, but native fungal attacks kill trees locally	Chronic, once they arrive	Introduced diseases impact once, then become a frequent presence	Major impacts on individual species when they arrive (elm, ash?)
Steep slope	Natural	Chronic	Constant	Patchy; operates on mature trees, but these 'knock on'
Snowfall	Natural	Episodic	Rare	Diffuse, more on weak trees and crown branches
Large herbivores	Native herbivores long since extinct; fallow and muntjac deer are naturalised	Chronic, but populations fluctuate and reserve has been fenced	Constant	Pervasive and selective for some tree and shrub species
Small mammals	Natural	Chronic with episodic outbreaks	Constant at low level, infrequent outbreaks	Pervasive
Grey squirrels	Introduced 1943	Chronic, with episodic outbreaks	Constant	Pervasive and selective, with slight patchy elements
Wild boar	Entered fenced reserve early in 2015	Let's see	Let's see	Patchy
Felling	By people	Episodic	Rare	Mostly upper and lower margins

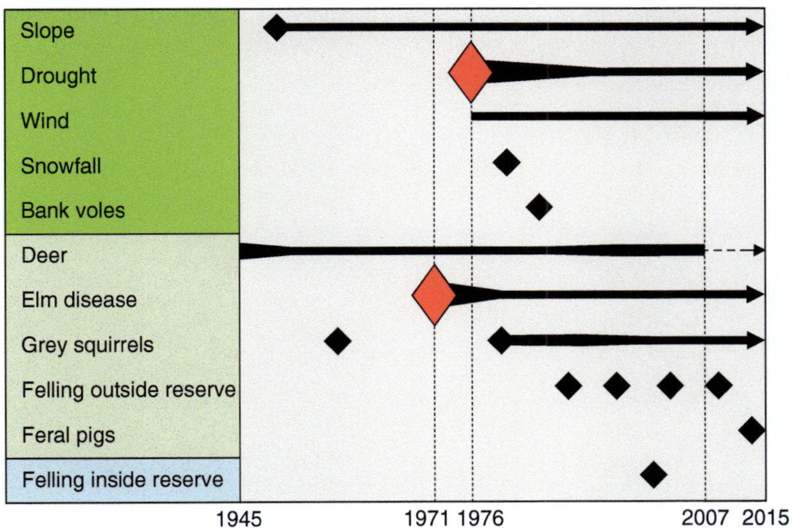

Fig. 5.10. A simplified history of disturbances in Lady Park Wood from 1945 to 2015 that separates natural disturbances (dark green) from semi-natural disturbances (pale green) and a management operation (blue). Events shown as diamonds; major events shown as red diamonds; chronic conditions shown as lines.

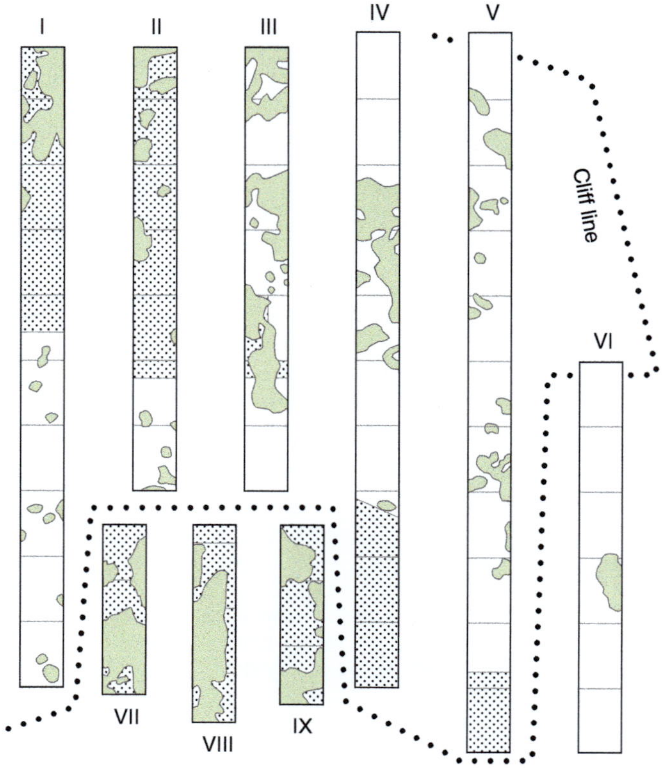

Fig. 5.11. Distribution of canopy gaps (green) as recorded in 1983 and 1985. The transects are arranged in relation to a 'London Underground' diagram of the cliff line. Old-growth stands are indicated by stippled shading.

ground before the felling, patches with minimal bramble and brash and patches disturbed by the forestry operations. Clusters established at the outset are remarkably persistent: one can still, for example, identify ash groups in mature stands.

The combination of mortality and recruitment results in population change. Two examples illustrate how the rate of change itself changes with time. In the old-growth (Table 5.6), stem density declined steadily until about 1977 (stand age 107 years), after which the density fluctuated, with a rough balance between mortality and recruitment, and mortality rates fluctuated around 2% per annum. By 2010, 333 trees in the old-growth remained from those alive in 1945; that is, 71% of all the individuals in the stand when it was 75 years old had died by the time the stand was 140 years old. There were four defining episodes:

- Intense competition between 1955 and 1977, when mortality greatly exceeded recruitment.
- High turnover in 1977–1992, when mortality and recruitment accelerated in the immediate aftermath of the 1976 drought. Recruitment was delayed until 1986–1992, when regeneration following the drought actually reached 1.3 m.
- Renewed high mortality, 1992–2010, when the recruits of 1986–1992 were excluded by shade or grazed out by deer.
- Since the last recording in 2010, many base sprouts that grew up after the perimeter fence was completed in 2007 have been recruited.

In contrast, the mortality and recruitment rates in the young-growth were generally higher (Table 5.7). Recruitment accelerated enormously after 2010 as post-2007 base sprouts reached above 1.3 m. Approximately 49% of the 1977 trees were surviving in 2002.

The long-term pattern of population change is reasonably clear. Mortality rates tend to be more steady and less episodic than the recruitment rates. When the stand is not disturbed, the steady background process of stem exclusion dominates and regeneration has difficulty becoming established. When the stand is disturbed, there is a small spike

Table 5.6. Changes in the population of all trees in the old-growth transects.

Year	Seasons	N at start of period	Density (N/ha)	Losses	Gains	Mortality rate (%/year)	Recruitment rate (N/ha/year)
1945–1955	10	1158	1287	−186	+137	1.74	15.2
1955–1977	23	1109	1233	−495	+23	2.54	1.1
1977–1983	6	637	708	−109	+72	3.08	13.3
1983–1986	2.5	600	666	−31	+18	2.10	8.0
1986–1992	5.5	587	652	−53	+242	1.71	48.9
1992–2000	9	776	863	−191	+74	3.09	9.2
2000–2010	9	659	732	−166	+113	3.17	14.0
2010		606	674				

Table 5.7. Changes in the population of all trees in the young-growth stands.

Year	Seasons	N at start of period	Density (N/ha)	Losses	Gains	Mortality rate (%/year)	Recruitment rate (N/ha/year)
1976–1977[a]	1	7993		−1085		13.57	
1977–1983	6	6908	3370	−1075	+99	2.78	8.05
1983–1993	9.5	5932	2894	−1237	+163	2.43	8.37
1993–2002	9	4858	2370	−1326	+125	3.48	6.10
2002–2013/15	11.5	3657	1784	−1313	+1889	3.79	80.14
2013/15		4233	2065				

Note: [a]Mortality rates have been calculated by adding the birch snags recorded in 1977 to the live trees and assuming that all the birch had died in 1976/7.

in mortality (gap formers are mostly few and large) and generally a larger spike in regeneration and eventually recruitment. Recruitment dominates at stand initiation, falls close to zero during the thicket stage (largely unrecorded here) and resumes when gaps form and when the canopy becomes structurally diverse enough to allow more light to penetrate below.

This pattern has been obscured to some degree by deer. It was a surprise when saplings failed to develop in the gaps created after 1976. Seedlings were common enough, but several factors seemed to have suppressed them – not just deer but also developing bramble thickets, voles, drought and crown expansion of mature trees, probably in that order of importance. After 1993, deer browsing was so intense that the bramble was largely eliminated, but this was no benefit to tree seedlings. After 2007, the fence eliminated deer browsing, more seedlings survived, but the bramble spread. In 2015, the race between sapling growth and bramble spread seemed likely to be won by the saplings, for many 5- to 10-year-old ash and beech, together with a scatter of wych elms, limes, oaks and maples, were well established and approaching the 1.3 m recruitment threshold, but it will be several years before they grow out of the reach of deer.

By 2015, it was also becoming obvious that vegetative regeneration could also be important. Throughout the stand, hazel, beech and especially limes were generating new base sprouts. Moreover, lime, ash, oak and beech trees blown down in gaps often survived to generate colonnades of new shoots from their trunks. The relative importance of seedling and vegetative regeneration in the next generation of trees will only become clear by 2050 or later.

5.10 Dead Wood

The feature that most often impresses visitors is the amount of dead wood. Not only must they pick their way round fallen trees but also gaze apprehensively at snags and dead trees standing leaning on living trees. Less often noticed are the many dead branches in the crowns of living trees and the scatter of smaller dead branches hidden among leaf litter and ground vegetation. Dead material increased substantially in the 1970s following elm disease and the 1976 drought, and by 1997 the volumes in the old-growth had built up to levels commonly found in so-called virgin forests. This is discussed further in Chapter 12, this volume.

5.11 Composition and Diversity

Diversity depends on the stage of succession and the frequency of disturbance, but the degree of any change is scale dependent. Take the young-growth stands: at the scale of a quarter section, diversity climbs quickly to a peak as new trees and shrubs colonise the vacant ground, but it descends rapidly through the thicket stage (Table 5.8). At larger scales, the peak is later and less pronounced (Table 5.9). Combining all the young-growth transects, the only species to have been lost entirely were: (i) alder, which regenerated in very small numbers on transect VI near the flood plain; (ii) 'bird cherry' – probably a mistake for gean – which was rare in the early thicket; (iii) broom, a common, but short-lived pioneer; (iv) the single wayfaring tree; and (v) the only privet to grow tall enough to qualify.

There is also a marked change in the balance between species groups as the stands age (see Tables 5.2–5.4). Pioneer shrubs, such as broom, vanish within 15 years, but the fast-growing, light-demanding trees, notably birch, dominate for longer. Eventually,

Table 5.8. Numbers of individuals of each tree and shrub species recorded in transect V: 800–850N, a quarter section that developed as a birch stand but which is now virtually an opening with numerous birch stumps and well-rotted, fallen logs.

	1944	1948	1958	1977	1983	1985	1993	2002	2013
Ash	A	111	6	2	2	2	2	1	1
Beech		1	1						
Maple		2							
Oak		3							
Aspen		87	9						
Birch	A	317	72	58[a]	40	31	17	13	3
Sallow		2							
Broom		36							
Dogwood		21	6	2					
Hawthorn		5		3	3	2	2	2	1
Hazel		2		1					[b]
Holly		1							
Individuals, N		588	94	65	42	39	21	16	5
Density N/ha		38,356	6132	4240	2740	2544	1370	1044	326
Number of species	2	12	5	5	3	3	3	3	3
Basal area m²/ha	0			22.8[a]	19.2	18.6	17.3	16.4	11.9

Notes: A = recorded as abundant. [a]By adding in the ten birch snags recorded in 1977, the basal area in 1976 was 24.5 m²/ha; [b]a hazel had fallen into the plot and produced six stems reaching above 1.3 m from the prostrate trunk, but none had rooted within the plot.

Table 5.9. Changes in the number of tree and shrub species recorded in the young-growth sections of transects since 1944.

Transect	Area (ha)	1944[a]	1948[a]	1955[a]	1977	1983	1993	2002	2013/15
I	0.33	12	12	18	13	12	12	12	9
II	0.10	11	11	8	12	12	12	10	8
III	0.30	12	15	16	17	17	17	17	14
IV	0.43	13	14	15	17	16	15	14	12
V	0.53	11	12	16	16	16	15	14	13
VI	0.37	11	19	19	17	13	13	12	8
ALL YOUNG-GROWTH ON TRANSECTS	2.06	21	23	25	24	21	20	20	20

Note: [a]All species recorded as seedlings, saplings and coppice growth, irrespective of height.

these pioneer trees drop out and the long-lived trees assume dominance: this transition took place from about 30–40 years in Lady Park, undoubtedly brought forward by the drought. Thereafter, as the old-growth demonstrates, the long-lived canopy trees dominate utterly, while small trees grow slowly below, pioneer trees are reduced to a small rump and shrubs vanish almost entirely. (The young-growth, where beech are squirrel damaged and less dense, may retain more underwood in the coming decades.) The naturally more disturbed stand below the cliff followed the same course, but perpetual tree falls limited the dominance of the main canopy trees and permitted a greater representation of small trees and underwood below. Most of the species that vanished from the transects remained in very small numbers elsewhere in the stands.

Time has also brought marked changes in the balance between the 'principal trees' in the canopy. Wych elm has lost out, of course, while ash and limes have gained. In the old-growth, beech has held on to its dominance despite the effects of drought and grey squirrels. Oak has just about held its own.

The above generalisations conceal a great deal of diversity. Thus, parts of the young-growth are lime woods, in the sense that small-leaved lime dominates a patch, whereas other patches are variously ash woods, hazel woods, beech woods and even, still, birch woods. Such differences have emerged since 1976 from an apparently homogeneous birch wood. Below the cliff, parts are now emerging as large-leaved lime woodland.

5.12 Looking Ahead: Prospects for the Unmanaged Stands

How will the wood develop in the future? Can we in fact make any predictions at all?

Until 1976, we would have predicted that beech would eventually become the overwhelming dominant, except below the cliff, for it was growing faster, surviving better and regenerating earlier than other canopy trees. However, the combination of the 1976 drought, voles and grey squirrels killed mature trees and established saplings, permanently reduced the growth rate of surviving drought-damaged beech and reduced many pole-stage beech to shrubs. Since ash and limes were unaffected and both could regenerate, we expected these species to become more prominent, i.e. the successional trajectory had changed. Likewise below the cliff: the trend to wych elm–ash dominance was halted by DED, leaving ash and limes in possession.

By 2015, the basal area of the old-growth had reached a plateau from which it seemed unlikely to increase further. Stem density and small-scale diversity, however, appeared to have 'bottomed out'. Large numbers of saplings were established but had not grown through the 1.3 m recruitment threshold. Most were ash, but there were many beech and a scatter of oak, wych elm, birch, sallow and both lime species. In addition, base sprouts of beech, hazel and both limes were common and widespread, and some of the shrubs were also regenerating in small quantities. A new ash-dominated generation seemed likely to develop; birches seemed likely to be reduced to small groups on thin soils; stem density looked likely to increase dramatically and small-scale diversity seemed set to increase.

Past experience suggests that, due to disturbances, none of the major forest trees can achieve long-term dominance, save perhaps for small-leaved lime above the cliff and large-leaved lime below. Sure, limes are vulnerable to deer and have been slow to spread, but oak (inability to regenerate in shade), beech (vulnerable to drought, wind, grey squirrels), ash (limited ability to recruit in shade) and wych elm (chronic disease) each have greater Achilles heels.

The problem is the known unknowns and the unknown unknowns. We know that disturbances can accelerate or change the successional trajectory, but we cannot predict when and how severely they will occur. Climate is predicted to become less stable and more extreme, but we can only place probabilities against the arrival of the next drought or storm. It is also becoming warmer, which will affect growth and regeneration.

Diseases seem more immediately predictable. In particular, ash disease is on the doorstep and promises to be a major disturbance. If it kills most of the saplings as well as the canopy ash, we anticipate that the stands will open up, wind falls of other canopy tree species will increase, hazel will be rejuvenated and the saplings of beech will face less competition. Birches, sallows and various shrub species may prosper, as well as bramble. In general, tree diseases and pests are being spread further and faster, but we do not know which ones will arrive, nor which will evolve into something new.

Deer are a further source of uncertainty. We know that they can prevent recruitment, but cannot be sure how long the fence will last or whether a change in local sentiment and management resources will allow effective control. At the time of writing, the perimeter fence still shuts out the fallow deer, but it is at risk from feral pigs, which not only have accessed the reserve themselves but also could let in the fallow deer. The newly arrived muntjac seem resident within the fence, and roe deer are in the offing. If the deer numbers build up within the fence, most prospective recruits will be browsed to oblivion.

Unless a 'Dutch lime disease' hoves into view or a grey squirrel pox allows the beech to grow once more into splendid trees, the two limes may well be 'the last trees standing'. The existing oaks will last another two centuries or more, and it would only take one favourable combination of a mast year, a major blowdown and freedom from excessive deer to renew the population. But, who knows? The fate of predictions is to seem absurd to our successors.

5.13 Comparisons with Other Woods

Have developments at Lady Park been typical of 'natural' temperate broadleaved woodland or exceptional? Other studies of tree populations at the individual tree level in unmanaged deciduous woodland show both similarities and differences, but the overwhelming impression is that each wood is a distinctive and individual combination of circumstances, composition and history.

Two American studies based on long-term plots in unmanaged mixed deciduous stands illustrate the point. In 1927, plots were established in 20- to 40-year-old mixed woodland of red maple, white oak, black birch, northern red oak, yellow birch and others in Meshomasic State Forest in eastern Connecticut and recorded every decade (Stephens and Ward, 1992; Ward, 2013). Like Lady Park, the study inherited a disturbed woodland, for the wood had formerly been grazed (though it had always retained some wooded cover) and blight had killed most chestnuts. Subsequently, the plots were defoliated by gipsy moth and canker worm in 1957–1967 and 1981–1982; disease killed elms in 1937–1957; dogwoods were killed by disease in 1977–1987; drought struck in 1957–1967; and fires burned locally in 1932. Like the young-growth in Lady Park, tree density decreased, while basal area and tree size increased. However, population turnover was much higher: mortality rates were steadier and generally higher (2.7–3.8%/year), while recruitment rates were much higher but fluctuated widely (0.5–5.7%/year or 17–73 stems/ha/year). In a sense, though the disturbances were different, this study indicates how Lady Park might have performed without the 1976 drought and in the absence of excessive deer browsing.

The other study was in the Davis-Purdue old-growth reserve in Indiana, where all the trees above 10 cm DBH in a 5.8 ha plot were measured and charted in 1926 and again in 1976 (Parker *et al.*, 1985). This stand, too, had been grazed in the past and had sustained a bout of elm disease, but many oaks of 150–200 years were present. The mortality rate (0.94%/year) was lower than Lady Park, but this was a minimal value because trees that were recruited after 1927 and died before 1987 could not be recorded. The recruitment rate (2.81%/year) was very much higher, again underlining the impact of deer in Lady Park. Very substantial differences were recorded between species, some of which were matched at Lady Park Wood, notably the low recruitment rate of oaks and the loss of elms to DED and their subsequent rapid regeneration. Sugar maple regenerated prolifically: this was a shade succession that would have had its counterpart at Lady Park if the beech had been allowed to regenerate. The Indiana study also shows that mortality tends to be higher in both the smallest size classes (due to competition) and the largest (due to senescence), which is broadly what happens in Lady Park. Radial growth rates differed markedly between species and between individuals of the same initial diameters, but were marginally fastest in the 50–80 cm DBH size classes.

Closer to home, we have ourselves followed the fortunes of an ancient beech–pedunculate oak–holly wood pasture at Denny Wood in the New Forest (Mountford *et al.*, 1999). By 1956, when the study started, this stand had developed a three-generation structure as described by Peterken and Tubbs (1965), in which the oldest beech and oak were over 300 years old. In the 40 years to 1996, it was subjected to alternating very wet and very dry summers in the 1950s, severe drought in 1976, lesser droughts in the 1980s and 1990s, and storms in 1987 and 1990. It was also grazed perpetually by deer and latterly ponies, and had been colonised by grey squirrels. Oak had ceased regenerating some time before 1956, but beech and holly regenerated strongly until the early 1960s – a classic shade succession.

The growth and mortality of beech dominated stand dynamics. Until 1976, most mortality was in the smaller trees, but thereafter rates for the older trees accelerated, due more to the 1976 drought than to later storms. Mortality of the 18th century generation rose from 1%/year to 3.8%/year in 1984–1988, concentrated in the slower-growing trees and often delayed, i.e. trees were initially weakened rather than killed outright, much like Lady Park. Mortality in the mid-19th century generation was steady and concentrated in the smaller trees, indicating that competition dominated but drought weakened some fast-growing trees (which were eventually killed by Honey fungus) and grey squirrels severely debarked many others. Mortality of the 20th century generation was initially high and concentrated in smaller trees, even though the canopy had opened: debarking by ponies and squirrels killed many. The older generations grew slower during the drought, but the survivors accelerated thereafter.

Oak mortality was generally slower at around 1%/year, but trebled briefly in 1984–1988. It was concentrated in the smaller trees, but was supplemented by windthrows and knock-ons from falling beech. Drought may also have contributed. Deer and ponies killed most of the holly underwood by browsing and debarking, and prevented regeneration, even below large canopy gaps.

Together with Lady Park, these examples illustrate two important general points. First, population changes in any wood are complicated (and difficult to explain clearly!), varying between species and in time and space. Second, each wood that becomes a 'natural' reserve is different, possessing its particular combination of species, formed by its own unique developmental history and impacted by its particular sequence and mix of disturbances after it becomes a reserve. We can recognise many common elements, but we would need far more examples to generalise about stand development. Accordingly, when we examine a wood in depth, it is the individuality that most impresses – just like people, in fact.

5.14 Stands in the Managed Compartment

The managed compartment was brought within the reserve under the agreement of 1984. It had been retained as native broadleaved woodland in the more accessible parts of which beech and a small group of larch had been underplanted. It has never been fenced, so fallow deer and, latterly, wild boar have had free access. It has also been more open to the public, though the impacts have been limited to an unofficial narrow, off-road motorcycle trail and an even more unofficial travellers' camp in the summer of 2012.

Five kinds of stand can be defined within the compartment, not all of which have actually been 'managed'. Apart from a fragment of unthinned young-growth, they all add to our understanding of local woodland development.

5.14.1 Old-growth

On and below the main cliff, the stand is still visibly grown-out coppice with standards, with both oak and beech standards and coppice of limes (mainly small leaved), beech, sessile oak, ash and hazel. Limes were probably logged selectively in 1939–1945, and large ash and dead elms at the base of the slope were felled recently in the interests of safety. It occupies the same topographical position as the stand below the cliff in the fenced reserve, but the slope is less steep, trees have fallen only when old stools have split; so, despite the loss of elms to DED, the basal area is much higher. In 1984, it reached 32 m^2/ha on transect X and over 40 m^2/ha locally where large oaks had been spared on the crags. Stem density decreased from 2219 in 1984 to 862 trees/ha in 2014. Of the 404 stems present in 1984, 130 survived to 2014. Structurally, it is more like the fenced old-growth above the cliff than the stand below the cliff.

5.14.2 Cliff woodland

The managed stand also includes limited areas of cliff woodland on both the lower (main) cliffs and in the dry valley near the top of the compartment, where yew, limes, beech and other species grow on the rock outcrops and sometimes out of crevices in the cliff face. Neither cliff is as large as the main cliff, but it is accessible enough to be able to measure trees where transect X crosses the cliff. Other measurements of trees on cliffs are available from the base of transects II–V and the top of transect V and VI, but these are fragmentary. The true cliff woodland is simply inaccessible.

Cliff woodland is remarkably varied. At their most sheer, cliffs are covered in little more than small shrubs or sheets of ivy, but on their fringes and on lesser outcrops, closed woodland can develop and large trees have generally been spared the axe. Site conditions favour yew, beech, whitebeams,

Fig. 5.12. Thinned young-growth in the managed compartment shortly after the 1984 thinning, showing strong coppice regrowth from hazel and lime stools.

large-leaved lime and small shrubs such as rose and privet, which can be surprisingly vigorous. If they remain firmly rooted, the trees may develop to great ages and sizes.

5.14.3 Thinned stand

The thinned stands are a form of young-growth. They were felled in 1942–1944, leaving a shelterwood of beech and oak standards, and later they appeared to have been patchily underplanted with beech. In 1982–1983, most parts were thinned and some standards were harvested. Regeneration of seedlings and coppice shoots was reasonably prolific, and the underwood developed into a thicket (Fig. 5.12). However, in 1992–1994, most of the stands were thinned again, and in order to do so, the woodmen cut and crushed virtually all the 10-year-old regeneration. This left an open stand of mainly beech, ash and limes, with some oak, birches and aspen. Unfortunately, by 1994, fallow deer had become so numerous that regeneration from seedlings failed completely and much coppice growth was browsed out. The stand remains open, but within the 1999 exclosure, beech, hazel, dogwood and small-leaved lime (from layers) have regenerated.

We can track changes in structure and composition from the Williams plots, established in 1979, and transect X, established in 1984, but only approximately. In 1944, the basal area would have been low, due to wartime fellings. It increased rapidly to about 35 m²/ha in 1983, of which about half was removed in the first thinning. It increased again until 1993, when the stand was thinned down to about 16 m²/ha. Since then, the retained trees have grown enough to return it to around 23 m²/ha.

5.15 Commentary

Growth in Lady Park and similar 'natural' reserves balances predictable changes against a sequence of unpredictable events. The predictable changes have been growth and thinning; the unpredictable factors have been natural impacts (notably drought), supplemented by semi-natural perturbations (notably elm disease). Between the two are such factors as tree falls resulting from slope, which are both predictable and, at the small scale and over a short timescale, unpredictable. In the very long run, however, one supposes that even the unpredictable storms, droughts and other natural phenomena are predictable, in the sense that they are bound to happen sometime. They become a matter of probability in any given period.

Clearly, one cannot predict what will happen next, merely state probabilities. One can describe how the stand has been changing up to the present (the recent trajectory) and then extrapolate forward to predict a future state, but there can be no guarantee that this will be realised. Accordingly, there can be no single 'natural' structure. In Lady Park, the predictable processes and unpredictable events have combined to generate an old-growth stand, but a storm blowing from the north-west could fell the lot and eventually generate a new, young stand, and both would be just as natural as the existing old-growth. Further, much the same can be said about the composition: there is no single 'right' composition, although we do see a place for all the extant species. At Lady Park, the trend until 1976 was towards dominance by beech, but the drought, squirrels and voles caused a change of trajectory towards a greater contribution by ash and limes. Soon, if ash disease arrives, there would be another change, presumably towards lime–beech dominance. If elm disease abates, as it has done in the past, wych elm may reassert itself.

In theory, growth of the managed stand should be much more predictable, but that is a moot point. The foresters of the 1980s and 1990s presumably thought they would achieve regeneration after the thinnings and a fast-growing overstorey in which beech would be prominent. In fact, they have been unable to control fallow deer and grey squirrels, so regeneration has failed and the beech timber has been wrecked. Elsewhere, the 1987 storm levelled both unmanaged and carefully managed stands equally, and disease outbreaks make no distinction either. Add to that unforeseen changes in the public sentiment and economic circumstances, and forest management plans must allow for unpredictability, however much the written plans may purport to dictate the outcome.

6 Ash: The Tree in the Spotlight

6.1 Trees in General

Changes in the stand show how trees develop and change en masse, but, like any human community, the 33 tree and shrub species found in Lady Park also coexist with and compete against each other. They are like characters in a play staged on the slopes and cliffs of the Wye Gorge, the course of which is determined not just by the collective dynamics of the community but also by their individual personalities and how they react to circumstances, events and each other. Some are dominant and others subordinate; some occupy all arboreal walks of life, while others stick to their particular, if narrow, niche. Some are opportunists enjoying a fast life, but a short one, while others are slow and persistent late developers, and a few manage to be both. The race does not always go to the strong, but to those with well-developed survival instincts. Like us, each may fall prey to chance and the unexpected, and react in different ways. Not that there is a conclusion to a race: it continues indefinitely, with endless twists and turns.

Trees can be grouped by their various characters. Beech, sessile oak, ash, wych elm and the two lime species are (or were, in the case of elm) the long-lived canopy trees. These 'make the weather', physically dominate the stand, create the environment for lesser species and, through their responses to adversity, determine the opportunities for and limitations on others. Then there is a group of short-lived canopy trees, alder, aspen, sallow, the two birches and gean, which can dominate the stand for a while but, in order to survive among the long-lived dominants, must be quick to take their chances for regeneration and grow rapidly when they get the opportunity. Third, there are subordinate trees – hazel, holly, hawthorn, field maple, rowan, service and whitebeam – that rarely grow tall enough to occupy the canopy but which adapt to circumstances, either by growing in shade or by concentrating their efforts where the dominants grow poorly. Then there are truly subordinate shrubs – dogwood, wayfaring tree, guelder rose, spindle, rose – that survive by various means in the undergrowth and on the margins. Finally, there are the naturalised immigrants, sweet chestnut and sycamore, which could eventually mix with the long-lived canopy trees.

At Lady Park, we have an opportunity to follow the fortunes of native tree species in near-natural mixtures, competing with each other in young and mature stands. This contrasts with the growth in plantations and other managed stands, which is carefully regulated and moderately predictable, except when overwhelmed by storms or other calamities. We will illustrate the detail by concentrating on ash, a reasonably long-lived canopy tree that regenerates well on all the local soil types and whose population is not dominated by individuals inherited from the former coppice, and then consider other trees in less detail in later chapters. We will find that the population of ash and other tree species develop like the stand as a whole, but each shows distinctive features.

6.2 Ash in Britain

'Ash groves are graceful' says the song, and indeed ash trees are tall, slim, pale, light and airy. Left to itself on good soils and with a moderate amount of room, ash will grow fast and straight, forming a lollipop-shaped crown with perhaps a few weak lower branches. These are the trees foresters seek to grow in managed woods. The tallest, according to

Fig. 6.1. A tall, slender ash growing below the cliff. Photographed in the mid-1980s, it was left isolated after the co-dominant wych elms had been killed by disease.

the Tree Register (2011), reached 40 m, with a girth of 280 cm. Up close, it is easily recognised as a fat-twigged tree with pale, finely fissured bark, black buds in pairs and leaves divided into pairs of leaflets.

In hedges and other open positions, ash still grows fast, but it forms a deeper and wider crown and does not develop such a tall, clean trunk. Many old field ash are pollards, whose crowns have regenerated with immense vigour after lopping, and whose trunks have become much wider and more hollow than they ever grow in woods, reaching an astonishing girth of 1078 cm (www.treeregister.org/champion-trees.shtml). In ancient woods, most ash were coppiced a short distance above the ground, the stumps re-sprouting vigorously from the outside edge, and after centuries of coppicing, they developed into immense, hollow, almost ring-like stools. Both pollards and coppice were neglected during the 20th century, so they developed larger, heavier, but fewer trunks, until the pollards become great spreading trees and the coppice developed into multi-stemmed individuals that sometimes seemed to be two or three separate trees. Now, sadly, many are simply falling apart under their own weight.

That is ash as a managed tree, but how does it perform when it is left to grow naturally? Many ash woods have regenerated naturally on abandoned pasture and arable, including the Dovedale ash woods (in Derbyshire and Staffordshire) and other famous stands. Most such secondary ash woods are the product of one flush of regeneration that has developed almost like an ash plantation. Other ash-dominated woods are ancient. Here, the ash has long been resident, but normally coexists with other species in mixed stands. These, too, are often even-aged, having been coppiced well into the 20th century – or, rather, they have strong even-aged elements, for the former coppice usually grows with older standards.

In Lady Park, we can follow an ash population through almost 150 years of near-natural development in competition principally with beech, sessile oak, small-leaved lime, large-leaved lime, wych elm and birch. In 1897, ash was more abundant as coppice than oak. No ash standards were mentioned, either then or in 1608, but two remain in the wood, so evidently there were a few. Otherwise, they appear to have regrown or regenerated after the post-1870 ash were thinned out in 1902 and *c.*1930: certainly, an ash that fell near transect I in 1999 had 85 rings at 7 m and could well have been a post-1902 recruit. Below the cliff, the origins could be more complex: of five trees near the base of the slope, two originated as coppice in 1864 and 1879 and three as maidens in 1880, 1920 and 1931.

Today, and presumably also in the past, ash is uncommon on the acidic soils in the core of the wood, but common and patchily dominant on the thin limestone soils at the top and the moist calcareous loams below the cliff.

6.3 Regeneration after Felling

Ash fruits copiously and often, and thereby maintains a substantial population of seedlings (Fig. 6.2). Like the poor, supposedly, ash seedlings are always with us, so much so that Hill reported in 1897 that seedling ash were very numerous in the then 26-year-old coppice. In some recent springs and early summers, they were so thick on the ground under the mature canopy that they covered half the ground surface, perhaps up to 100 seedlings/m^2 (equivalent to 1 million/ha). Seedling populations peak in spring with the addition of new germinants, then decrease steadily through the year, but huge numbers still survive and can be found as clusters of black-budded bristles among the winter leaf litter. The density of small seedlings is probably greater near seed trees and where ground vegetation is otherwise sparse.

Ash ranked second only to birch in the thicket of saplings that developed after the wartime felling. In late September 1944, ash seedlings were generally abundant, but patchy in the southern half of the reserve. Four growing seasons later, in October 1948 and March 1949, seedlings and small saplings were still frequent to abundant throughout, and a few had reached 7–10 feet (2–3 m). Some ashlings had a head start, having germinated a few years before 1942, and these tended to become the tallest and strongest in the developing thicket, though very few overtopped the other vegetation at this stage. Summarising his impressions, Dr Eustace Jones observed that:

> the ash evidently starts slowly; seedlings which are probably 5 years old are only 6–18in. tall over much of the area, though on the moist soils of the lower slopes they may be 3–6ft tall by this age. Many of the saplings which are 2–4ft tall are no thicker than a pencil, and are supported by surrounding vegetation, and there are very many small saplings only 1–2ft tall.

By 1955, the successful saplings were well above the height at which they could be damaged by deer. For example, transect IV: 200–300 was:

Fig. 6.2. Ash and wych elm seedlings.

> almost continuous birch with patches of ash. There has been dense continuous blackberry up to 3–6ft tall throughout, mostly now dead. Ash saplings (are) very abundant amongst it; many are exceedingly drawn up and 8–10ft tall but no thicker than a pencil.

However, many ash saplings had been pushed over by brambles and others had been 'fantastically contorted' by clematis and the whipping action of birch saplings. Mostly, they were clustered into pure, dense groups (Fig. 6.3). For example, in transect VI, one quarter section had none, while another had 9784 ash saplings/ha. Most were:

> very slender and whippy and have lost leaders several times and are obviously not going to last. Grazing is only light – indeed it is often not possible to be certain whether they have been bitten or merely died back and broken.

Moreover,

> Dense tangles of Rubus have pushed over saplings of ash. Ash saplings 8–12ft high may be little more than 0.25 inches diameter and worming their way through crowns of coppice shoots or birch. Ash seedlings under suppression die back repeatedly and produce fresh shoots.

Deer had certainly browsed some, for tooth marks were visible on larger saplings and the adjacent bramble had been eaten.

After counting the growth rings of a sample of ash saplings in 1958, it seemed fairly clear that the developing ash population arose from seedlings

Fig. 6.3. The clusters of ash saplings in regenerating stands persist until late in stand development. In this lightly thinned young-growth stand, the ash cluster stands out as pale trunks among the darker beech.

established just before and shortly after the wartime fellings; that opportunities for regeneration diminished as ground vegetation and coppice stools developed; but that new seedlings still entered the race in some patches after 1949. Standard beech and oak prevented regeneration locally, partly because deer were attracted to shaded ground and ate the ash seedlings among the grass. On unshaded ground, ash had to do battle with bramble thickets, mainly by extending rapidly at the expense of stem strength, and tended to succeed in clusters. Location determined their fates: ash growing by deer tracks were browsed; those growing on the margins of charcoal hearths were perhaps favoured; and any ash that reached above 5 feet (1.5 m) had effectively got away, albeit in some cases knocked out of shape by bramble, clematis and neighbouring birch saplings. Nevertheless, ash prospered as a population because potential dominants emerged early: well within the first 15 years, a few individuals were clearly growing much faster than the rest, probably older individuals that had survived the wartime felling. Sixty years later these were, with few exceptions, still the leaders.

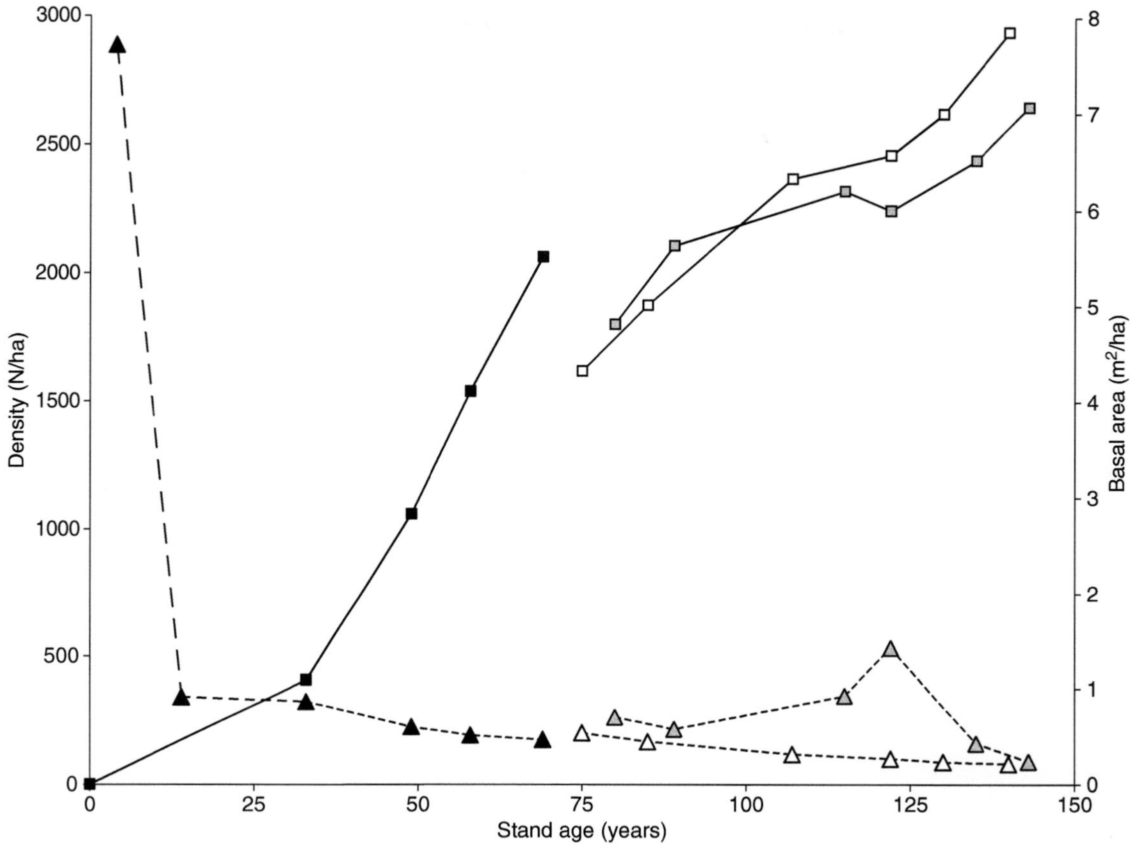

Fig. 6.4. Changes in the density (triangles) and basal area (squares) of ash in relation to age in unmanaged stands in Lady Park Wood. This is a chronosequence: black = young-growth; clear = old-growth; shaded = stands below the cliff.

6.4 Outline of Ash Population Changes over 150 Years

Recruitment ceased about 1950 and thereafter the new stands grew as a thicket of saplings, all more or less the same age. The changes that took place over the next century and a half can be described in broad terms by combining the young-growth and the old-growth into a single sequence (Fig. 6.4).

The total contribution ash makes to the stand can be measured by basal area (Tables 5.2–5.4). This climbs fairly steadily from zero at the outset to about 7–8 m²/ha at 143 years. Progress has been irregular. In the old-growth and below the cliff, ash has oscillated at 19–21% and 26–33% of stand basal area, respectively, while in the young-growth the steady loss of birch helped it to climb from 5% in 1976 to 23% in 2013. While its basal area is still increasing in all three stands, its relative contribution has remained much larger below the cliff.

Conversely, the density of ash declined throughout the 143 years. From 100,000s/ha in 1948, it declined steeply to thousands in 1958, about 300 by 33 years and about 60 by 143 years. The rate of decrease seems to be heading for a low point of 50–60 trees/ha. The old-growth at 75 years had a higher density than the young-growth at 68 years: evidently, the *c.*1930 thinning allowed a new cohort of ash to enter the stand.

With an increasing basal area being accumulated in ever fewer trees, the size of trees and the rate at which individual trees accumulate basal area have both increased. One measure of this is basal area/tree, the increase of which steadily accelerates.

These trends are unlikely to continue for much longer. The large canopy ash are increasingly vulnerable to wind, disease and loss of vigour, while gaps are – deer permitting – likely to be filled with large numbers of ash saplings. Basal area may plateau, but density would increase and average tree size would decrease.

6.5 Population Turnover: Mortality and Recruitment

By the time the young-growth stand had reached 32 years in 1977, 95% of all the ash recorded in 1948 had long since died. The population density was still 324 ash/ha (Table 6.1), but it continued to decline and had halved again by 2013. The mortality rate was high in 1977–1983, partly because several smaller ash were struck by falling birches killed by the drought. Thereafter, it decreased steadily, probably because competition from birch had been reduced after the drought. Recruitment was always very limited, but it picked up a little after the drought. Comprising both regeneration from seed and base sprouts from established trees, most of it failed due to deer browsing and debarking, but six of the 1983–1993 seedling recruits survived to 2013, and one was so well placed in a gap on the colluvial fan that by 2013 it had almost reached the canopy, potentially a future elite tree.

In 1977, ash was common on the steep and alkaline ground, especially the upper plateau, and sparse on acid soils above the cliff (Table 6.2): this pattern was still clear in 2013, despite large differences between zones in survival rate. Small ash greatly outnumbered larger ash throughout, but especially on the upper plateau and steep slopes. Some 38% survived to 2013, but mortality was particularly heavy on the lower steep slopes and colluvial fan, where they met severe competition from hazel, beech standards and other ash. All the large ash and most of the medium ash survived. The latter fared worst on the upper plateau, where competition between ash was severe, and on the colluvial fan.

Much the same story emerges from the old-growth (Table 6.3). In 1945, when the stand was 75 years old, the density of 200 trees/ha was slightly above that of the young-growth in 2013 (when the stand was 68 years old). By 2010, it had fallen steadily to 64 trees/ha. The mortality rate was higher in

Table 6.1. Changes in the ash population in the young-growth stands.

Year	Initial density (N/ha)	Live at start	Losses	Gains	Mortality rate (%/year)	Recruitment rate (N/ha/year)
1977–1983	324	664	–128	+1	3.51	0.08
1983–1993	262	537	–95	+18	2.03	0.92
1993–2002	224	460	–69	+1	1.79	0.05
2002–2013	191	392	–44	+8	1.03	0.34
2013	174	356				

Table 6.2. Survival of ash in young-growth stands on different topographical zones, 1977–2013.

	All ash, 1977	Number of ash stems in 1977 in three size classes (DBH)			Survival to 2013 (%)		
Zone	Density N/ha	Small 1–6 cm	Medium 7–12 cm	Large >12 cm	Small	Medium	Large
Upper plateau	839	58	18	1	27	78	(100)
Upper steep slopes	484	67	21	1	52	95	(100)
Mid-slopes	509	154	26	19	43	88	100
Lower slopes	220	139	23	17	24	83	100
Lower plateau	135	14	5	3	36	80	(100)
Plateau above cliff	0	–	–	–	–	–	–
Lower steep slopes	473	34	12	12	2	83	100
Colluvial fan	163	20	12	8	20	67	100
TOTAL	324				38	84	100

1977–1983, due primarily to wind felling trees on the upper steep slopes, and in 1992–2000, due to a combination of wind damage and deer browsing. It rose to its highest level in 2010–2013, when several recruits from 1986–1992 failed. Recruitment was very slow, but continued throughout, even though saplings were severely browsed by deer until 2007. It picked up enough in 1983–1992 to exceed mortality when post-drought recruits reached over 1.3 m. When deer were particularly numerous, only weak basal sprouts recruited, but several clusters of saplings recruited within the fence on transect IV. Just two leaderless stump sprouts that originated on an older tree after the drought survived in 2013. A new feature developed after 1992: as some canopy ash weakened, they grew new sprouts from the base and lower trunk.

Population change of ash in the young and old-growth can be read as different stages in a single story – a chronosequence – but the stand below the cliff gives a contrasting perspective (Table 6.4). In 1950, the population was still recruiting after the wartime selective felling. Recruitment must have slowed in the 1960s as the canopy closed, but in 1971 Dutch elm disease (DED) suddenly relieved ash of competition from wych elm, and the canopy opened again (Fig. 6.1: see p. 76). Regeneration from seed was substantial, and by 1985 many clusters of ash saplings were well established in gaps and more followed before 1992. Since this is the classic pattern for ash regeneration, we fully expected recruitment to continue and the best ash in each group to develop into new canopy trees, but recruitment rate fell and few of the saplings prospered. The most likely causes were deer and increasing shade, brought about by vigorous crown expansion in the canopy and strong hazel growth in the underwood. While the mortality rate of the oldest trees fell steadily after 1985, mortality of the younger generations accelerated. By 2013, ash had declined to about one-third of the 1950 population, a density of 89 trees/ha, only slightly higher than the old-growth above the cliff.

What happened? All the original underwood and subcanopy ash died, and the smaller canopy trees also died after first retreating down to the subcanopy. The only original trees that survived were larger trees, and even some of these fell or were knocked down when other trees fell against them, having remained robust, fast-growing, large-crowned

Table 6.3. Changes in the ash population in the old-growth stands above the cliff.

Recording year	Initial density (N/ha)	Live at start	Losses	Gains	Mortality rate (%/year)	Recruitment rate (N/ha/year)
1945–1955	200	180	–36	+5	2.21	0.56
1955–1977	166	149	–51	+1	1.81	0.05
1977–1983	117	99	–16	+2	2.90	0.37
1983–1986	97	85	–7	+2	3.38	0.89
1986–1992	96	80	–6	+14	1.41	2.83
1992–2000	100	88	–21	+7	2.98	0.86
2000–2010	87	74	–7	+5	1.10	0.62
2010–2013	80	72	–15	+3	7.49	1.11
2013	64	60				

Table 6.4. Changes in the ash population in the stand below the cliff.

Period	Initial density (N/ha)	Live at start	Losses	Gains	Mortality rate (%/year)	Recruitment rate (N/ha/year)	Mortality rate of trees present in 1950 (%/year)
1950–1959	263	130	–32	+9	2.98	2.07	2.98
1959–1985	215	107	–76	+137	4.65	10.90	4.87
1985–1992	348	168	–51	+141	5.04	41.69	2.88
1992–2005	534	258	–184	+4	9.16	0.64	1.96
2005–2013	159	77	–41	+7	9.07	1.81	0.75
2013	89	43					

trees until they toppled. By 2013, just 16 of the 53 canopy trees on the transects in 1950–1959 were still standing, the largest being 255 cm GBH (cf. 144 cm GBH in 1950), and one of these had fallen and re-sprouted in 2005–2013. Few later recruits lasted: the survivors by 2013 had been recruited in 1950–1959 (1), 1959–1985 (11), 1985–1992 (9) and 2005–2013 (5).

Viewed together (see Fig. 6.4), the decline of ash density above the cliff is largely negatively exponential, as one would expect in a single cohort, whereas the stand below the cliff exemplifies the wild fluctuations associated with pulses of recruitment. In the absence of disease, similar pulses of regeneration would eventually be expected in the old and young-growth too.

6.6 What Limits Regeneration and Recruitment?

The curious history of the stand below the cliff prompts questions about the factors that restrict the success of regeneration. The great majority of seedlings must either be eaten by voles and slugs, 'suffocated' by shade or parched before their roots grow deep enough to tap subsoil water, and those that survive beyond the first year become increasingly vulnerable to deer browsing. However, such losses are routine: like early Victorian children, so we are told, it is the fate of those that survive for, say, 5 years or more that matters.

Recruits (regeneration reaching 1.3 m) were found mostly in gaps (Fig. 6.5), but even there they grew slowly at first, save for one or two leading individuals. And some were swamped by bramble growth. Above the cliff, their progress has been limited by deer browsing and the increasing shade cast by the expanding crowns of trees on the gap margins. However, deer and shade from the canopy were not solely to blame. When we erected small exclosures in the late 1980s and included several small ash saplings within them, these were ringbarked by voles in 1991 and 1992. For example, on 15 May 1992 we noted that a 2-m-tall ash sapling growing among bramble within one exclosure had been healthy until the previous summer, had then been ringbarked up to 8 cm above the soil surface and had tried to re-sprout in 1992, but had, nevertheless, died.

But why did the substantial 1985 and 1992 cohort below the cliff fail to prosper, despite in some cases having what appeared to be plenty of canopy space to grow into? The answer seems to be a combination of factors. Even though the broken topography restricts deer locally, most saplings remain vulnerable to browsing. The canopy may open frequently, but tree growth is vigorous and crowns expand rapidly. Further, underwood hazels and yews grow vigorously and outcompete saplings. And several well-established saplings were crushed by falling trees or branches before they could reach the canopy. Even so, clusters survive, including one round a rock outcrop on transect IX, where the saplings have had to fight their way up through the branches of well-established yews. Later accounts will show that other species have suffered too, so the factors restricting ash regeneration are general.

Three features stand out. First, disturbance facilitated recruitment. The main regeneration opportunities above the cliff came after the wartime felling and, less so, after drought in 33- to 48-year-old and 107- to 122-year-old stands. Tree falls and elm disease promoted regeneration below the cliff. Second, established saplings, apparently with good prospects, do not necessarily generate new canopy trees. Third, regeneration opportunities increase as the stand ages, although until 2007 they were almost nullified in Lady Park by the deer.

Today, within the protection of the perimeter fence, some basal sprouts have developed from some mature ash, trunk sprouts are growing from recently fallen canopy trees and droves of small saplings have become established in much of the wood. The saplings are growing slowly, but groups are accelerating and now a new cohort of saplings is starting to recruit, especially on the colluvial fan and within new canopy gaps. At the same time, muntjac deer have arrived and feral pigs bid to let in the fallow deer, so the prospects for a substantial new generation are in the balance. If it succeeds, the population density in the 143-year-old-growth of 2013 will prove to be the nadir at which the surviving population is still largely made up of trees initiated at the outset. This old generation represents less than 1% of the population at 3 years of growth, and a truly minute fraction of all the ash that germinated.

6.7 Growth and Survival within Established Stands

Death constantly stalks the ash population. How does it relate to the size, vigour and location of individual trees?

Fig. 6.5. Ash regenerating in a gap on transect V, created by the fall of a particularly large hazel.

6.7.1 Young-growth

In 1977, the live ash in the young-growth ranged from 1 to 26 cm DBH (3–82 cm GBH). Their size distribution was strongly skewed: the vast majority was small, but there was a long 'tail' of larger trees. Dominant individuals had long since identified themselves.

Survival was largely determined by size in 1977, which meant growth rate before 1977 (Fig. 6.6), since this was a nearly even-aged population. Between 1977 and 2013, almost all the trees above 13 cm DBH survived, most of those below 7 cm died. By 2013, the largest tree had grown into a fine emergent tree of 172 cm GBH at a radial growth rate of 5.4 mm/year – almost as fast as the 7 mm/year achieved by shoots from vigorous pollards at Beechwood House (in the lower Wye Valley, 10 miles from Lady Park) in their second and third year – but most were very much slower. Growth rates of those that survived increased with size, but the shape depended on whether it was expressed as basal area increment (concave) or ring width (convex) (Fig. 6.7). In short, over 36 years, the smallest were doomed by slow growth, while almost all the larger ash grew well enough to survive.

Canopy position strongly influenced survival and growth rate (Tables 6.5 and 6.6). In 1993, the position of individual ash was recorded in four classes: canopy, subcanopy (trees that occupied the upper stratum, but more than 25% of their crown was overshaded by canopy trees), underwood (greater

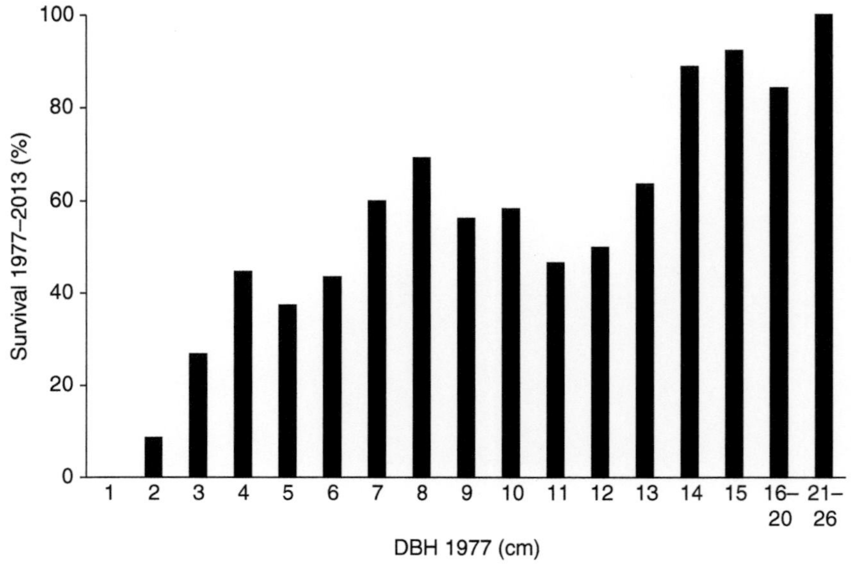

Fig. 6.6. Survival of ash in the young-growth stands, 1977–2013, related to initial size.

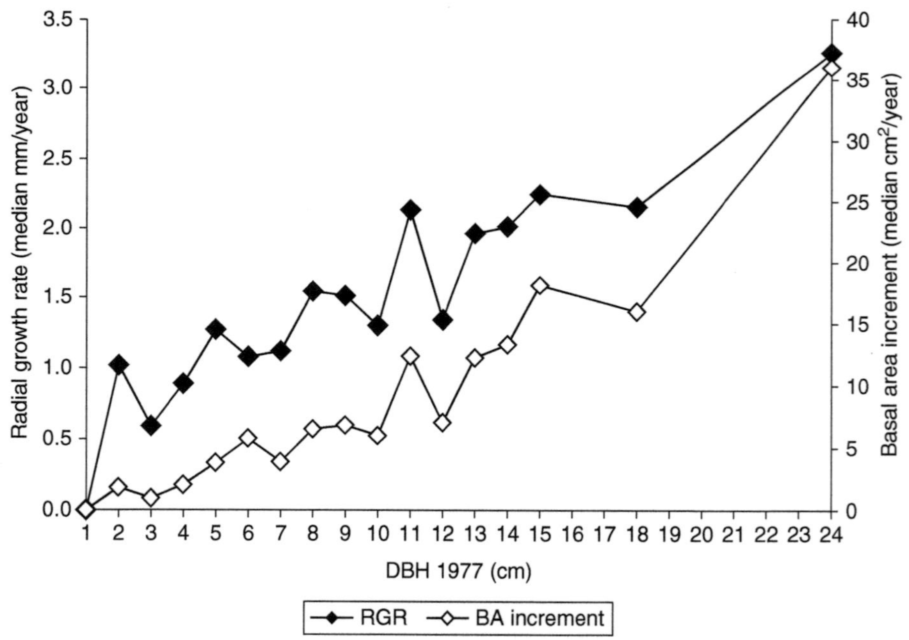

Fig. 6.7. Growth rate of ash in young-growth stands that survived from 1977 to 2013 related to initial size. Growth is given as medians of radial growth rate in mm/year (filled symbols) and basal area increment in cm^2/year (open symbols).

than 2 m, but their foliage remained well below the upper stratum) and 'ground' (individuals less than 2 m tall). Most trees maintained their station in the vertical structure over the next 20 years. Some apparently ascended from the underwood into the subcanopy, and from the subcanopy into the canopy, but some of this was more apparent than real, because different definitions were used in

Table 6.5. Fate of ash, 1993–2013, according to canopy position.

		Canopy position in 1993			
		Ground	Underwood	Subcanopy	Canopy
	N[a]	34	169	122	144
Fate by 2013[a]	Survived	2	94	112	142
	Alive <1.3 m	2	6	1	1
	Fallen			1	
	Dead	30	69	8	1
Canopy position 2013[a]	Canopy	–	5	62	135
	Subcanopy	2	51	48	7
	Underwood	2	44	4	1
Crown size 2013[a]	Large and medium-large		3	12	80
	Medium and medium-small		21	50	44
	Small	2	50	42	17
	Tiny	1	24	9	
	Base sprouts	1	2	1	1
	Broken				1
Radial growth rate 1993–2013 (mm/year)					
	Range	1.31–1.84	0–2.86	0–3.65	0.12–5.47
	Median	1.63	0.39	1.16	1.75
Basal area increment (cm^2/year)					
	Range	1.6–2.4	0–15.2	0–27.9	0.5–75.7
	Median	2.0	0.9	4.8	11.0

Note: [a]All trees living, including fallen trees and those alive only below 1.3 m.

Table. 6.6. Median radial growth rates (mm/year) and basal area increments (cm^2/year) of ash in the young-growth in relation to crown size and vertical position.

		Crown size in 2013			
	Stratum in 2002	Tiny	Small + medium-small	Medium + medium-large	Large
Radial growth rate 2002–2013 (mm/year)	Canopy	0.62	1.09	1.55	2.29
	Subcanopy	0.57	0.78	1.31	
	Underwood		0.37	0.65	
Basal area increment 2002–2013 (cm^2/year)	Canopy	1.7	4.6	8.3	19.1
	Subcanopy	1.2	2.6	5.8	
	Underwood		0.9	1.6	

2013. The few that dropped back from the canopy to the subcanopy really did drop back. Growth rates were largely determined by canopy position. By 2013 – and no doubt also in 1993 – most underwood trees usually had small crowns, most canopy trees had medium and large crowns and subcanopy trees were intermediate.

All very predictable, one might say, but this straightforward expression of the self-thinning rule obscures some interesting features. Thus, although growth rates were strongly determined by original size and position in the canopy, there was immense variation around this relationship; that is, during any given period, similar trees could be growing at substantially different rates depending, presumably, on the performance of neighbours. While most canopy trees had large crowns, grew fast and maintained their position in the canopy, a few had smaller crowns and were slower growing. These are likely to be the first to fail. Conversely, while most ash that had not become established in the canopy by 1977 were doomed by 2013, a few prospered and even developed medium-large crowns.

The growth rates of many individual trees changed considerably from one period to the next, but in general a tree's role in the stand remained unchanged. By and large, the fast growers before 1993 were still the fast growers after 2002, and vice versa, although there were a few individuals that accelerated substantially and others that braked hard. At all periods there were static trees – those just hanging on. In general, radial growth slowed considerably between 1977 and 2013, whereas basal area increment declined after 2002 (Table 6.7), but in every period some trees were 'out of step'. Thus, contrary to trend, 14 trees accelerated after the millennium. Acceleration also allowed a few individuals as small as 13 cm GBH in 1977 to prosper, but smaller individuals showed no ability to take advantage of opportunities, such as new gaps in the canopy.

Walking round the young-growth, one's impression is that ash growth has been stronger on the colluvial fan, but the records paint a more complex picture (Table 6.8). Vigorous and nearly static trees can be found in all zones, and slow-growing individuals outnumber the fast growers everywhere, but if we confine our attention to the fastest 25% of ash in each stand – just those trees that foresters would select for a final crop if the stand were managed for timber – the ash on the colluvial fan stand out. Not only have they grown faster and taller but also their trunks are straight and clean and their crowns emerge from the canopy. On the dry limestone at the top, the largest also have large crowns, but they are squat by comparison and their growth has been consistently slower. Elsewhere, there is little to choose between the best trees of the mid- and lower slopes. The best ash on the thin soils of the lower steep slopes have consistently grown better than those on the upper plateau and steep slopes, perhaps because the soil remains moist.

Summarising, the golden rule for success was to grow into the canopy, and to do so quickly. In a nutshell, free market economics prevailed. The ash population has been subjected to a feedback whereby, in a developing thicket, a fast-growing tree can expand its crown and shade its neighbours, thus increasing its growth rate at the same time as it depresses theirs. Large crowns and faster growth promotes survival, while smaller crowns and slower growth leads to even slower growth. With few exceptions, the big got bigger, differentials increased and the small went to the wall. Most failing individuals first slowed down while still appearing healthy and vigorous, before growing very slowly just before they expired. They usually fought back for a

Table 6.7. Changes in growth rate of ash as the young-growth stand developed: radial growth rate and basal area increment of the 333 trees that survived from 1977 to 2013 and were measured on all dates.

	Radial growth rate (mm/year)			Basal area (cm²/year)		
Young-growth	Mean	Median	Range	Mean	Median	Range
1950–1977	1.41	1.11	0.4–4.8	2.29	1.05	0.1–19.6
1977–1993	1.71	1.54	0.1–5.6	6.45	4.45	0.2–43.6
1993–2002	1.52	1.33	0–5.3	8.52	5.70	0–63.7
2002–2013	1.16	0.94	0–5.4	7.79	4.91	0–82.2

Table 6.8. Influence of site on growth: radial growth rates (mm/year) between 1977 and 2013 of ash in the young-growth stands in each topographical zone.

	N	Range	Median	Top quartile mean	Range	Median	Top quartile mean	Largest (cm GBH)
Upper plateau	35	0.14–2.94	1.50	2.39	0.2–19.2	6.1	9.7	100.5
Upper steep slopes	52	0–3.36	1.11	2.26	0–24.0	3.7	12.2	107.5
Mid-slopes	125	0.03–4.16	1.40	2.58	0–38.3	5.0	15.5	137
Lower slopes	67	0.23–3.66	1.39	2.57	0.2–32.5	5.8	16.7	130
Lower plateau	13	0.44–3.15	1.29	2.55	0.8–17.2	5.9	14.7	100
Lower steep slopes	23	0.25–2.92	1.88	2.64	0.8–22.0	11.8	18.9	109
Colluvial fan	20	0.16–5.38	1.66	3.87	0.2–61.5	7.0	43.3	172

while by producing new shoots from low on their trunk and, as a last resort, generating new sprouts from the base, but usually to no avail.

6.7.2 Old-growth

The old-growth ashes continue the processes established in the young-growth. Just 27% of the individuals recorded in 1945 survived to 2010, mostly the larger trees in the canopy. Over the entire 69-year span, the vigorous ash expanded by up to 2.6 mm/year, while others survived on just 0.2 mm/year. The growth rates of survivors accelerated, especially when they were released from competition after the drought, but in all periods there were trees that hardly grew at all (Table 6.9). Whereas radial growth rates were much the same as in the young-growth, basal area increments were larger. Most trees that died had a premonition of death – they grew slowly before they perished (Table 6.10) – but a few – mostly trees that were broken or tipped over – were growing well. Mid-slope trees grew slightly faster than those on the upper plateau.

6.7.3 Below the cliff

The growth of survivors below the cliff was generally faster than those above, due, one supposes, both to the moist alkaline soils being more suitable for ash than the dry and acid soils above and the repeated opening of the canopy due to wartime felling, elm disease and the higher rate of tree falls. In fact, an ash near the base of transect VII recorded the fastest sustained radial growth rate anywhere (4.53 mm/year). Several of the larger, fast-growing trees fell, and all ash smaller than 30 cm GBH in 1950 grew slowly and died before 2005. Radial and basal area growth of survivors slowed in 1985–2005 (Table 6.11), just when one might have expected them to speed up after the canopy was opened up by elm disease. Both measures, but especially basal area increment, increased sharply in 2005–2013, possibly because the ash stood almost free of competition.

Growing below the cliff in a relatively open stand, the tall ash seem to be limited only by their ability to stay upright. They are increasingly likely to fall as they get larger, especially those growing immediately below the cliff, which developed unbalanced crowns, and those on the lower margin, which are exposed to wind. At the base of the slope, ash have reached 255 cm, but on the slopes they generally fall at much smaller sizes, often 130–160 cm. Many have attracted great growths of ivy on which snow accumulates, thereby increasing the risk of bringing them down. Indeed, one such fall led to a Lady Park ash becoming possibly the most studied tree in Britain. One dark evening in late December

Table 6.9. Changes in the growth rate of ash as the old-growth stand develops: radial growth rate and basal area increment of the 45 trees in the old-growth stands that survived from 1945 to 2013 and were measured on all dates.

		Radial growth rate (mm/year)			Basal area (cm²/year)		
	Seasons	Mean	Median	Range	Mean	Median	Range
1945–1955	10	1.05	1.21	0–2.8	7.60	7.39	0–30.5
1955–1977	23	1.20	1.24	0.3–2.7	10.55	10.55	1.1–31.4
1977–1992	14	1.49	1.32	0–3.8	15.28	14.00	0–44.8
1992–2000	9	1.42	1.33	0–3.0	16.33	14.02	0–39.8
2000–2013	12	1.50	1.39	0–3.9	19.63	15.98	0–64.3

Table 6.10. Premonition of ash death in the old-growth stands: comparison between radial growth rates (mm/year) in 1945–1977 of trees that died and those that survived to 2013.

	N	Range	Median
Survivors: alive in 1945 and survived to 2013	49	0.31–2.50	1.26
Later mortality: alive in 1945, but died 1992–2013	20	0.03–1.63	0.59
Early mortality: alive in 1945, but died 1977–1992	32	0.14–2.60	0.32

1999, a local resident walking home along the old railway track in a snowstorm was crushed by an ivy-laden ash. He lay under the tree for several hours, but survived after his shouts were heard by someone checking the generator at Biblins Lodge. Investigations for the subsequent legal proceedings showed that a contributory factor had been the death of just one of the four main roots, which had evidently 'supplied' a subsidiary trunk that had died several years earlier.

6.7.4 Managed compartment

Thinning allows retained ash to grow faster, especially and unexpectedly on the deeper, more acid soils (Table 6.12), which implies that even the dominant ash in the unmanaged stands have not freed themselves from competition and that ash likes deeper soils more than it dislikes a moderate increase in acidity. In addition, the managed compartment supplements information from the reserve in three ways:

- The ash in the unthinned old-growth, which also lie 'below the cliff', were slower (median 1.17 mm/year, maximum 2.89) than the ash below the cliff within the reserve, probably because the stand was never opened by DED or tree falls and was on a gentler, drier slope.
- The one ash on the flood plain was fast growing (3.34 mm/year).
- And the two ash growing out of the cliff were not as slow as one might have expected (0.45 and 2.20 mm/year).

6.8 Survivors and Failures

The general rule in undisturbed stands is simple: large trees in the canopy tend to grow fast and survive, while the small trees generally grow slowly and die. Even in the canopy, however, some trees lack vigour, develop smaller crowns, become 'squeezed' by their neighbours and eventually perish, though many fight to the death. Thus, failing canopy trees develop new shoots on the trunks and sprouts from the base, and when they die they still send up sprouts from the base of the old trunk. If, by chance, a gap opens nearby, these sprouts may grow well and rejuvenate a hitherto condemned tree – tantamount to natural pollarding and coppicing.

Ash saplings and slow-growing poles often persist after being flattened by a falling neighbour. Those that are merely heavily shaded display the physiological equivalent of stubbornness by extending lateral shoots towards distant light, even as the leader dies, thereby forming misshapen underwood poles

Table 6.11. Changes in growth rate of ash in the stand below the cliff: radial growth rate and basal area increment of the 16 trees that survived from 1950 to 2013 and were measured on all dates.

		Radial growth rate (mm/year)			Basal area (cm²/year)		
	Seasons	Mean	Median	Range	Mean	Median	Range
1950–1959	9	2.45	1.85	0–8.3	20.47	17.52	0–54.4
1959–1985	25	2.00	2.12	0.1–3.6	20.79	21.89	0.5–41.9
1985–1992	7	1.52	1.36	0.5–2.3	19.43	18.74	2.6–46.6
1992–2005	13	1.63	1.47	1.0–3.8	22.41	20.03	6.9–72.3
2005–2013	9	2.31	1.99	0–6.5	38.58	28.12	0–154.8

Table 6.12. Comparing radial growth rates and basal area increments of surviving ash in unmanaged and thinned stands. The values given refer to all trees that survived from 1977 to 2013 in the young-growth (unthinned) stands and from 1984 to 2014 in the equivalent (thinned) stands in the managed compartment.

	Radial growth rate (mm/year) Median (top quartile)		Basal area increment (cm²/year) Median (top quartile)	
	Unthinned	Thinned	Unthinned	Thinned
Upper steep slopes	1.11 (2.26)	1.54 (2.76)	3.7 (10.0)	5.1 (19.4)
Mid-slopes	1.40 (2.58)	2.15 (2.71)	5.0 (9.6)	10.9 (20.3)
Lower slopes	1.39 (2.57)		5.1 (10.6)	

with few live twigs. Occasionally, small, moribund trees 'get lucky' when a gap opens up near them, but by 2013 (when they were about 68 years old), they were failing to respond: evidently, they had lost their capacity for recovery. Once entrenched in a particular stratum, trees tend to maintain station: whatever span of years one considers, smaller trees rarely overtake larger trees and large trees rarely fall behind small ones. Like the University Boat Race or the football leagues, once you are ahead, you tend to stay ahead.

This relentless and unforgiving rule is not, however, absolute, for some trees buck the trend. Individual trees vary in health, luck and circumstances, particularly the competitive power of their neighbours. The two small individuals that grew exceptionally fast were both new saplings growing strongly in new gaps, which evidently possessed the vigour of youth and the capacity to respond when gaps opened above them. These must be the harbingers of the next pulse of recruitment.

Conversely, large ash sometimes weakened and died. With some, the contributory factors were obvious. Thus, a vigorous ivy growing into an ash crown may eventually break a major branch; a forked tree may simply split; or rot may be admitted after a falling neighbour damages bark low on the trunk. Some large trees grow too close to an even more vigorous tree – usually a beech, lime, another ash or formerly a birch – and slowly but surely fall behind, develop an unbalanced crown and then either expire in heavy shade or lean at an increasing angle until they fall. These large casualties are trees that initially have competed strongly, but lose – sometimes narrowly – as competition intensifies in the later stages of stand growth. Ivy seems to pick on weak trees and may be partly responsible for the weakness.

But survival and failure are not just a matter of canopy position and space, supplemented by accidents of various kinds. Just as some small individuals have the luck to be given more space when a gap opens nearby, some larger individuals just decline gently when they already have enough space to grow better. Some of these are cankered (Fig. 6.8) and the rest may be under attack from fungi in the soil.

Smaller trees persisted in greater numbers on the upper slopes, where growth was slower and competition correspondingly less intense on the shallow limestone soils. Equally, the few canopy trees with tiny or small crowns are commoner on the upper plateau, upper steep slopes and mid-slopes, where growth is generally poorer and individuals that would have been eliminated in a more vigorous stand survive. One, for example, twists up to the canopy between hazel, whitebeam and other ash. Several grow in ash groups where larger and more vigorous ash dominate; and others have been outgrown by nearby birch, lime or beech. A few have been helped on their failing course by damage in the crown, heavy ivy growth competing in the crown or what appeared to be damage caused by deer gnawing. On the other hand, a few successfully replaced lost leaders by generating sprouts from their trunk (Figs. 6.9, 6.12). Several, indeed, were strong enough to return to the canopy.

So, at any given time, we have the largely unthreatened dominants of ash society; the vulnerable underclass; the losers and the unlucky who are about to be excluded, despite heroic resistance in many cases; and opportunists, the lucky and the plucky, who will live to make a sustained contribution to society.

6.9 Causes of Death

Most of the ash recorded at the outset have since died. In most cases, a meticulous coroner would record not one cause but a chain of causation, a combination of factors acting in sequence; for example, when competition forces a tree to lean towards a gap, but wind actually brings it down. Summarising, the main links are:

1. Competition. Individuals in the underwood and subcanopy are shaded by other individuals, grow slowly in both height and girth, languish and die. Competition also affects trees in the canopy, where crowns are restricted by other canopy trees. If one has the ill luck to grow next to a really vigorous tree, its crown will be seriously restricted and possibly distorted, thereby setting up a feedback that eventually leads to the exclusion of the weaker tree, sometimes via a stage where the weaker tree is left behind in the subcanopy.

2. Slope. Both canopy and underwood trees growing on slopes tend to develop unbalanced crowns, with strong growth on the downhill side and correspondingly weak growth uphill. Such trees lean increasingly as they grow and become progressively more vulnerable to wind and the instability of the ground. If leaders are still active, they develop into bow-shaped trees. Alternatively, the more they lean, the more likely they are to reiterate new, vertical leaders from down the trunk, leaving the old crown

Fig. 6.9. Resilience! This ash lost one leader, but developed a new one from a side branch. At some point it fell, but, as its new leader failed in shade, new potential leaders have grown from all points on the trunk.

to degenerate. Trees face particularly strong competition from trees growing on the slope above them, for the latter start with a height advantage. In fact, some trees grow at a lean from the start.

3. Wind. We have never seen an ash fall in high winds, but they clearly do get blown down (Fig. 6.10). They are especially vulnerable if they have developed a strong lean, or if a gap has opened near them, thereby removing both the breaking effect of adjacent trees on wind speed, increasing the amplitude of the swaying that precedes a fall, and reducing the possibility that a falling tree will lean on a neighbour and remain more or less upright and live. Wind-blown trees are usually uprooted, leaving a pit. They usually fall to the ground, but a minority rest against other trees. The outcome, however, may be much the same: both have weakened root systems, and even leaning trees find their crown shaded by the crown of the tree that supports them. If a healthy

Fig. 6.8. Although its crown expanded in the canopy, this cankered ash is destined to fail and be succeeded by the small-leaved limes below.

Fig. 6.10. Recently fallen canopy ash that excavated a substantial pit as it fell. Much of the soil and humus will eventually wash from the roots back into the pit.

ash falls beneath a beech or lime, new growth usually dies in the shade, but if it rests below birch, ash or a gap, it can reiterate strongly. Indeed, new shoots from the stool can exceptionally form a new, strong tree exactly where the previous tree grew (Fig. 6.11).

4. Tree falls. Ash are sometimes brought down by falling neighbours, less often by large branches torn from neighbours and just occasionally by ivy supplemented by snow. Some are not uprooted, just shattered, usually fairly low down (1–4 m). With intact root systems, they tend to grow again from the base. Like trees that tip over, they often try to grow again from the trunk.

5. Grazing and browsing. New growth from stumps and fallen trunks is vulnerable to deer browsing, and this has certainly killed some trees that might otherwise have been able to grow again. Rarely, too, deer have debarked parts of an ash trunk to the quick, and this has introduced rot. Seedlings and saplings also fall prey to slugs, snails and voles.

6. Disease. Just occasionally, a large canopy tree that ought to remain dominant starts to die back, develops cankers and numerous trunk and base sprouts and then starts to break up. More often, one suspects that trees that have already become vulnerable due to competition are finished off by a fungus. At the time of writing, Chalara has not been detected in the wood.

Singly or in combination, causes of death vary with age. Grazing affects saplings. Competition is

Fig. 6.11. While most ash die after they fall, this windthrown tree on the upper plateau has generated three strong new trees from near the rootstock. Other, weaker, stems growing from the former upper trunk and crown branches also survive for now.

dominant, especially in the thicket and pole stages, while slope, wind, tree falls and disease become more prominent in mature stands.

6.10 Decay: Snags and Logs

Ash that die on their feet, leaving a snag, are usually small. They rot and disintegrate rapidly and the stump eventually falls when all the roots have rotted. Large ash usually live until they fall, and some reiterate new shoots. Most die after falling, because either they are shaded or most of their root system has been broken in the fall. Their logs rot rapidly and will mostly be consumed within a decade. Occasionally, large trees die standing: the snag of a 58 cm GBH tree that died between 1992 and 2000 was still standing in 2013.

6.11 Review

Even without the looming ash dieback, the ash population in Lady Park Wood has reached an interesting stage. After nearly 150 years of relatively natural succession on a range of site types, we have reached the point where the largest trees are increasingly vulnerable to wind and the instability associated with slopes, while a new cohort is on the brink of recruiting in large numbers. Only the regenerants in gaps are likely to recruit, provided they can outpace the bramble. Voles may yet ringbark many, and the deer fence will have to hold up for another decade, but even that may be insufficient, as our recent experience below the cliff indicates. Nevertheless, the prospects for ash recruitment are improving, with ever-larger gaps appearing and browsing reasonably under control. They would increase further if storm, drought or any other disturbance afflicts Lady Park, or another species becomes diseased.

In the hierarchy of forest trees, ash stands between the long-lived dominants, oak, beech and limes, and the fast-growing, light-demanding opportunists, birch, gean, aspen and sallow. In natural woodland, it should in theory retreat when the long-lived dominants are not disturbed and advance with moderate disturbance, but its prospects after a stand-destroying disturbance are uncertain. While it might languish among thickets of birch and sallow, it is now more likely to be outpaced by regrowth from fallen limes. If the disturbance regime of the past 70 years continues, ash seems most likely to sustain dominance on the thin limestone soils at the top of the slope, whereas on deeper, base-poor soils it could be ousted by beech, oak and limes.

Speculation about the future of ash in Lady Park seems almost futile, and not just because we cannot predict the disturbance regime. Even now, its potential in Lady Park has not been fully tested, for we do not know its maximum size, how long it can live, nor whether it will live longer on the upper plateau, where it is slower growing. Seventy years of study enables us to understand its performance in increasingly natural woodland, but mortality and recruitment patterns keep changing in time and space, and these fluctuations cannot be explained fully. In any case, it is still growing out from its managed past, with the consequences of former management still recognisable as stools and even-aged cohorts. Moreover, the species has been advancing in the lower Wye Valley for the past century or more (Peterken, 2008a), so past experience may be a poor guide to the future.

Whatever we might have predicted if asked a few years ago, the prospects for ash have changed again. Chalara ('ash dieback') has reached the district but not the wood, though we expect it to arrive shortly. Further east, some ash populations have already been devastated, but genetic markers suggest slightly higher tolerance in the British population as a whole compared to Denmark (Sollars *et al.*, 2017), which might be because British ash are largely derived from a different glacial refuge, i.e. the Iberian Peninsula (Fraxigen, 2005).' Genetic heterogeneity affords some long-term hope, but hope recedes when we contemplate the emerald ash borer on the horizon (Straw *et al.*, 2013; Herms and McCullough, 2014). We await developments with interest and apprehension!

Fig. 6.12. A natural counterpart of David Nash's *Ash Dome* sculpture, formed by a cluster of ash in Lady Park that refused to give up the search for growing space. The new shoots at their base show that they are still under stress and still not giving up.

7 Beech and Oak, The Major Forest Trees

For centuries, beech and oak have dominated in Lady Park Wood. Not only have they formed the largest individual trees they have also been the main concern of woodmen and foresters. Yet they are competitors: in both managed and natural woodland, the fortunes of one declined as the fortunes of the other improved. Oak long dominated in both wood pasture and coppice, but beech was favoured in the conversion to high forest and 'natural' woodland, only to be brought up short by grey squirrels and the 1976 drought. Today, both their futures are in the balance.

7.1 Beech in Britain

Beech is the tree that makes the ecological weather. In theory, it has the characteristics of a natural winner: not only does it cast such deep shade that few trees can live in its shadow but also it bears shade when young, lives long, regenerates well and is mildly resistant to browsing. In undisturbed woodland, it ought to dominate, and, once dominant, it should have the capacity to stay dominant.

In practice, beech has feet of clay. Its shallow roots and thin bark make it vulnerable to drought, windthrow and the instability of steep slopes, so it has a well-deserved reputation for being accident-prone. When it falls or dies, it provides a clear opportunity for all other trees, for the ground it vacates is bare. Thus, when a beech wood suffers a major disturbance or a series of minor disturbances, other trees have their chance to regenerate and dominate, albeit while sheltering a new generation of beech saplings waiting to reclaim their place.

Beech is principally a tree of southern Britain, where it dominates woodland on both calcareous and strongly acid soils, but not usually on wet, neutral and poorly drained substrates. Regarded as a native in south-east England, the Cotswolds and south Wales, it has been so commonly planted elsewhere that it is now naturalised almost throughout Britain. In fact, there is a case for thinking that planting over the last three centuries has enabled it to reach its natural limits, having been prevented from doing so earlier by prehistoric woodland clearance in the English lowlands.

At its best, beech is a tall tree with a deep crown of simple, oval, but pointed leaves and heavy crown branches. In winter, it is easily recognised by its smooth, pale bark and long, sharply pointed buds. In spring, its bright young foliage forms a misty green haze, but the leaves quickly darken to cast shade deep enough to exclude other trees below. Characteristically, therefore, a large, spreading beech stands clear of other growth, forming a 'shaded clearing', free of underwood.

Within woodland, beech will grow into a tall, slender, graceful tree up to 25–35 m high. As it grows, the long twigs on the leading shoots may extend by well over 30 cm each year, though as it nears top height the apical growth of even healthy trees is reduced in dry summers (Power, 1994). In the open, beech, like most trees, forms a hemispherical crown supported by branches that spread wide and eventually lean down and touch the soil. In closed woodland, it usually maintains a clear leader. While a beech is still growing in height, the leading twigs have a spiky appearance that is lost when the tree matures – one can always tell whether a beech is still vigorous.

The early growth of beech is usually slow. If seedlings survive their first 2 years, they will sit in the undergrowth growing very slowly for years, reach no more than 2 m in 20 years and can survive for decades in this quiescent state, waiting for their chance to grow into a well-lit canopy gap. However,

Fig. 7.1. A tall beech that died in 1976, photographed in 1984, by which time the bark was flaking off and the smallest twigs and branches had fallen.

if they are released from this torpor by the fall of a large branch or a nearby canopy tree, they will become more vigorous. Like Norway spruce, which can also maintain a population of suppressed, but potentially vigorous saplings, released beech saplings can reach the canopy, a property that once made beech a favourite in Chiltern beechwoods and other selection forests. On the other hand, beech can grow really fast almost from the outset. In 'the finest beech stand in Ireland' (Elwes and Henry, 1906–13: Vol 1, p. 26), a plantation at Woodstock, County Kilkenny, radial growth rates (RGRs) from 1846 to 1901 – the main growth phase – were 1–7 mm/year (median 2.5), well in excess of the growth rates of old-growth canopy beech in Lady Park.

Traditionally, beech was coppiced and pollarded. However, growth after felling and lopping tended to be weak, so much so that beech was sometimes treated as selection coppice by retaining some stems at each felling as a ruse to stimulate vigorous sprouting from the stool or bolling. Nevertheless, beech developed giant hollow stools and massive pollards, and both could grow into substantial trees if they were left uncut. Beech has long been planted as specimen trees in parks, where it grows into huge globes of branches, twigs and foliage, marginal limbs brushing the ground. Grazing animals rarely allow low branches to grow freely, however, but where they do the branches sometimes take root. In Mark Ash in the New Forest, an old beech has leaned, rooted, grown a new tree, which also has leaned, rooted and grown a third tree, a real-life ent!

7.2 Beech in Lady Park Wood

Beech was part and parcel of the medieval Dean woods and was first recorded in Lady Park in 1604 as a sparse population of standards, possibly relicts of the earlier wood pasturage. The 1897 survey – and stools present in the wood (Fig. 7.2) – showed that it was common as coppice, but by then almost all the standards were oak. It owes its present abundance in the canopy to a period of nearly a century when it became the foresters' favourite. Hill recommended in 1897 that Lady Park be promoted to beech high forest; many of the promoted trees were retained as 'standards' when parts of the wood were felled in 1942–1944; and in 1982 and 1993, it was the species that the Forestry Commission sought to promote when the managed compartment was thinned. It was a different kind of favourite in the late 18th century, when large, spreading, prominent beeches were retained on the skyline (Fig. 7.3). Today, sadly, it demonstrates better than other trees the folly of releasing grey squirrels into Britain (Fig. 7.4).

Most of the canopy beech in the old-growth and the standards in the young-growth originated after 1870. We know this from six ring counts made in late 1944, which indicated origins in or near 1859, 1874 × 4 and 1884. One of the beech in transect III that died in 1977 was ring-counted at *c.*103 years in 1986, i.e. it also originated in 1874. A tree that fell across the track in May 1999 and was cut to permit access had about 85 growth rings at 7.3 m (34 cm DBH), of which the last 17 were markedly narrower than earlier rings. If 68 represents the number of rings in 1976, then the tree was 7.3 m high in 1908, which in turn indicates a late 19th century regenerant that was retained in the 1902 thinning. In September 2000, the large beech at the base of transect IX was felled for safety reasons: its *c.*96 rings indicated an origin after the 1902 fellings. Combined, and taken literally, these counts indicated that the old-growth beech were 113- to 156-years-old in 2015.

Apart from those on the skyline, most beech in Lady Park have grown in closed stands from seed, long-established coppice stools and even low pollards (Fig. 7.5). Beech was favoured in the 1902 thinning, when the foresters might well have removed both unusually strong ('wolf') trees along with weaker brethren, leaving the rest to grow evenly. Some beech were retained during the 1942–1944 felling and in effect became 'standards', but they had never been part of a coppice-with-standards system. Since 1944, the population has thinned out; regeneration has been episodic and patchy; differences have become apparent between trees of different origins; large beech have proved to be vulnerable to drought (Fig. 7.1: see p. 98); beech of all but the smallest sizes have suffered from grey squirrels; and we have observed differences relating to topography, marked changes in performance over decades and clear responses to thinning.

7.3 Changes in the Population of Beech

The populations in the three stands followed distinctive courses. By 1945, 40 years of promotion had ensured that beech dominated much of the old-growth. The 351 live individuals were roughly 75 and 42 years old and already well established in the canopy and subcanopy. Some 20% were maidens, and these were somewhat larger than the coppice

Fig. 7.2. An ancient and unusually vigorous coppice beech stool growing on a limestone outcrop at the north edge of the wood.

stems on average, probably because many of the latter had regrown after 1902.

The population decreased until 1986, but has since remained fairly steady (Table 7.1). Mortality rate was lowest in 1945–1955, when the stand was still fairly open after the *c.*1930 thinning, increased in 1955–1977 as competition intensified, and peaked after the 1976 drought. Recruitment was high initially, due mainly to the ingrowth of coppice stems, but the emphasis shifted to slow maiden recruitment in 1955–1986, then accelerated as coppice stems grew from established individuals in 1986–1992. The emphasis shifted again to maiden recruitment, especially in 2000–2010. Since 2010, a new cohort of both saplings and coppice regrowth from old trees has developed in the near absence of deer and is poised to recruit. Of the 135 recruits between 1945 and 2000, 57 remained alive in 2010.

By 1977, the young-growth population comprised 11 standards, 575 coppice stems, 201 trees of seedling origin, 4 trees alive only below 1.3 m, one fallen live tree, plus 39 dead stems of coppice origin and 18 of seedling origin, i.e. it had been thinning. The population declined thereafter, with a fluctuating recruitment falling far short of matching an accelerating mortality rate (Table 7.2). The 70-year-old population in 2013/15 stood at a much lower density than the 75-year-old population density in the old-growth, a measure of the degree to which beech had been promoted before 1945 and the later impact of grey squirrels.

Fig. 7.3. One of the skyline beeches, with wide-spreading crown. The 1976 drought killed most of the crown, but most crown branches produced new shoots. Several years later, a main branch fell under its own weight, the trunk having been weakened by rot.

The various components of the population performed differently. One standard died in the drought, but the other ten survived to 2015. Trees originating as seedlings after 1942 declined as a proportion of the population from 26% in 1977 to 17% in 2013/15. Their mortality rate was consistently higher than that of coppice stems (Table 7.3). Saplings were resilient: several fell and remained alive, one for at least 19 years; and a few 'died', but remained alive below 1.3 m. Recruitment was very limited, except after the drought: all the 2002–2013 recruits were new shoots from established trees. Trees of coppice origin formed by far the majority of the population. Their mortality rate accelerated steadily, especially after the millennium, when many large old stools died completely (Table 7.4), but recruitment of new stems on old stools also accelerated, especially after the fence was erected in 2007. Overall, beech individuals established long before 1942 were increasingly dominating the population 70 years later.

Below the cliff, the thinly scattered beech population lived an exciting life! The 1950 population was turning over, both literally and metaphorically. It comprised 19 trees (4 coppice, 15 maidens), all in the canopy, plus 1 sapling, 7 saplings just below 1.3 m, 1 fallen but live beech of 45.7 cm GBH and 2 dead trees. The largest beech (188 cm GBH), which grew at the top of transect VIII, fell days after the transects were recorded. The population has since fluctuated widely (Table 7.5) and, exceptionally, the mortality rate has varied as much as the recruitment

rate. In the absence of a record about 1970, one supposes that the population declined in the dense stand that preceded elm disease, and again since 2005, in common with other species. (Much of this recent decline has been due to heavy losses at the base of transect IX, where hazel regrowth is strong.) Beech was recruiting before 1959, when the canopy was still open after wartime thinning, and recruited again in 1985–1992 as regeneration after elm disease reached 1.3 m. Despite continuous recruitment, the population declined from 1950 to 2013.

7.4 Beech Regeneration

Beech seedlings have always been easy to find (Fig. 7.6), especially after mast years, but recruitment from seedlings largely failed until recently. After the 1942 felling, they were infrequent and saplings were overtaken rapidly by the faster-growing birch and ash, but they persisted in the shade, growing slowly. During the 1970s, a scatter of later saplings became established, mainly in the old-growth, which grew into small, but strong, plants waiting the chance to enter the stand, but in the event, they failed when a plague of voles in the summer of 1985 killed most of them by ringbarking at or just below the litter surface. Older saplings and poles were debarked by grey squirrels, which killed their leaders, distorted their growth and reduced them to large shrubs. Worse, deer browsed out almost all regeneration until the reserve was fenced, except within the exclosures. Their impact can be appreciated in the safety exclosure (erected 1990), where there is now a thicket of well-established beech saplings (Fig. 7.7), and throughout the wood following the fencing of the whole wood. The post-2007 cohort of saplings has been growing slowly and still mostly stands well below 1.3 m, but by 2016 some were starting to accelerate and recruit.

The old coppice stools also sprouted again after 1944, and a few produced growth strong enough to grow into the canopy, where they are now scarred by grey squirrels. The rest developed into contorted underwood stems, many of which were eventually crushed by falling trees or undermined by rot in their ancient stools. Many eventually died; others survive only as weak stems on well-rotted stools,

Fig. 7.4. A branch from an old Wye Valley beech with a long scar on what was its upper side. Originally formed when a grey squirrel tore off a long strip of bark, the scar has been partly healed by new bark growth.

Table 7.1. Population changes of beech in the old-growth stands.

	Initial density (N/ha)	Live at start	Losses	Gains	Mortality rate (%/year)	Recruitment rate (N/ha/year)
1945–1955	390	351	−27	+39	0.80	4.33
1955–1977	403	363	−135	+14	2.00	0.68
1977–1983	269	242	−48	+13	3.62	2.40
1983–1986	230	207	−13	+1	2.56	0.44
1986–1992	217	195	−17	+47	1.64	9.50
1992–2000	250	225	−36	+21	1.92	2.59
2000–2010	233	210	−34	+28	1.80	3.46
2010	227	204				

Table 7.2. Population change in beech in young-growth stands.

	Initial density (N/ha)	Live at start	Losses	Gains	Mortality rate (%/year)	Recruitment rate (N/ha/year)
1977–1983	383	787	−80	+4	1.77	0.33
1983–1993	347	711	−117	+31	1.87	1.59
1993–2002	305	625	−119	+7	2.32	0.34
2002–2013/15	250	513	−162	+62	3.13	2.63
2013/15	202	413				

Table 7.3. Population change in beech of seedling origin in young-growth stands.

	Initial density (N/ha)	Live at start	Losses	Gains	Mortality rate (%/year)	Recruitment rate (N/ha/year)
1977–1983	98	201	−35	+4	3.14	0.33
1983–1993	83	170	−51	+29	3.68	1.49
1993–2002	72	148	−46	+2	4.05	0.10
2002–2013/15	51	104	−39	+7	4.00	0.30
2013/15	35	72				

Table 7.4. Population change in beech of coppice origin in young-growth stands.

	Initial density (N/ha)	Live at start	Losses	Gains	Mortality rate (%/year)	Recruitment rate (N/ha/year)
1977–1983	98	575	−44	0	1.32	0
1983–1993	83	531	−66	+2	1.39	0.10
1993–2002	72	467	−73	+5	1.87	0.24
2002–2013/15	51	399	−118	+55	3.00	2.33
2013/15	35	336				

many broken at squirrel scars. After withstanding this for half a century, their mortality rate seems to be increasing.

Third, a less expected form of regeneration. The Lady Park beeches generate basal sprouts from mature trees and main roots close to the trunk in a fashion that is unusual in Britain and far better developed in *Fagus americana* (which is virtually a self-coppicing species). Heavily shaded, most of these sprouts never develop beyond weak, contorted stems that lose their leaders and retain only wisps of foliage towards the base, but some grow

Fig. 7.5. Beech pollard on the upper plateau.

into leaning underwood trees. Persistent though they are, they usually wither into a fuzz of stumps with short side shoots and eventually die, or they are undermined when the principal tree is blown over or killed by drought. They never, as far as we know, become large canopy trees.

Fourth, beech can layer from low, spreading branches and reiterate from fallen trees. We have no examples of the former in Lady Park, but several of the latter. One fine tree that was blown over near the top of transect V has generated several trees from points on the trunk and former crown, some of which have certainly rooted (Fig 7.8) (Fig. 7.20).

Summarising, the amount and timing of regeneration has varied according to the state of the stand, while recruitment has been limited by shade, deer and voles. In the young-growth, there was substantial recruitment from both seedlings and coppice stools before it formed a thicket, but few thereafter. In the old-growth, limited recruitment from sprouts and seedlings eventually followed the 1976 drought. In both stands, potential recruitment was browsed by deer and killed by voles. Below the cliff, the stand was still recruiting after 1950, but that stopped until elm disease and the tree falls associated with a maturing stand allowed in a long pulse of regeneration and recruitment – shade-tolerant though it is, beech regeneration still benefits from more light. In the managed stand, deer totally excluded recruitment, despite beech's relative unpalatability.

7.5 Beech Growth and Mortality

Growth and mortality are strongly interlinked. Just like populations of ash and other tree species, smaller beech generally grow slower and are more likely to die. Distinctively, however, small beech can survive deeper shade than most, while large beech are more vulnerable to 'events'. The best demonstration of the former is found on the margins of the old-growth, where a group of 'large saplings' have remained static for decades, hardly growing, but surviving (Case study 7.1). Large beech suffer the consequences of developing only a shallow root

Table 7.5. Population changes of beech below the cliff.

	Initial density (N/ha)	Live at start	Losses	Gains	Mortality rate (%/year)	Recruitment rate (N/ha/year)
1950–1959	41	20	0	+6	0	1.38
1959–1985	54	26	−16	+6	3.75	0.48
1985–1992	33	16	−1	+26	0.92	7.69
1992–2005	85	41	−8	+9	1.66	1.43
2005–2013	87	42	−28	+2	11.88	0.46
2013	33	16				

Case study 7.1. The geriatric beech 'saplings'

Several tiny beech in the old-growth just above the main track have hardly changed in decades (Fig. 7.9). In the 1980s, they looked like established saplings, but their true status was revealed by one that was included within transect II. First recorded in 1977 at 9.4 cm GBH, its girth had grown slowly to 15.5 cm by 2010. In 1983, it was 4 m tall, 'growing well' mainly on the downslope side. By 1992, epicormics were developing on the trunk, and by 2000 it had several dead branches. In 2010, we recorded a 'thin, spindly sapling with leader snapped off. Weak lateral branches, well suppressed. Trunk epicormics.' Radial growth from 1977 to 2010 was just under 0.3 mm/year, but growth before 1977 had been double that, assuming that it reached 1.3 m after 10–15 years.

Clearly, this is no 'sapling' but a 65-year-old veteran that must have grown from seed shortly after the adjacent stand had been felled in 1942. Having become established in the temporarily well-lit margins of the old-growth, it has persisted, healthy, but clearly under stress.

These must have become established soon after the 1942–1944 felling on the other side of the track, but they did not keep up with the rapid regeneration in the young-growth. Heavily shaded from the mid-1950s, they have survived with thin foliage, neither gaining in height (about 4 m) nor increasing their girth.

Fig. 7.6. Beech seedlings and small saplings are widely scattered through the wood, but these have no future unless the old canopy tree above them dies.

Fig. 7.7. Beech regeneration within the 'safety' exclosure, made in 1990. This hints at how the beech might have regenerated if deer had been excluded before 2007.

system: they are vulnerable to wind, unbalanced growth on slopes and particularly to drought.

The great 1976 drought bore most heavily on the old-growth beech (Peterken and Mountford, 1996; see Fig. 7.1). Canopy beech had been growing well, but decelerated during 1955–1977, presumably due to increasing canopy competition (Table 7.6), but in 1976 they lost their leaves in August, many died immediately (Fig. 7.10), either desiccated or scorched, and the survivors almost stopped growing. In the following years, great scars developed where bark had been killed, and these developed into rot hollows that ate their way into the trunk (Fig. 7.11). Up to 75% of crown branches died back from the growing points, leaving jagged ends and dead branches hanging in the remaining crowns. Splits in the bark up to 3 m tall developed at the base of otherwise undamaged trees, and trunk necroses developed, i.e. patches of bark simply died and later sloughed away, leaving the wood exposed.

The trees responded by attempting to heal the bark splits and necroses, but only the smaller bark splits healed completely. Most developed into visible rot, which expanded in some cases to 80% of the circumference in trees that somehow remained upright. Crown branches generated new shoots from within the crown and by epicormics developing on the trunk below the crown, and to some degree recovered, but they never again developed the spear-like tips of fast-growing crown branches. Increment rates recovered, but stabilised at rates well below the rates achieved during the previous

Fig. 7.8. Fallen beech near the plateau that lived to regenerate several trees from the prostrate trunk and crown.

half century (Fig. 7.12), though, remarkably, even after the drought, the fastest growing trees maintained their growth rate and the best placed of the underwood beech actually accelerated. Several survivors eventually succumbed to the rot and loss of vitality. For example, a 264 cm GBH beech, which probably survived the 1870 coppicing as a small sapling, fell in January 2016, by which time only 13% of its basal area was still alive at 1.3 m, 72% was rotten and 15% was a hollow that tapered from the ground to 3.5 m. In effect, this tree had been dying for 39 years.

Today, even the trees that have lived for 40 years since the great drought are in a poor condition, even though many look vigorous (Case study 7.2). Many are scarred and rotten in the lower trunk. Few have vigorous crowns. Indeed, crown branches keep falling, invariably broken at points where grey squirrels have stripped the bark from their upper sides. Whole trees also keep falling, either breaking where rot has weakened the trunk or simply by losing balance on a slope, and several large spreading trees have split, shedding substantial parts of their crown. Nevertheless, the species is still a major component of the old-growth stands. The largest beech had reached 387 cm GBH in 2014, but now grows slowly (Fig. 7.13).

The arrival of grey squirrels in about 1943 introduced a new limitation on beech growth. Until then, red squirrels had allowed them to grow unhindered into splendid trees, but grey squirrels tear off strips of bark for reasons known only to grey squirrels, thereby ringbarking some and admitting rot to the trunks and branches of the rest (Fig. 7.15). As we have found in Lady Park, they leave saplings alone, but attack pole-stage growth, especially trees growing fast, and thinning merely incites them to even more vigorous stripping (Mountford, 1997, 2006; Mountford and Peterken, 1999). They also debark the crown branches of mature trees, especially on the upper sides where the damage is invisible from below, leaving them to rot on the upper side, sometimes forming a trough in which water collects. Inevitably, the branches

Case study 7.2. Recovery of the skyline beech at transect V: 100 from the 1976 drought

This was and is a splendid large, spreading beech on the upper steep slopes, one of several that may have been retained on the skyline to mitigate the visual impact of coppicing when the Wye Tour was at its height (Fig. 7.14). In 1977, however, our records described its crown was 'half dead' and by 1983 'all the main branches had died far back', yet by 1992 major crown branches were 'now re-shooting'; by 2000 its crown had 'regrown from sprouts at ends of branches'; and by 2010, when its girth had reached 306 cm at breast height, it had 'recovered well with a massive, vigorous crown'. Radial growth rates reflected this near-death experience: 2.28 mm/year before 1977, assuming that it was 200 years old in 1976, 0.24 mm/y in 1977–1992, 2.03 mm/y in 1992–2000 and 1.83 mm/y in 2000–2010. However, the drought also killed a large patch of bark at the base of the trunk, leaving a scar 4 m high and introducing rot. By 2016, the bark had grown back over much of the scar, but the damage was still visible.

Table 7.6. Radial growth rates (mm/year) and basal area increments (cm^2/year) of the 92 beeches recorded in the old-growth at all dates from 1945 to 2010.

	Radial growth rates		Basal area increment	
	Median	Maximum	Median	Maximum
1945–1955	1.62	7.28	11.0	104.0
1955–1977	1.07	3.29	7.0	51.5
1977–1983	0.00	7.08	0.0	136.3
1983–1992	1.26	4.62	10.8	100.5
1992–2000	0.88	4.51	10.7	77.7
2000–2010	0.80	4.06	9.6	64.9
1945–2010	1.11	2.75	10.0	53.1

Note: In 1983, diameter was measured with a calliper as the average of two diameters at right angles.

break easily in storms. Indeed, most branches that fall are found to have snapped at squirrel scars. Squirrel damage is patchy – perhaps reflecting territorial individuals with different enthusiasms for stripping – and slow-growing poles are more likely to be ignored. The overall effect is to reduce beech poles to contorted bushes, leave more fortunate poles with scars on their trunks, and accelerate dieback in the crowns of canopy trees. Unless squirrels can be controlled – and the Forestry Commission has tried – beech seems very unlikely to form towering, well-formed trees in the future.

Beech in the young-growth were less affected by drought but were damaged more by squirrels: in 2002, half the trees attaining 20 cm GBH had been severely debarked. Trees originating from seed grew at least as well as coppice stems (Table 7.7), and both tended to slow down. While competition must have been a factor, debarking and the increasing age of the stools must have contributed. In fact, many stools died completely. In both old-growth and young-growth, many small trees of both sapling and coppice origin grew very slowly, yet survived. Thinning of similar stands in the managed compartment did increase radial growth rates, despite the concomitant increase in bark stripping by squirrels.

All but two of the standards grew well. The exceptions were the largest, which perished in the drought, and one that consistently grew slowly, even though its crown looked healthy. Surprisingly, most of the old spreading skyline beech also survived in reasonably good condition, though two were badly damaged in 1976 and have since shed major branches (see Fig. 7.3). They are now growing slowly: the large beech at the top of transect V (see Case study 7.2) has 'JR 1922' etched into its bark, and it has hardly expanded. Growing from birth on thin, dry soils, perhaps they have become accustomed to drought.

Below the cliff, radial growth rates were generally faster and more stable than those above the cliff. They suffered only a short slowdown after 1985, which implied that they remained unaffected by the drought, or that drought impacts were counterbalanced by relief from elm competition. Three faster growing trees in 1950–1959, growing at 3.03–4.04 mm/year, continued to grow fast until they fell. Individual trees on the same stool differed: on one stool, one tree accelerated gently while the other braked hard and recovered; on another, one tree slowly gained dominance over the other. Some trees were consistently fast and others consistently slow. The largest and better-growing trees leaned increasingly and eventually fell (or were felled for safety). The upper size limit imposed by the steep

Fig. 7.9. Susan Peterken measuring one of the geriatric beech 'saplings'.

Fig. 7.10. An even-aged group of beech in the old-growth stands on transect III, photographed in 1983, which must have perished as a group in, or immediately after, the 1976 drought.

Fig. 7.11. From the other side, this canopy beech looks healthy and vigorous, but since 1976 a great scar has developed almost up to the crown. Photographed 40 years after the drought, this is one of several trees that could split at any time.

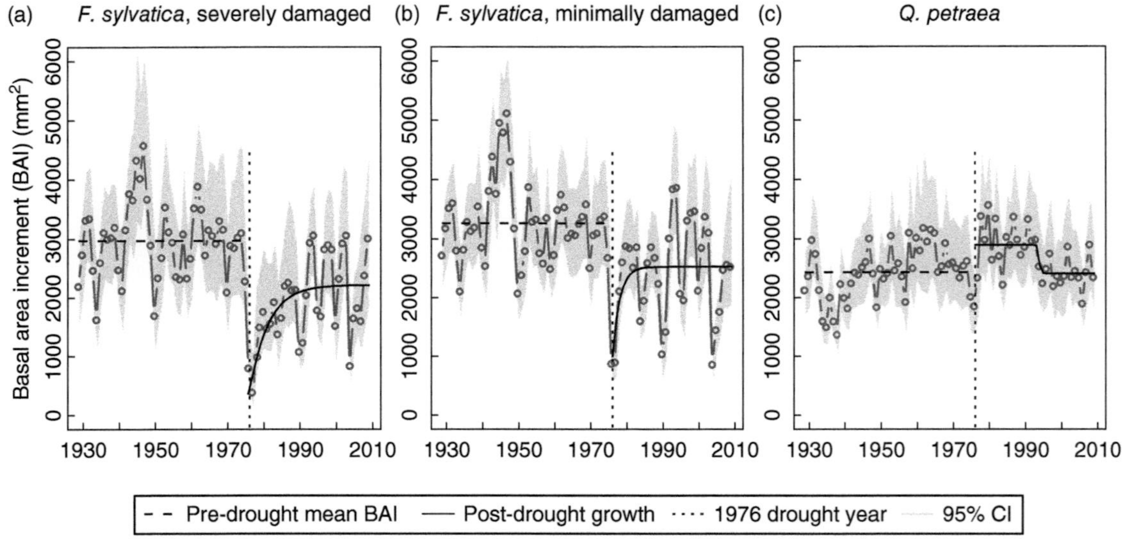

Fig. 7.12. Annual growth of beech and oak in old-growth stands as revealed by ring widths measured in cores. Growth of both oak (c) and severely affected beech (a) was depressed by the 1976 drought, but oak recovered quickly and prospered for a few years, whereas beech never recovered its earlier growth rates. Beech that were mildly affected (b), as judged by external symptoms, showed the same response in muted form. (From Cavin *et al.*, 2013.)

slope must be above 200 cm, but not far above: a large, rotted beech at the base of the slope just north of transect IX reached 268 cm, way above any others.

Comparisons with growth rates at the opposite edge of beech's European range show that Lady Park's trees are far from disgraced. In virgin forest at Rajka, in the mountains of Albania, beech grew at 0.37–2.64 mm/year (average 0.96 mm) (Tegel *et al.*, 2014), whereas in Lady Park the 109 beeches in the canopy and subcanopy in 1945 that survived to 2010 grew at 0–2.75 mm/year (average 1.13 mm). Comparable averages for 'dominant' and 'suppressed' trees in Rajka were 0.99 mm and 0.73 mm, respectively, whereas in Lady Park averages for canopy and subcanopy trees were 1.13 mm and 0.51 mm, respectively. The dendrochronological study from which the Rajka values come appears not to have selected truly suppressed individuals.

7.6 What Kills Beech?

Many agencies kill beech (Table 7.8), but determining which is the most important can be difficult. In most cases, several causes contribute, some of which develop slowly over years, others of which are events that happen in seconds. Case study 7.3 illustrates the point: the beech that died was weakened by drought, ivy and butt rot, but it was probably grey squirrels that finally enabled decay to gain the upper hand over growth.

This death, like so many others, was the outcome of a chain of causation, not a single cause. Several factors combine in parallel or in sequence to weaken trees before some other factor kills them. Weakening factors include debarking by squirrels; ivy growth competing in the crown; a particularly vigorous neighbour in the canopy; physical weaknesses, such as a forked trunk; physical blows, such as a stone rolling down the slope and damaging the bark of the trunk enough to let in fungi; a partial windthrow which ends with a tree leaning on another, with its crown below its supporter; and the unbalanced growth and distorted form that often results from growing on a steep slope. A coppice-origin tree may be growing well, but the old stool from which it grows rots, weakens and may eventually split, cutting the ground from under the tree's feet, so to speak. Any factor could be the initiator and any other factor could administer the *coup de grâce*. A tree on a steep slope develops a lean, but its fall may be due to wind, another tree upslope that falls on it, or an upslope root that breaks. In human terms, many beech are just unlucky.

Fig. 7.13. In 1963, Andy and Toot etched their names into a beech growing on the lower slopes above the cliff. Photographed in 2014, their bid for immortality demonstrates just how slowly the tree has been growing; they should just have celebrated their golden wedding anniversary.

Case study 7.3. Contrasting fates of two canopy beech on transect II: 450–500

This is the story of two nearly identical trees that were growing near each other on similar ground, initially with equal vigour.

Beech 156 was a canopy tree that had grown to 176 cm GBH by 1977, having grown at 3.23 mm/year and 2.68 mm/year in 1945–1955 and 1955–1977, respectively (basal area increments of 41 cm^2 and 42 cm^2/year). However, it was severely affected by the 1976 drought, and by 1983 the top of the crown was dying and bark was falling from large trunk necroses. By 1986, it was recorded as 'dying': most of the crown was dead, a few branches remained alive in the lower crown and even more bark was missing. By 1992, the upper crown had disintegrated; bark was splitting and pealing from the base to 4 m; rot had set in; but parts of the crown were regenerating and epicormics were springing from the trunk just below the crown. By 2000, the trunk epicormics were dead, but the crown still contained some vigorous new shoots. These, however, were being debarked by squirrels. By 2008, it was dead, very rotted at the base, but still standing as an ivy-clad snag. In 2013, the snag fell and the prostrate trunk was so rotted that it looked like a dugout canoe.

Beech 153 was a slightly less vigorous canopy tree, having grown to 160 cm GBH in 1977 at 3.23 and 1.07 mm/year, respectively, but this one survived. No damage to the crown was recorded, and from 1986 onwards it was described as healthy and expanding towards a small gap. It was, however, growing very slowly at 0.09, 0.62 and 0.56 mm/year in 1977–1992, 1992–2000 and 2000–2010, respectively, reaching 183 cm GBH in 2010.

Fig. 7.15. Cut stump of a vigorous beech near Lady Park that appears to have been repeatedly debarked by grey squirrels, but has tried to heal the wounds. The scars and subsequent expansion of surviving bark can be seen on both the cut surface and the root visible top right.

At most times and in most populations, small beech are more likely to die than large beech. This 'steady-state' or chronic mortality is the outcome of competition. The underlying mechanism must be multifarious: smaller beech are more likely than larger beech to be shaded, suffer from root competition, grow slowly or be flattened by falling trees and branches.

Fig. 7.14. The apparently fine beech at the top of transect V with a drought-induced scar that it has been unable to heal in 40 years.

Table 7.7. Post-drought radial growth rates of beech.

Stand	Tree origin	N	Radial growth rates (mm/year) Range	Median	Girth in 1977/85 (cm) Range	Median
Old-growth	Saplings	48	0–4.04	1.13	9–226	119
	Coppice	102	0–3.59	0.40	9–198	35
Young-growth	Saplings	59	0–2.29	0.96	3–44	13
	Coppice	279	0–2.84	0.71	6–100	19
	Standards	10	0.10–3.36	2.05	145–242	198
Below the cliff	All trees	10	0.69–2.03	1.59	61–185	122
Old-growth below the cliff, not thinned	All	36	0–3.00	0.27	6–59 (201)	18
Younger growth, not thinned	All	20	0.05–2.39	1.06	9–35	18
Upper cliff woodland	All	25	0–1.59	0.32	16–146	45
Thinned stands on limestone	All	19	0.19–3.82	1.43	9–71 (262)	40
Thinned stands on deeper soils	All	42	0.16–5.76	2.28	4–91	32

Notes: Old-growth, 1977–2010; young-growth, 1977–2013/15; below the cliff, 1985–2013; managed stand, 1984–2014. Tree origins not distinguished in the managed stand.

Table 7.8. Causes of death of beech.

	Impacts on small and underwood beech	Impacts on canopy beech
Competition	Despite bearing deep shade, most beech die small after growing slowly in shade.	Canopy beech are restricted by vigorous neighbours. Subcanopy beech can be overtopped, restricted, then killed by taller oak and lime.
Wind	Very little, except when other trees fall on them.	Capable of uprooting mature, healthy trees, a process made easier by the shallow root system. Also breaks debarked trees and trees with rotting cores.
Drought	Very little.	Can kill trees outright, but usually weakens them by killing crown branches and patches of bark, especially near the base of trunks.
Grey squirrels	Ringbark poles, destroying leaders repeatedly, especially fast-growing individuals.	Debark crown branches, mainly on the upper side, base of trunks and exposed roots, allowing in rot.
Deer	Browse saplings and coppice growth, but other species are preferred.	None.
Slope	None.	Develop unbalanced crowns and severe leans.
Other falling trees	Flattens some saplings, breaks poles; but may also leave an established pole in a gap.	Even large beech can be knocked over or broken by trees falling against them from above.
Ivy	None.	Can compete with crown and break crown branches. Ivy 'trunks' reach 50 cm GBH on beech.
Clematis	Can smother and break poles.	None.
Weak coppice stools	None.	Often break under the weight of large trees.
Fungi	Very little.	Veteran and damaged trees admit wood-rotting fungi. Root systems of ageing trees may be weakened by fungi.
Damage to trunk bark	Can be frayed by deer or debarked by squirrels.	Rarely grazed by another falling tree or by a rolling boulder.
Physical weaknesses	Very little.	Large forked trees often split; crown branches may spread too widely; wide-spreading veteran trees may break up.

This steady exclusion of smaller beech is supplemented by the irregular, often dramatic and sometimes inexplicable deaths of larger trees. Some die standing, as they did during and after the 1976 drought. Others split and fall, a common fate of trees that are weakened by drought-induced trunk rot. Still others – mostly tall beech growing on steep slopes – fall over, tear up roots and deposit soil and rocks into a mound, leaving a complementary pit where the tree had been growing. The frequency, periodicity and pattern of large-beech mortality depend substantially on stand age, site conditions (soil depth, slope) and the incidence of droughts and storms. Death comes in clusters, not only when groups of large trees knock each other over but also perhaps due to variation in soil depth, or even the incidence of particular soil fungi. There are also inexplicable falls, such as the large canopy beech with a balanced healthy crown and an intact root system that fell with a prolonged, rending crash one calm, sunny afternoon in October 2013 when we were measuring trees nearby (Case study 7.4).

Mortality patterns have differed between stands. In the upslope old-growth, where the 1976 drought hit hardest, the death of large standing trees from 1977–1986 coincided with much reduced growth rates and almost outpaced the losses of small trees. On deeper soils in the old-growth above the cliff, beech in the canopy also grew much slower, but survived the drought. Below the cliff, the steady attrition of large beech has been associated with fast growth and destructive falls of live trees that have grown fast to the point when the roothold failed. In the young-growth, mortality has been dominated by the conventional exclusion of smaller beech and losses caused by debarking, reinforced by the group failure of stems on aged and rotten stools.

7.7 Decaying Beech

The beech killed by the 1976 drought remained standing as snags in tree-sized gaps, eerie reminders of the fine trees they once were. Initially, their main branches and much of their crown remained intact, but as the years passed the snags 'contracted' as twigs, crown branches and then main limbs successively broke away. Eventually, their trunks remained as a pillar of increasingly rotten, bark-free wood until either they snapped, leaving a short stump, or their roots rotted and the trunk tipped over without excavating a pit. In the end, each tree was reduced to a conical pile of rotten wood.

When live canopy beech fall over, the trunk and branches lie spread over 30 m of woodland floor, the butt rotates, excavating a pit and exposing the root plate encased in mingled soil, rocks and torn root. Fallen trees steadily lose their bark, but remain intact until rot becomes so advanced that

Case study 7.4. The falls of two large beech from the upslope old-growth

A massive tree fall occurred on 2 August 2013 in the old-growth of transect I, just above the main ride. A large beech that had been damaged in the 1976 drought snapped off at about 1 m in an evening storm, which briefly brought heavy hail and strong gusts. It fell past a small canopy oak, stripping off the crown branches, brought down a spindly canopy ash and fell against another drought-damaged large beech. This, too, snapped off at *c.*1 m and fell between other beeches, to lie across the path. Both beeches were live in about 25% of the circumference, and the top beech was hollow. In fact, it was amazing that it stood as long as it did.

While recording on transect III on 17 October 2013, we heard a prolonged crashing of a large tree fall from the stand above and to the south. Investigating, we found a large beech had fallen from the canopy of the upslope stand midway between transects II and I, roughly in mid-slope (Fig. 7.16). The tree was healthy, with a full crown and several leafy branches down the trunk, and there were no visible trunk lesions. It fell downslope, of course, missing other trees within range, save for an adjacent tall ash, from which one of three large crown branches had been torn (but which was left against the crown of a canopy oak, ready to be a widow-maker). The large pit revealed a healthy root system, but with no large roots upslope. Since this was a calm, sunny, warm day, as balmy and quiet as it gets in October, and the weather recently has been devoid of any extremes, one would normally expect to be safe. Right by it but slightly above was a slender canopy beech that looks more than half rotted through the trunk, and which ought to be a prime candidate for falling. So, we can talk about the causes of tree falls all we like, but this one fell for no discernible reason at an extremely unlikely moment. If it was just too heavy in the crown for its root system, why leave it to a calm day?

Fig. 7.16. The fallen beech, photographed 30 minutes after it fell.

the trunk fragments into a line of rotten logs. Eventually, the butt rots down, the soil and root remnants fall away and the base of the trunk eventually decays completely, leaving a mound beside the pit. The many branches that fall from still-living trees follow the same course, but quicker.

Small beech lose their identity within a few years after they die, though they can sometimes be found as bark-free sticks leaning against living trees. Likewise, branches that fall in full leaf lose their leaves within a year and break down quickly, mostly within a decade. It is the larger trees that take longer.

The drought killed several groups of mature beech, one of which happened to be crossed by transect III (Case study 7.5). Here, the canopy trees that eventually died survived for a year or two, but were clearly dead by 1983. Ten years after the drought, one still stood as a snag in the canopy and even retained some crown branches, but 16 years after the drought, all snags had either fallen or remained only as short and largely bark-free stumps. Short snags were still visible after 23 years, but the remains had completely rotted after 34 years. Predictably, the smaller trees disappeared faster than the larger trees.

Much the same happened elsewhere in the old-growth. The large beech that fell from transect VIII in 1950 could not be detected in 1985, so it evidently decayed completely in less than 35 years. Drought victims in transect I were still mostly standing in February 1989. In another group of drought victims north of transect II, five snags were still standing in May 1992, to a height of 14 m, but with 95% bark loss. In the young-growth, two standards that died in 1977 still had snags of 15 m and 10 m in 2002, but they had fallen by 2013.

Many drought victims took decades to die, having been weakened by, for example, substantial bark necroses that developed into very rotten trunks and large bracket fungi. They died eventually, by which time the main trunk was half decayed. One example on transect II, which took over 20 years to die, is given as Case study 7.3. Delayed mortality of drought victims probably explains the presence of well-rotted snags in 2015, 39 years after the drought.

7.8 Oak in Britain

Oak was the heart of both Olde England and early British woodland ecology, and it remains

Case study 7.5. Response of canopy beech growing in the old-growth on the mid-slopes, transect III: 400–600

The beech here dominated a tall, densely stocked stand, but in 1976 the drought killed several, leaving a large, irregular gap. If we included a tree that was very close to the transect, there were 15 beech of 69–217 cm GBH that had been in the canopy since at least 1945. All were recorded as living in 1977, one year after the drought, but by 1983 eight were dead. The remaining seven were damaged to varying degrees, but all recovered and survived to 2010. In general, it was the less vigorous trees (as measured by radial growth rates in 1945–1977) that died. In one telling example, of two trees of 138 cm GBH growing together, the trees that had been growing faster before 1976 survived, but the slow-growing tree died.

The dead trees had all vanished by 2010, 34 years after the drought. In fact, two smaller trees had vanished by 1992, 16 years after. In 1986, several still stood as tall snags up to 27 m, some with branches still intact, but all with some loss of bark. One had fallen after its roots had rotted and another had snapped at 0.5 m. By 1992, the snags were much shorter, 1.5–7.5 m, and bark had been reduced to 10–50%. Two snags were still visible in 2000, but had gone by 2010.

The survivors were all recorded as damaged in 1983. In fact, one was 75% crown-dead and others were almost as bad. But, by 1986, and especially by 1992, epicormics were growing vigorously in the crowns and on the trunks, and they continued to grow vigorously thereafter, expanding into the gaps. In fact, by 2000, the record showed that the drought damage was no longer obvious and trees that had been confined before the drought were taking advantage of the additional space to grow larger crowns. Some of this new crown growth was badly debarked by grey squirrels, and a few new branches were actually killed, but the trees were still prospering.

the symbol of both the National Trust and the Forestry Commission. As many old houses and barns testify, it was the main indigenous timber tree, growing prominently in wood pastures, coppice and farmland, where some oaks achieved great age, astonishing girths and immense crown spreads. Ecologically, under the terminology adopted by Tansley (1939) and other founding ecologists, much of the variety of British woodland was concealed within an all-embracing class of 'oakwoods', a term that made sense then because oaks were generally the tallest trees in the ancient coppice woods.

Despite its symbolic importance, oak has been on the retreat for years. It is not just that huge numbers were felled during the 20th century wars, but also, along with other broadleaves, it was supplanted by European larch, Norway spruce, Douglas fir and several other imported conifers as state timber trees. Further, its ecological resilience has been impaired: never a true shade-bearer, it was sustained well enough for centuries by the open conditions in wood pastures and the repeated cutting of coppice woods, but around 1908 the silver-leaf oak mildew *Microsphaera alphitoides* spread across Europe, and this has further reduced its ability to regenerate from acorns within woods (Rackham, 2003). At the same time, the mature oaks left in former wood pastures and coppices have faced severe competition for the first time in their lives as new regeneration has infilled wood pastures and coppices have grown tall with neglect. Further, the huge post-war reduction in farmland trees meant that vast numbers of oaks were lost from fields and boundaries. The change has not been all one way – some former pastures have developed into oak-dominated woodland – but, in most places, oak in the 21st century is a species facing a new reality.

Two species are native to Britain. Pedunculate oak combines stalked acorns with leaves possessing short stalks and auricles at the base. Sessile oak has short-stalked acorns, stalked, broad leaves and is generally a more clean-limbed tree. Both are found throughout Britain with their hybrids, but pedunculate oak dominates the lowlands, while sessile oak is associated more with the uplands. This pattern is not rigid. Pedunculate oakwoods are scattered through the uplands, from Loch A'Mhuilinn (Sutherland) to the famous Wistman's Wood on Dartmoor, and lowland sessile oakwoods are found in both ancient coppices, such as Swanton Novers (Norfolk) and wood pastures, such as the New Forest. Underlying this pattern is a preference for (but not a restriction to) deep, base-rich soils and tolerance of waterlogging on the part of pedunculate oak, and a predilection for thin soils and free drainage on the part of sessile oak. In the western uplands, lowlands and south-east lowlands, this places most sessile oaks on strongly acid soils, but in the borderlands, sessile oak is frequently the main oak on limestone, as it is in Lady Park Wood. This pattern seems to be natural, but it has been modified by planting.

Oaks are chunky trees whose bark changes from smooth, almost silvery in youth to finely fissured in maturity. The lobed leaves remain late and conspicuous on the tree in autumn. The foliage of pedunculate oak is clustered, whereas the leaves of sessile oak tend to be broader, more spread and altogether more graceful. The largest oaks project an overwhelming sense of mass and strength, but deep in the woods they can be surprisingly difficult to spot because they cast less shade than beech and generally stand among a thicket of lesser growth.

7.9 Oaks in Lady Park Wood

After the detailed account of beech, this account of oak will seem like a mere tailpiece, but this accurately reflects the difference in their importance for Lady Park's dynamics. Whereas beech exerts great influence on other species, oak is more like a genial big brother, physically substantial, but allowing other species to get on with their own lives.

The oaks in Lady Park Wood are all sessile, save for one pedunculate oak on the Wye flood plain, one on the colluvial fan and another on the lower slopes of the managed compartment. Before 1870, oaks were both the main standard trees and a component of the coppice, and they must have been prominent in the preceding wood pasture. Coppice oaks were concentrated on the acid soils above the cliff, infrequent elsewhere and hardly present at all below the cliff. As standards, they were evenly distributed throughout the stand, due, we assume, to the careful retention of promising oak saplings during coppicing. They were retained in 1902 when foresters initiated the promotion to beech high forest, but many were felled in the war, leaving only a thin scatter as a shelterwood (three per hectare in the young-growth).

How old are they? Growth-ring counts gave an origin between 1746 and 1842, which suggested that new standards were regularly promoted after each coppicing. However, there was a notable concentration in 1800–1810, which hinted at planting, or at least a marked effort to recruit timber trees. There is no record of planting, and many oaks in Lady Park have been promoted from coppice stools, but the remaining stands of old timber elsewhere in the former Hadnock Wood include a generation of fine, mostly maiden, sessile oaks dating from around 1800, which must be the survivors of the late 18th century generation that persuaded the Crown to buy Highmeadow Woods in 1817 (see Chapter 2, this volume). By 2015, the largest oak had reached 425 cm GBH, another on a crag above the Slaughter reached over 400 cm (Fig. 7.17), and several others scattered throughout the wood had achieved 320–390 cm. Most look sound and healthy. Several must have originated well before 1746, for they are much larger than the ring-counted trees would have been.

The Lady Park oaks, like the oaks in so many ancient coppices, were sustained carefully for centuries by promoting coppice shoots and nurturing selected saplings originating from acorns. Now they have lost their privileged position and must fight for survival against the species that had formerly been subjugated by coppicing.

7.10 Changes in the Population of Oaks

Oaks effectively failed to recruit after the 1940s, either on the transects or elsewhere – the few actual recruits quickly died. Oak seedlings, however, are seen often enough, and some have survived a few years before perishing, but even good mast years such as 1995, and the abundant gaps present after 1976, have not initiated new oaks. In 2016, however, there were real prospects of limited recruitment. Strong, small saplings have become established in and near canopy gaps towards the lower end of transect III and in a large tree-fall gap close to transect VI on the colluvial fan. If the deer let them get away, they will still be vulnerable to deformation or worse by grey squirrels.

Population change is thus a simple story of steady thinning out of most small and slow-growing stems, as well as a few large ones. In the old-growth, the population decreased steadily from 75 trees in 1945 to just 24 in 2010, of which 6 had been standards (Table 7.9). Seedling oaks were seen occasionally, but they all perished before they exceeded 10 cm. Shoots were still being recruited from coppice in 1945–1955, and two more grew in after 1986, but all died before 2010.

There was little doubt about what was happening. As the underwood grew into the canopy, more and more oaks were weakened by shade. As the shade increased, most lost so much of their crown that their growth slowed, they became completely static, then died. Even the survivors lost large crown branches:

Fig. 7.17. Fine standard oaks retained on the steeply sloping ground overlooking the Whippington Brook.

indeed, two of the larger oaks in 1945 died before 2010. In 2000, descriptions of the survivors record that most had lost much of their crown, all were 'squeezed' by neighbours, and a few were becoming overtopped by beech, lime or ash. Only the 1976 drought provided a decade of mild relief.

The conditions for oak in the young-growth were different. Oak coppice was felled with the rest of the underwood in 1942–1944, and so were many standards. This relieved the remaining standards of competition, rejuvenated the coppice and made regeneration from acorns possible. Despite this, seedling regeneration was limited and quickly became shaded by birch. Most coppice regrowth was also weak, perhaps because many stools were large, old shells that had been robbed of their former substance by shade after 1902 (Fig. 7.18). By the autumn of 1948, the leading sprouts had grown to 1.5–4 m, with the very tallest at 5.5 m, while even the strongest seedlings reached only 60–90 cm (LPW EWJ A8). Some of the felled standards also sprouted, but their growth was also weak. In addition, even the stronger stools were browsed by deer and quickly subjected to competition by fast-growing birch.

After canopy closure in about 1950, both coppice stems and seedling regeneration thinned out. By 1977 (Table 7.10), 177 were left alive, with 12 dead coppice shoots and 6 dead maidens still standing. The trees that originated as seedlings after 1944 were generally smaller than the coppice stems. Thereafter, predictably, the remaining standards

Table 7.9. Population change of oak in the old-growth stands.

	Initial density (N/ha)	Live at start	Losses	Gains	Mortality rate (%/year)	Recruitment rate (N/ha/year)
1945–1955	83	75	−10	+7	1.42	0.78
1955–1977	80	72	−35	0	2.85	0
1977–1986	41	37	−2	0	0.65	0
1986–1992	39	35	−2	+2	1.06	0.40
1992–2000	39	35	−8	0	2.84	0
2000–2010	30	27	−3	0	1.30	0
2010	27	24				
1945–2010		75	−51	0	1.74	0

survived better than those in the old-growth, and coppice stems survived better than maidens. Two standards with prominent crowns sickened and died. One was healthy in 1983, dying back in 1993 and dead by 2002. The other took longer to die: it was dying back slightly in 1983, but perished after 2002. Elsewhere within the young-growth, most standard oaks have remained healthy.

7.11 Growth of Oak

In all stands and periods, mortality rates were greatest in the smaller stems, but the size–survival correlation was weak. Much depended on neighbours. Some small stems competing mainly with birch survived, whereas some initially large stems growing near larger and more vigorous neighbours perished.

The 24 survivors in the old-growth grew steadily from 43–221 cm GBH (median 81) in 1945 to 83–295 cm (median 155) in 2010 (Table 7.11). At all times, growth rates varied enormously between individuals, but generally grew slowly before 1977, when stand density was highest, and accelerated in 1977–1992, when the drought relieved them of competition from beech. They slowed again in 1992–2000 as beech reasserted itself, but accelerated after 2000.

The changing circumstances were clearly visible on the trees. Before 1977, competition with beech and the up-growing lime coppice killed the lower branches of the former standards. In the 1980s, many were again free to expand their crowns, but instead they developed a sheath of epicormic growths on the well-lit side of their trunks. After a decade or so of respite, the oaks were again subjected to increasing competition as the crowns of beech, lime, ash and others expanded and increased in height. Mortality rates picked up, growth rates fell and the sheaths of epicormics died. Most large oaks survived, but with increasingly restricted crowns. Latterly, gaps created by tree falls have had a mixed effect: while the mortality rate declined, growth rates increased slightly and one or two oaks have visibly benefited and others have been knocked down, including one of the very large trees. In 2015, the fallen trees were generating new sprouts, but they looked unlikely to keep pace with other new growth.

In the young-growth, the surviving standards grew at 0.54–3.38 mm/year (median 1.92) from 1977 to 2015, the largest reaching 342 cm GBH. The slowest was chloritic and dying back in 1993, but its health has since improved. Some coppice shoots and maidens were even faster, but most were much slower. Among the survivors, there was surprisingly little difference between coppice stems (median 1.14) and maidens (median 1.32). By 2002–2015, just five trees of seedling origin and eight trees of coppice origin were bidding to become elite trees by growing at over 1.5 mm/year and not decelerating, but the rest were small, slow growing and/or decelerating. Here and in the old-growth, the wheat was increasingly being sorted from the chaff.

The only oak growth rates available from below the cliff were those in the unthinned part of the managed compartment. The eight trees, including one former standard, grew at 0.27–1.38 mm/year (median 0.96), slower that those within the reserve. In the thinned parts of the managed compartment (Table 7.12), oaks grew well at 0.48–3.82 mm/year (median 1.77) from 1984 to 2014, which indicated only a limited response to thinning once one allowed for the removal of small stems. One oak in a Williams plot achieved the phenomenally fast rate

Fig. 7.18. This standard oak was felled in 1942 and re-sprouted. By 2015, two sprouts had survived to generate trees in the young-growth. The standard oak had been promoted from an old and already hollow coppice stool, probably in the early 19th century.

Table 7.10. Population change of oak in the young-growth stands.

	Initial density (N/ha)	Live stems at start			Mortality rate (%/year)		
		Standards	Coppice	Maidens	Standards	Coppice	Maidens
1977–1983	86	6	138	33	0	2.30	2.70
1983–1993	75	6	120	28	0	1.20	2.05
1993–2002	66	6	107	23	2.45	3.45	5.37
2002–2013/15	47	5	78	14	2.47	4.49	4.75
2013/15	28	4	46	8			
1977–2013/15					1.09	2.93	3.76

Note: There were no recruits between 1977 and 2015.

of 4.71 mm/year from 1979 to 2015, which showed what was possible.

Growth rates collected by Eustace Jones (1959) when he was still involved at Lady Park showed that the best Lady Park oaks were as vigorous as any elsewhere. Dominant oaks in plantations on 'moderately good' sites average 2.3 mm/year until they reach 150 years. The best sessile oaks in the

Table 7.11. Radial growth rates and basal area increments of the 22 oaks in the old-growth that survived from 1945 to 2010.

	Radial growth rates (mm/year)		Basal area increments (cm^2/year)	
	Median	Range	Median	Range
1945–1955	0.81	0–3.64	6.7	0–37.3
1955–1977	0.99	−0.14–3.19	9.6	−2.6–39.8
1977–1992	1.42	−0.73–3.71	13.9	−8.4–50.4
1992–2000	1.15	0.18–3.54	12.2	1.6–96.3
2000–2010	1.46	0–6.37	15.9	0–115.3
1945–2010	1.14	0.35–3.20	11.0	5.0–46.8

Table 7.12. Radial growth rate of oaks in thinned and unthinned stands, 1977/84–2010/15 (mm/year).

Zone	N	Range	Median	GBH in 1984 (cm)
Not thinned				
Old-growth, trees of all origins	24	0.06–3.84	1.33	57–245
Old-growth below the cliff in the managed compartment	8[a]	0.27–1.38	0.96	12–217
Young-growth, standards	4	0.54–3.38	1.72	170–276
Young-growth, coppice origin	47	0.08–3.96	1.14	13–82
Young-growth, seedling origin	8	0.20–3.60	1.32	9–66
Young-growth in the managed compartment	13	0.11–2.84	1.09	18–53
Thinned (managed compartment)				
Young-growth on limestone	7	0.58–2.97	1.91	12–72
Young-growth oung growth on deeper soils	8	0.48–3.82	1.63	24–60

Notes: [a]Includes one pedunculate oak. Old-growth, 1977–2010; young-growth, 1977–2013/15; managed stand, 1984–2014.

great oak forests of Bellême and Reno-Valdieu (France) average 2.6–5.9 mm/year until they are 240 years old. Standard oaks in British coppices achieve 1.6–3.8 mm/year for 100 years, but field oaks regularly grow at 5–6 mm/year or even faster. A little to the south of Lady Park, the Newland oak, once Britain's largest tree at 13.3 m GBH, must have grown at an average of over 2 mm/year for over a millennium.

7.12 Oak Decay

The tannin-filled wood of oak decays slowly. Stumps of oaks felled in 1942 were still solid enough to sit on 73 years later (Fig. 7.19). Piles of branch wood left by wartime loggers were still visible until the 1980s. Coppice stools decay at the core, leaving only a rough ring of living fragments on the larger stools, but even dead stools remain as shells for decades. Trees that die standing remain as snags for several years until their roots rot. And, remarkably, a few of the oaken posts with which Eustace Jones and his students marked the transects in 1944 remain in position and erect, albeit fragile with rings standing proud from the cut ends like the cheekbones of mummies.

7.13 The Relationship between Beech and Oak

Beech and oak are both long-lived trees that have jointly formed the largest trees in Lady Park since at least the 13th century, but their relationship has been complementary. Oaks are dominant historically, and the largest oaks are still the most impressive trees, but they are ecologically subordinate in near-natural woodland, relying on their longevity to survive long enough to find the circumstances in which they can regenerate. Beech is the true natural dominant in the sense that its fate determines whether these circumstances arise.

The 1976 drought illustrated this perfectly. Throughout southern Britain and central Europe beech growth rates slowed sharply (Power, 1994; Zimmermann *et al.*, 2015), and in the old-growth

Fig. 7.19. Seventy-three years of decay: felled in 1942, this stump of an oak standard was still firm enough to sit on in 2015. Such stumps proved to be a magnet for wild boar, foraging in the thick litter.

of Lady Park its hitherto relentless advance ceased abruptly and the oaks immediately took advantage. As Liam Cavin's examination of annual growth increments clearly showed, oaks slowed only a little after the drought and quickly recovered to grow faster than before, whereas the beech slowed severely, took longer to bounce back and never really recovered (see Fig. 7.12). The growth of the oak slowed again as competition intensified, but by then lime and ash were more influential. In Lady Park, the greater ability of oak to withstand drought can also be seen in the concentration and stronger growth of oaks on the dry slopes immediately above the cliff. Sampled by plot VW10.1 at a point where former beech–oak coppice had apparently been singled before 1940, the RGR of oaks was 0.68–2.58 mm/year (median 1.76) between 1979 and 2016, whereas the beech grew at 0–0.98 mm/year (median 0.15). On the transects in 1977–1992, the median RGR of oaks (1.42) was well above beech (0.63).

The other side of the relationship is regeneration. Beech generally has little problem, but since the early 20th century oaks have almost ceased to regenerate within woodland, although it still regenerates well enough in open dry grassland and grass heaths. After heavy mast years, oak seedlings may be abundant in woods elsewhere, but they survive mainly on the margins. Lady Park, despite the deer, has conformed to this pattern: beech has recruited from seed when and where it has had the chance, whereas oak seedlings have been uncommon and have failed completely since 1945.

Beech grows with one or both oaks almost throughout Britain. It was regarded as native only in the south, but there were early outposts in north Wales and northern England; it has long been planted throughout Britain, and it has naturalised successfully in the north. Within its undoubted native range, it associates with pedunculate oak on a very wide range of soil types, including some of the most acid (e.g. the New Forest) and the most alkaline (e.g. chalk scarp hangers). The Lady Park combination of beech with sessile oak is less common: outside the southern Welsh borderlands, it is found mainly on strongly acid soils in the wood pastures of the south east lowlands (e.g. the New Forest, Hampshire; The Mens and The Cut, Sussex) and a scatter of coppices (e.g. Perry and Winterbourne Woods, Kent). In the southern Welsh borderland, exceptionally, they coexist on the full soil range from alkaline to strongly acid. Following planting and naturalisation, beech and sessile oak also combine throughout the western uplands, mainly on acid soils, but occasionally on hard limestones.

In classical succession on acid soils, beech follows birch and then oak (Tansley, 1939). Oak readily colonises grassland and scrub, but beech is slower. Beech, however, bears shade, so enters under oak, eventually replaces it and can then maintain its dominance by regenerating under and around itself. Nevertheless, oak (and birch) can persist because beech excludes other species from the underwood and is subject to catastrophes, a combination that provides wide open spaces in which light demanders can regenerate. The natural balance between beech and oak at any particular time is determined largely by the history of catastrophic disturbances to beech and their timing in relation to mast years, but the most likely outcome is probably coexistence with beech as dominant. On calcareous soils, analogous relationships lead to beech dominance over ash.

Does this apply to natural woodland? In future-natural woodland, the balance must have shifted in favour of other species, for beech growth is impaired by grey squirrels and increasing vulnerability to drought (Zimmermann *et al.*, 2015), oak regeneration is inhibited by silver-leaf fungus and ash is about to be blighted. In original-natural woodland, limes were dominant in the lowlands and beech was absent or localised. If such woods had been allowed to develop naturally to the present time, one could assume that beech would have become more widespread, but it would have been contending with the two limes, which is the circumstance that Lady Park allows us to examine.

The combination of beech, sessile oak and one or both limes is particularly characteristic of the lower Wye Valley and its immediate vicinity, but it is not unknown elsewhere. For example, the two species grow with small-leaved lime on alkaline soils in Covert Wood, Kent, and Chudleigh Rocks, Devon; and on acid soils at Buckland Woods and Holne Chase, both on the fringes of Dartmoor, Devon. Outside the lower Wye Valley, the only ancient wood we know where these three are also joined by large-leaved lime is Kings Wood, South Yorkshire, which occupies alkaline soils over Magnesian Limestone, but the combination might well have been widespread on calcareous loams in the south-east lowlands if woodland had been allowed to develop naturally during the past few thousand years.

With limes in the mixture, beech faces serious shade-bearing competitors whose longevity is superior and whose ability to regenerate within woodland is almost as strong. Nevertheless, in central Europe, the heart of their ranges, beech typically dominates limes, except on shady slopes or ravines where the soil is stony and humus-rich, and on deep, moist soils at the foot of slopes (Ellenberg, 1988), and in these locations both limes co-dominate with ash and wych elm. In Britain, and especially in the borderlands, where limes are close to the edge of their ranges and ash and wych elm have been relatively stronger competitors, the relationships may be different. In particular, vegetative reproduction may be more important, which would favour the limes. Certainly, this is the direction in which experience so far at Lady Park points.

The other consideration is herbivory. If Vera (2000) was right and pre-Neolithic woodland included large open areas maintained by large herbivores, the balance would go against highly palatable limes and in favour of beech, and especially oak, whose relationship would then matter more. But, lime was abundant in much of southern Britain, and it was only later as woodland pasturage developed that beech replaced lime in the New Forest and in Epping Forest (Baker *et al.*, 1978).

The beech–oak relationship has become important because we have evicted limes and spread beech and oak by 5000 years of woodland clearance, pasturage, planting and perhaps soil impoverishment. In Lady Park and other ancient woods where limes have survived, we may conclude that, droughts notwithstanding, beech still seems to

Fig. 7.20. A fallen canopy beech in the upslope old-growth stand which retained enough of its root system to re-spout vigorously.

have a clear long-term advantage over oak, principally because it regenerates copiously when it has the chance and oak now has to battle against the imported silver-leaf fungus (Rackham, 2003), but in the long run, their relationship may be unimportant. Limes are increasing, and crucially they are also free from imported ailments, though to say that may tempt providence. Were it not for the anticipated Chalara outbreak, one would predict a future more as lime–ash–beech woodland, but shortly the trend may be towards lime–beech dominance, perhaps the very condition that would have developed if pre-Neolithic woodland had been allowed to develop naturally.

Finally, this is the best place to mention that a few sweet chestnuts can be found in Lady Park. None, however, grow in the transects and plots, so we have little information on their performance. The largest, a marginal tree at the bottom of the colluvial fan that was not felled in 1942, is 192 cm GBH. Another, a maiden growing in the young-growth near transect I, is 187 cm GBH and manifestly growing fast (RGR *c.*0.4 mm/year sustained for 70 years). Seedling chestnuts also have occasionally been found. This relative of the oak has grown in the district since at least the 12th century (Hart, 1966), may have been introduced long before, and is well naturalised.

8 Limes and Wych Elm

The remaining long-lived canopy trees in Lady Park Wood are small-leaved lime, large-leaved lime and wych elm. Field maple might have been added, but its performance in the wood relegates it to a subordinate tree. Wych elm, too, ought to be relegated, but it was an important canopy tree in parts of the wood for the first 30 years of this study.

Both limes and wych elm have played fourth fiddle to oak, beech and ash as timber, so even now they are seen more often as slender subcanopy trees with smoothly arching branches than as towering trees in the canopy. Both can dominate the underwood in ancient coppices, but deer and domestic stock browse them avidly – they have been important sources of foliage fodder in the past – so they are uncommon in pasture woods and are usually restricted in the uplands to less grazed, rugged ground. Their fates diverge today: while numerous wych elms have succumbed to disease, the limes grow serenely on and even seem to be benefiting from climate change.

8.1 Limes in Britain

Limes are the ecologists' trees. Having only limited attractions as timber and fuel, they have been little planted in woodland, so their distribution is explained more in terms of natural factors than the history of forestry. Pollen studies show that they were once dominant in the pre-Neolithic woods of southern, especially lowland, Britain (Greig, 1982), and their present distribution, scattered mainly in ancient woods and hedges, bears the hallmark of a relic population. Their graceful, ascending form has made them favourites in parks, though some of the cultivar hybrids are encased in unsightly base sprouts.

Two lime species are native (Pigott, 1991, 2012). Small-leaved or winter lime (SLL) is widespread but patchy in the ancient woods of England and Wales north to Cumbria, being particularly prominent in central Lincolnshire – a county named after lime – and the southern Marches. Large-leaved or summer lime (LLL) (Fig. 8.1: see p. 130) is much more restricted. Unlike SLL, which grows on a wide range of soils, including strongly acid sandy loams, LLL is concentrated on limestones. LLL, however, is rare. It may once have been dominant on just those base-rich, well-drained loams that have been cleared for agriculture (Pigott, 1981). The two species are distinguished on leaf size, shape, hairiness and the orientation of inflorescences, ascending in SLL, descending in LLL. Less obviously, they differ subtly in growth form, SLL being more ascending, LLL having a greater tendency to generate untidy base sprouts. Where both species occur together in limestone woods, intermediate forms can be found, though not in Lady Park.

8.2 Limes in Lady Park Wood

Lady Park is one of the few woods where both species are common. First recorded in 1897, limes have certainly been present much longer. Those on the deeper, acid soils above the cliff are mostly SLL, but on the limestone outcrops and the slopes below the cliff most are LLL. As coppice, limes were usually cut close to the soil, but some were cut high as stubs and miniature pollards (Fig. 8.2), a common practice around the lower Wye. Either way, they would not have flowered, so regeneration from seed must have been rare. Nevertheless, they reproduced somehow, for the wide range of stool sizes and clusters now present, including several giants whose age is great but can only be guessed, implies a range of ages.

Clustered and concentrated on the outcrops and steeper slopes in Lady Park, but rare in the northern

Fig. 8.1. Large-leaved lime in the upslope old-growth stand. This is clearly a single individual contributing two trees to the canopy.

part of the reserve and the rest of the former Hadnock Wood, their distribution may be a relict of wood pasturage, maintained by the difficulty of spreading under the later coppicing. This antipathy to wood pasture is clear elsewhere in the lower Wye Valley, where limes are conspicuously rare in the remnants of the former Wyeswood Common, but common in former coppices on the steep slopes.

Most lime stems are no more than 90 years old, exceptionally 113 years, but how old are the stools? Seedling regeneration must have been exceptionally rare in the past, which implies that the large stool clusters are clones, some of which must be at least several hundred years old, and this belief is reinforced by leafing dates, leaf form and growth form (Case study 8.1; Fig. 8.3).

8.3 Population Changes

The initial populations differed subtly between stands. In the old-growth, where a small and scattered population of LLL grows among the predominant SLL, limes had been cut in 1902 and *c.*1930 to promote beech and oak, but in the young-growth, which happens to be all SLL on the transects, they had been cut again in 1942–1944. Below the cliff, where the majority are LLL, they may have been spared in 1902 and *c.*1930, but some were cut during the Second World War for lightweight timber to make aircraft frames. Now they are fully-fledged members of the canopy, 70 years old or more. The great majority have grown from low-cut stools.

The striking feature of subsequent population changes has been the differences between stands. In the young-growth, the old stools generated dense clusters of shoots, which must have thinned out strongly before 1977 after canopy closure, presumably slowed in 1977–1983, before accelerating steadily thereafter (Table 8.1). Recruitment was negligible until the fence allowed a strong flush of base sprouts to develop after 2007 (Fig. 8.4). In the old-growth, the population was still recruiting in 1945–1955, but it has declined steadily, mitigated only by a flush of new sprouts from established trees after the drought (Table 8.2). Below the cliff, the records show a massive increase after 1985, followed by an almost equally rapid decrease, but pre-1985 'fluctuations' may be an artefact of recording (Table 8.3). The 1971 enumeration revealed that a few large limes had been spared during the wartime fellings, but none on the transects.

8.4 Regeneration and Recruitment

Although limes have a predilection for vegetative regeneration, they regularly produce fertile seed and sometimes recruit through the germination and growth of new seedlings. Eustace Jones maintained that lime had last recruited from seed in 1942, and one can still see the legacy: in 1977, 98 ex 227 individuals in the young-growth were maidens, plus 12 in the old-growth. In recent years, seedling limes have been found in the old and the young-growth (but not on the moss-covered limestone rubble below the cliff) and sometimes, as in 2015, they are numerous. Most disappeared immediately, but a few persisted, though any that germinated before 2007 were heavily browsed. Since 2007, a few saplings of both species have grown above 1.3 m, including a

Case study 8.1

This, the most impressive lime in the wood, is a small-leaved lime growing close to the northern boundary, isolated from other limes (Fig. 8.3). It has the uniform leafing and form typical of a clone and currently takes the form of a tight cluster of eight stools, four of which are certainly linked. In February 2015, this comprised 49 live stems ranging from 15 to 162 cm GBH (median 36) and including 19 over 92 cm that were collectively destined to be the long-term dominants. Assuming that they started growth in 1942, the fastest had sustained radial growth of 3.58 mm/year, well above the maximum within the old-growth. The only downside was that the largest lateral trees had grown with a strongly oval cross section, enlarged 'inside to outside' to form strengthening buttresses. Trees are clearly polarising into large and small, with 18 trees above 100 cm GBH, 27 below 50 cm, but only 4 in the mid-range of 50–100 cm. This clone is the seed source for a dense cluster of maiden saplings developing in a large clearing created by the fall of a standard beech nearby.

Fig. 8.2. Small-leaved lime treated as a pollard, a remnant of the pre-1870 coppice management.

Fig. 8.3. The giant small-leaved lime clone near the northern edge of the reserve. Bud-break, leaf-drop, leaf shape and stem form are uniform throughout.

Table 8.1. Changes in the population of limes in the young-growth.

	Initial density (N/ha)	Live at start	Losses	Gains	Mortality rate (%/year)	Recruitment rate (N/ha/year)
1977–1983	434	889	−52	0	1.00	0
1983–1993	408	837	−133	+7	1.81	0.36
1993–2002	347	711	−168	+3	2.95	0.16
2002–2013/15	266	546	−171	+125	3.21	5.30
2013	244	500				

cluster of over 40 SLL saplings in a gap on the colluvial fan near transect VI (Fig. 8.5, 8.15).

Vegetative regeneration, in contrast, has been prolific, and it has taken several forms:

1. Regrowth from felled stools, stubs and very near surface roots. The main form of regeneration when the wood was coppiced and the source of most trees present in 1945.

Fig. 8.4. Down but not out: a small-leaved lime regenerates after falling.

Table 8.2. Changes in the population of limes in the old-growth.

	Initial density (N/ha)	Live at start	Losses	Gains	Mortality rate (%/year)	Recruitment rate (N/ha/year)
1945–1955	166	149	−15	+30	1.06	3.33
1955–1977	182	164	−61	+1	2.00	0.05
1977–1983	116	104	−13	+1	2.20	0.19
1983–1992	102	92	−15	+45	2.20	6.25
1992–2000	136	122	−27	+5	2.74	0.62
2000–2010	111	100	−22	+8	2.45	0.89
2010	96	86				

2. Layering from the tips of flattened and pendant branches. Being shade bearers, the subordinate shoots on stools and other slow-growing stems tend to survive as weak, arching or leaning stems. Many are flattened by falling branches, or fall to the ground as old stools decay, and others simply lean far enough to touch the ground. Once in contact with the soil, they are capable of developing roots and generating new shoots, though the shoots are vigorous only if they are well lit. The most spectacular instance of

Table 8.3. Changes in the populations of both limes in the stand below the cliff.

	Initial density (N/ha)	Live at start	Losses	Gains	Mortality rate (%/year)	Recruitment rate (N/ha/year)
1950–1959	180	87[a]	–4	+14	0.52	3.21
1959–1985	201	97[a]	–35	+167	1.77	13.91
1985–1992	476	229	–17	+96	1.09	28.38
1992–2005	639	308	–119	+26	3.67	4.14
2005–2013	447	215	–88	+25	6.33	6.46
2013	317	153				

Note: [a]If we allow 15 stems for individuals recorded as 'over 10 stems' and 6 for individuals recorded as 2–10 stems, the number of stems in 1950 and 1959 would be 192 and 146, respectively.

this form of regeneration should have followed the late snowfall in April 1981, when fairly strong, but still subordinate, stems were bent to the ground and pinned under disintegrating birch that had died in 1976–1977 (Fig. 8.6, 8.14). In fact, none rooted, because the new shoots were browsed by deer.

3. As the stands have aged, so tall limes have fallen, especially in the old-growth and below the cliff. Most retain enough of their root system to survive and produce colonnades of strong sprouts from base to crown of the original trees, especially when they fall into clearings (Fig. 8.7). In practice, the strongest and most lasting growth usually springs from near the base of the old trunk, close to the roots and where the shoots are best lit (Fig. 8.8) – similar, in fact, to funding in the Civil Service, which is most assured if you are close to the Treasury. Until 2007, much of this potential new growth was browsed to oblivion by deer (Fig. 8.9).

4. Healthy trees constantly generate new shoots from the base of the trunk and near points on surface roots, especially LLL trees (Fig 8.1). Until 2007, these were invariably browsed by deer, but since 2007 they have been allowed to grow. By 2015 there was an immense crop of new shoots, which will remain subordinate unless the main tree falls or breaks, in which case they can grow into trees.

The three stands show some interesting contrasts, but all confirm the importance of vegetative reproduction. In the young-growth, the very limited recruitment in 1983–2002 was entirely from layers, save for one new shoot from a tree that had retreated below 1.3 m. Accelerated recruitment in 2002–2015 was entirely due to new basal sprouts from existing individuals: most were thriving, but by 2015 three had fallen, one had retreated below 1.3 m and four were pseudorecruits, i.e. older shoots growing so slowly that they were actually recorded as dead in 2002. Until recently, there were far more recruits in the old-growth. The stand was still recruiting in 1945–1955, and recruited again when the canopy was more open and competition from beech was much reduced in the wake of the drought. Not one certainly came from seed. Most were coppice sprouts, ten were layers, a few were sprouts from below 1.3 m on trunks, and just occasionally a new shoot grew up from a fallen tree. They grew poorly but persisted: of the 90 recruits, 35 were still alive in 2010.

Below the cliff, recruitment was substantial but fluctuating. It must have been rapid in the 1940s and 1950s as a reaction to felling, but cannot be quantified. The size distribution in 1985 implied that recruitment accelerated rapidly when disease opened the canopy and eliminated competition from wych elm. Here, seedling regeneration was successful: ten recruits grew from seed, of which two survived in 2013, and several other saplings persisted for several years without reaching 1.3 m. Even so, most recruitment originated as new coppice shoots from standing trees, supplemented by several from layers and occasional regrowth from fallen trees that generated short colonnades.

Although both limes bear considerable shade, recruitment progresses better in gaps and other well-lit positions. Vegetative growth from layers and fallen trees grew strongly under, or on the edge of, gaps. It is also stronger from near the rootstock of tipped-over trees, because the original top usually falls into the shade of a neighbour, while the rootstock is exactly under the gap created by itself. Regrowth from still-standing trees rarely gets far because it is in shade. Layers have been more successful below the cliff, where the rubble slopes are less accessible to deer, the canopy is constantly opened by tree falls, and LLL – the main species – generates

Fig. 8.5. Small-leaved lime regenerating from seed in semi-shade on the colluvial fan.

more vegetative sprouts than SLL. Layers can grow well enough to form trees themselves, which may later lean and generate more layers, almost like a glacially slow looper caterpillar. Seedlings are far weaker than vegetative growth, so they normally fail, but there are now signs that they can succeed, either as clusters in gaps or as single individuals fortunate enough to grow below a light canopy. Deer can, however, negate all forms of regeneration, even when the canopy is open, as the managed stand is currently demonstrating.

8.5 Growth

Small-leaved limes grow as clean-limbed trees to over 35 m when they have the room, but suppressed stems persist as weak and usually leaning growth. On large coppice stools, new shoots quickly develop into a cone of fresh growth in which some of the central stems become dominant and eventually one or two develop into the main tree, while the rest languish in its shade, often reduced to almost leafless stumps. On large, fragmented stools, a dominant individual often develops on each fragment until, after 50–70 years, it must compete for canopy space with its neighbouring dominants. Eventually, stems on individual stools – and in the population as a whole – polarise into dominant stems which become the long-term successful trees and subordinate stems that persist but eventually fail, and they do this even when they have room to develop several dominant trees. A superficially similar size distribution developed when deer were excluded and base sprouts on mature trees were allowed to grow.

Fig. 8.6. Subordinate stem of small-leaved lime, arched over by a late April snowfall in 1981, which would have rooted at the tips if the wood had not been heavily grazed.

However it comes about, the size and growth-rate distribution is severely skewed. All stands contained individuals that survived for decades with almost no increase in their radius – a tribute to their shade-bearing capacity – so medians and ranges form the best basis for comparing growth in different stands (Table 8.4). On this basis: (i) growth within the fenced reserve was slower in the young-growth, where there were many small stems, than in the old-growth and (ii) faster still below the cliff; (iii) thinning enabled the best SLL to grow marginally faster than the best in the unthinned young-growth; and (iv) the growth of trees on the apparently inhospitable cliff was fast, although there were no outstandingly vigorous individuals.

Competition between the stems arising from individual stools is much the same as competition between separate individuals in stands. A feedback operates in which sprouts that were more or less equal initially steadily differentiate into winners and losers. Eventually, the winners reach the canopy, but continue to compete. Small differences in crown position and size combine with weaknesses in the remains of the stool to determine which form the major trees. Limes are unusual in retaining several losers as weak, arching, marginal stems that can grow vigorously again if they get the chance.

Growth of both limes appears to be sensitive to competition and available light. This explains the benefits of thinning and also the slower growth in the old-growth below the cliff in the managed compartment, which has never been thinned by large tree falls or by elm disease. It also explains fluctuations in growth rates. Thus, limes in the old-growth grew fastest when the canopy was still open from the *c.*1930 thinning, slowest when competition from beech was most severe before 1977, and fast again in 1977–1992, while beech competition was reduced by the drought (Table 8.5). Below the cliff, it was conspicuously slower before elm disease

Fig. 8.7. Colonnade of regrowth from a small-leaved lime that fell below a group of ash in the young-growth.

opened the stand, when the trees were generally smaller and less likely to topple (Table 8.6). Perhaps surprisingly, the few stems of seedling origin in the young-growth that survived to 2013/15 grew at much the same range and median rates as surviving stems of coppice origin, but the latter included many small suppressed stems.

Elsewhere, a few measurements made by Pigott (1991) suggest that radial growth rates of SLL vary with circumstances and hint at a decline with age. A large, maiden tree growing on limestone below Chudleigh Rocks, Devon, grew at 0.7 mm/year for between 200 and 300 years, which contrasted with parkland trees that grew at 6.4 mm/year initially but decelerated to 3.0 mm/year after 200 years. In the southern Lake District (Pigott, 2012), SLL had grown at 3.6 mm/year on deep, moist soil at the base of a slope, but at 1.25 mm/year when rooted in fissures of a limestone cliff. The rate Pigott gives for top-quality lime plantations in Russia is 2.5 mm/year.

8.6 Mortality and Decay

Mortality in the young-growth followed the usual self-thinning rules, but with interesting departures. By 1977, stems of seedling origin were generally smaller than those of coppice origin, but their subsequent growth and mortality patterns were much the same. Most of the very small stems (below 3 cm DBH) had died by 2015; most large stems (above 12 cm) lived, and there was a gradation of survival in between. Survival was correlated with median radial growth rates, though there was a

Fig. 8.8. Windthrown small-leaved lime stub forming a pit and mound, which regenerated from rootstock to crown. Only the new growth from near the roots actually survived.

sharp increase in growth rates between 10 and 12 cm DBH. Huge disparities in growth could be found at all sizes. Thus, many individuals, some as large as 14 cm in 1977, hardly grew at all in 38 years, and fast-growing trees could be found across almost the whole initial size range. Canopy position was crucial: median radial growth rates (RGRs) of the stems that survived from 1977 to 2013/15 were 1.65, 0.23 and 0.12 mm/year for stems in, respectively, the canopy, subcanopy and underwood in 1993. The probability that a stem would survive depended on how large it was in 1977 and the canopy position it achieved.

Mortality rates varied. In the young-growth, they were low after competition was reduced by the drought, but picked up since, partly because some stems were debarked by deer or brought down by falling trees. Below the cliff, it increased after 1992, but the reasons were not clear. Its apparent increase in the old-growth may have been distorted by under-recording of small stems in 1945 and 1955. The more recent increase was linked to increased deer browsing and fraying before 2007 and accelerating loss of weak, marginal stems on stools.

Although smaller stems are always more likely to die, they are remarkably tenacious. As they fail, they retreat: their leaders weaken and die and continue to sprout weakly from the base. Small side shoots on large stools commonly lean and produce new shoots from their upper sides as they weaken.

Fig. 8.9. This small-leaved lime was singled in 1983, after which the stumps sprouted vigorously, but the new shoots were cut back during a subsequent thinning operation. Most still try to spout again, but for the past two decades have been browsed to oblivion.

At any stage, if circumstances change, weak stems can spring back to vigorous life.

Larger lime stems *do* sometimes perish. Most topple when the stool becomes too weak to support their weight. Others, being slender, suffer 'misfortune' when other trees fall on them. Crushed trees often linger, growing slowly for 10–20 years. Still others are blown against other trees, where they survive as weakened trees, their crowns shaded and their root systems damaged. Below the cliff, a large pollard with seven large branches collapsed. Groups of tall, vigorous trees topple together if they have grown from one stool. If large trees manage to hang on to their roots as they fall, they will sprout vigorously, forming colonnades of new trees. If their trunks break or their roots are all torn out, even large trees rot quickly once they die.

8.7 Lime Populations in Terms of Plants, not Trees

The mortality and recruitment rates given in Tables 8.1–8.3, which refer to stems, are influenced strongly by the minority of large stools that can produce dozens of stems. An alternative would be to express population turnover in terms of genetic individuals, but this could be equally misleading because large, fragmented clones cannot be easily identified as such. The nearest one can approach is to express populations in terms of separate plants.

Table 8.4. Radial growth rates (mm/year) of limes in different stands that survived throughout the period indicated.

Stand	Period	Species	N	Radial growth rate range	Radial growth rate median
Stands not thinned					
Young-growth within fence	1977–2013	Mainly SLL	132	0–3.30	0.44
Old-growth within fence	1945–2010	Mainly SLL	37	0.03–3.24	0.77
Below the cliff within fence	1950–2013	Mainly LLL	28	0.03–4.74	1.13
Old-growth below cliff in managed compartment	1984–2014	Mainly SLL	49	0–2.57	0.64
Cliff in managed compartment	1984–2014	Mainly LLL	8	0–2.41	1.53
Thinned stands					
Young-growth on limestone	1984–2014	SLL	20	0.20–3.66	1.74
		LLL	15	0.20–2.98	0.94
Young-growth on deeper soils	1984–2014	SLL	7	0.68–3.40	2.55

Table 8.5. Growth rates of the 39 lime trees surviving from 1945 to 2010 in the old-growth.

	Radial growth rate (mm/year)		Basal area increment (cm²/year)	
	Range	Median	Range	Median
1945–1955	0–8.09	1.21	0–76.0	6.0
1955–1977	0–1.78	0.51	0–19.8	3.3
1977–1992	0–4.16	0.99	0–66.5	8.1
1992–2000	0–2.83	0.53	0–50.9	4.3
2000–2010	0–3.02	0.71	0–42.5	5.4
1945–2010	0.03–3.36	0.85	0.1–44.7	5.5

Table 8.6. Growth rates of the 27 lime trees surviving from 1950 to 2013 in the stand below the cliff.

	Radial growth rate (mm/year)		Basal area increment (cm²/year)	
	Range	Median	Range	Median
1950–1959	0–5.95	0.45	0–46.9	1.1
1959–1985	0–3.91	0.69	0–37.6	2.9
1985–1992	0–6.37	1.59	0–67.0	6.3
1992–2005	0–5.02	1.47	0–77.4	7.1
2005–2013	0–5.87	1.59	0–116.4	7.9
1950–2013	0.03–4.74	1.11	0.1–56.5	5.3

In practice, much the same picture emerges in the old-growth and the young-growth, where the principal process is thinning with negligible recruitment. The young-growth transects include 56 plants that have produced more than five stems between 1977 and 2013/15. In the old-growth, in contrast, most limes have just one or two stems, probably because thinning in 1902 and 1930 weakened larger stools and restricted their ability to re-form large trees under the beech–oak canopy. Here, 78 plants were recorded on the transects in 1945, of which 43 were dead by 2010, a mortality rate of 1.2%/year, well below the rate for stems. They were temporarily augmented by 23 new individuals, all but one of which was dead by 2010. The few recent recruits came from layers, but since most first appeared in 1955, when they were recorded as multi-stemmed, we suspected that they had been omitted from the record in 1945.

The LLL below the cliff is more revealing. It comprised 24 separate individuals in 1950, of which 19 survived to 2013 and another fell but remains alive. One large coppice individual died when its entire root system was broken in a fall, and three small individuals died before 1985, apparently shaded out. Over the same period, 22 new individuals were recruited, but only 7 lived in 2013. The population thus comprises a limited number of large individuals with multiple stems, some of which have developed into large trees that dominate not just the stand but also all other stems on their stools, while a transient population of new individuals is maintained in their vicinity by means of layers and seedling regeneration.

8.8 Wych Elm in Britain

Elms are the tragic trees of Britain's woods and farmland. Repeatedly diseased down the millennia,

Fig. 8.10. One of many small wych elms growing on the upper plateau below an ash.

they were struck again in the early 1970s, and with them went a defining feature of lowland farmland, memorialised by a Plant a Tree Campaign, a book entitled *Epitaph for the Elm* (Wilkinson, 1978) and many columns of newsprint. Wych elm, the species in Lady Park, is one of several species of elm in Britain, but the only one that is agreed by everyone to be a native. In form, it is 'a magnificent widely branching and lofty tree when well grown' (Tansley, 1939), but it was met more often as coppice and pollards. From a distance, it is sometimes confused with lime, perhaps for no better reason than both species frequently grow as multi-stemmed trees with arching branches, and both are obviously not oak, ash, beech or birch. Close up, wych elms are distinguished by broad, shallowly toothed leaves, some – like Neptune's trident – with three points (Fig. 8.10).

Wych elm can grow very rapidly in ideal circumstances. For example, Elwes and Henry (1906–13) recorded a tree at Hildenhay, Yorkshire, which reached 32 m by 4 m girth after 80 years, and an instance from Ireland of prolific and rapid growth from layers: after 26 years, six layers had reached 1.2–1.5 m girth at 1.5 m. Pollarding and coppicing enabled individual elms to live for many centuries and to grow to great sizes. Then, as country estates developed their ornamental parks, maiden elms were allowed to grow to full height. The individual wych elms that most impressed Elwes, Henry and their informants included the 'finest' wych elm at Studley Royal, Yorkshire, which was 32 m high by 7 m girth at 1.5 m; a tree on Middle Hill, Broadway (Worcestershire), with very wide spreading branches, that measured 26 m high by 7.5 m girth; and the largest wych elm in Scotland, 38 m high by 9 m girth at

1.2 m. Even these were dwarfed by a huge wych elm recorded in Robert Plot's *The Natural History of Staffordshire* (1636), which was 36 m high by 5.2 m in diameter (16.3 m girth). When felled, apart from an immense amount of sawn timber, it yielded 61 loads of firewood, each requiring six oxen to draw it to the house at Field (a Staffordshire village), and we know this because eight local worthies signed a document to that effect.

Elms – presumably wych elms – were one of the major trees of pre-Neolithic forests and the defining component of Birks *et al.*'s predominantly western 'Hazel–Elm Province' (Birks *et al.*, 1975), but at the start of the Neolithic, disease most likely (Rackham, 2006) caused a sharp drop in the amount of elm pollen falling into mires – the Elm Decline – and it has remained low ever since. Wych elm is, nevertheless, a common species of base-rich woods in the western uplands (median pH 5.5) and borderlands (median pH 7.0), but infrequent in the lowlands. While essentially moisture loving – it is most frequent on moist sites and in a moist climate – it paradoxically also grows out of rock outcrops and on apparently dry sites in the borderlands, lowland and south-east lowlands. Throughout its British range, it is associated most strongly with ash and hazel.

8.9 Wych Elm in Lady Park Wood

In Lady Park Wood, wych elm was only a minor component of the coppice-with-standards system before 1870 and was entirely confined to the underwood, where it was sometimes cut as stubs. Its distribution within the underwood must have been much as it is today, concentrated on the deep, alkaline soils below the cliff, with lesser concentrations on the shallow limestone soils of the top edge and around the main cliff. Within the body of the wood on acid soils it is rare.

The key event was, of course, the arrival of Dutch Elm Disease (DED) in 1971, and by 1977, most of the larger trees were dead or dying. By 1992, 65% of all the elms present below the cliff in 1950 had died, but, remarkably, the population had actually increased by 40% through seedling regeneration (see Fig. 6.2) and stump regrowth from the base of old trees. Suitably surprised, and recognising that our records enabled us to relate 22 years of response to the disease to 25 years of disease-free growth before it struck, we analysed the changes and published our findings in the *Journal of Ecology* (Peterken and Mountford, 1998).

We found that disease had split the elm population into two fairly distinct elements, a large, fast-turnover population and a small, slow-turnover population. Trees in the former grew fast, reproduced copiously, but died young, a 'hedonistic' lifestyle that involved: copious regeneration from seed, much of which was in and near gaps; fast growth of some saplings and stump sprouts; early fruit production from these new poles; but reinfection with DED once these poles reached 8–9 m (Fig. 8.11). In contrast, trees in the slow-turnover population grew slowly, reproduced occasionally and lived long, an 'ascetic' lifestyle characteristic of elms concentrated just below the cliff and at the top of the upper slope. Growing on shallow soils over limestone outcrops, hidden in the underwood and subcanopy, and subsisting on a limited water supply, they grew slowly and remained small. This evidently saved them, for they would have only laid down small narrow xylem vessels which appeared to be unaffected by DED and thus continued to conduct sap (Solla and Gil, 2002). Put another way, if an elm had anticipated the arrival of DED and had wanted to live, its best strategy would have been to grow on a site near the limits of its tolerance, not on its 'best' soils.

We concluded that wych elm was resilient in the face of DED. It was flexible enough to persist in shade, but outgrew most other trees when gaps appeared above saplings. Patchy and irregularly distributed though it was in 1950, subsequent regeneration showed that it could grow virtually anywhere below the cliff and that its pattern in any year was determined by chance patterns of mortality, survival and regeneration.

Subsequent experience has forced us to temper our optimism. By 2013, and in possession of another 21 years of observations, we could ask whether our 1992 conclusions had been right, and the short answer was that we were right only in part.

8.10 Population Changes

Before reviewing the conclusions in the light of new observations, we should point out that quantifying changes in the population has been a nightmare. Wych elms grow fast; are frequently knocked over, crushed or distorted; reiterate easily from stumps,

Fig. 8.11. A group of exposed wych elms below the cliff infected with elm disease. Lower parts of the trees remain alive.

bent trunks and fallen trunks; generate new individuals by layering; regenerate copiously from seed; bear shade to the degree that crushed saplings can remain hidden in the mess of fallen trees and still emerge later; and frequently change size and shape profoundly under the influence of DED. To follow their progress properly, we really needed more frequent observations, particularly observations for 1971. In order to make something of the records we have, we elected to interpret change as conservatively as possible: thus, for example, when in doubt about which trees corresponded across the interval between 1959 and 1985, we assumed least change. Since also some saplings will have come and gone between recordings, the actual change will have been greater than the change recorded. We have also used the elm records from the whole-wood enumeration of 1971, even though these cannot be directly related to the transect records.

Since 1950, the elms below the cliff have endured a roller-coaster ride (Table 8.7). A massive increase in density has been followed by a massive decrease, with a high turnover throughout. Recruitment before 1985 may be exaggerated by recording small, multi-stemmed individuals as single stems but there was no doubt that recruitment was high in 1985–1992 in the wake of DED (and might have been high before 1985) and has since decreased.

The mortality rate in 1959–1985 was not as high as might have been expected, but that was an artefact of the long period between observations. In 1950–1959, the stand was still open after wartime felling and only a few small individuals had died. Thereafter, the mortality rate probably increased gradually, but it was still the smaller elms that perished, though larger elms occasionally toppled. After 1971, large elms died in droves, but smaller elms may have thrived, though many must have been crushed as canopy elms disintegrated and other canopy trees toppled. At the same time, some large trees generated new stems and regeneration from seed was common. After 1985, crushing continued, shade increased and browsing by deer was substantial – many small survivors were debarked or browsed in 2005, but many new saplings and shoots were recruited. However, the mortality rate increased steadily thereafter until it became rapid in 2005–2013. There was also a decline in the vigour of small stems growing from old stools. After 1992, the elms with the best chance of survival were underwood trees of medium size and moderate growth rates (Table 8.8).

Radial growth rates in 1950–1959 were generally high in the medium and large elms at the base of the slope, three medium-sized trees exceeding 5 mm/year in 1985–1992, but these headline figures disguise the fact that most elms were small and grew slowly. The fate and performance of the top ten elms by girth at each recording date (Table 8.9) showed that many survived to the next recording date, even after 1971, but that their vigour was extremely variable. The fastest-growing trees in 1950–1959 all died before 1992, but survival of the most vigorous trees of later periods was somewhat better, no doubt because they were less prominent in the canopy (Table 8.10). Just seven of the 1959 elms survived the turmoil. Two represent the scatter of slow-growing underwood trees by the cliff, which have escaped the disease altogether, and another is an individual that has fallen repeatedly, thereby remaining 'out of sight' of DED's beetle vectors. The rest are stump sprouts from once-large trees that have grown slowly. The largest elm in 2013 was 122 cm GBH, but this was exceptional and well below the 154 cm GBH achieved by the largest elm before DED.

Growth, mortality and regeneration have combined to alter the elm size distribution throughout the 65 years of observations. The profound changes wrought by DED are demonstrated by including the 1971 enumeration in the sequence (Fig. 8.12): although the changes are somewhat exaggerated because the transects did not include the very largest elms, this

Table 8.7. Changes in the populations of wych elm in the stand below the cliff.

	Initial density (N/ha)	Live at start	Losses	Gains	Mortality rate (%/year)	Recruitment rate (N/ha/year)
1950–1959	114	55	−2	+17	0.41	3.91
1959–1985	145	70	−41	+128	3.46	10.59
1985–1992	325	157	−20	+99	1.93	29.27
1992–2005	488	236	−114	+25	4.95	3.98
2005–2013	304	147	−99	+21	13.06	5.43
2013	143	69				

Table 8.8. Mortality of elms below the cliff, 1992–2013, in relation to preceding growth rate.

Radial growth rate (mm/year) in 1985–1992	N	Mortality (% of 1985 population)
0–1	76	76
1–2	30	67
2–3	12	58
3–4	12	42
4–5	4	50
5–8	3	100

Table 8.9. Survival and growth rates of the ten largest elms over each recording interval in the stand below the cliff.

	Trees that survived to next recording date (out of 10)	Radial growth rate of trees that survived each period (mm/year)	
		Range	Median
1950–1959	10	0.45–3.14	1.63
1959–1985	4	0.31–2.38	1.30
1985–1992	8	0–4.77	0.67
1992–2005	5	0.49–2.08	1.22
2005–2013	8	0.20–5.57	1.25

Table 8.10. Survival of the ten most vigorous elms in each period to subsequent recording dates in the stand below the cliff.

	1959	1985	1992	2005	2013
1950–1959	10	3	0	0	0
1959–1985		10	8	6	2
1985–1992			10	5	4
1992–2005				10	8

sequence correctly shows the rapid growth of canopy elms to 1971 and their complete loss thereafter. The sequence shows a population growing out before DED struck from past coppicing into a negative exponential density–size distribution said to be characteristic of continuously regenerating populations.

The elms above the cliff add little to the story. There is still a scatter of slow-growing underwood trees on the slopes at the top and just above the cliff, including one old stub from the 19th century coppice, but the largest trees have died. The 17 elms in the old-growth in 1945 were reduced to a rump of three by 2010 via a switchback ride through high points in 1955 (18) and 2000 (13) and a low point in 1983 (5). The four largest trees died of DED, but many more small elms died, some due to DED, many after severe debarking by deer. Three individuals (and one just outside the transects) survived from 1945 to 2010, all fairly high on transects I and II, growing slowly at 0.15–1.08 cm/year. The 50 elms in the young-growth of 1977 were also reduced to three by 2010. Mortality rate was high in 1977–1983 (5.3 %/year), when larger trees died disproportionately of DED, and again in 2002–2015 (7.6 %/year), when smaller trees died disproportionately. In the intervening periods, mortality shifted steadily towards smaller trees.

8.11 Long-term Impact of Elm Disease

Reassessing our 1998 conclusions, but now with 45 years of post-disease experience, we can still recognise the small population of slow-growing survivors on and near the cliff and on thin limestone soils, and confirm that its rate of regeneration has remained very limited. The fast-growing population on the deeper soils below the cliff has not, however, been anywhere near as resilient as it seemed in 1992. Far from growing strongly and regenerating prolifically, growth and regeneration has slowed and mortality of seemingly established saplings has been high, even though the canopy remains open. DED remains a chronic risk for tall, fast-growing and prominent elms, but shade, deer and falling trees have been more lethal to saplings and small trees than we expected. Further, the prolific regeneration witnessed before 1992 must have been based on a one-off shower of seed from dying canopy trees and shoots from still-vigorous root systems. Substantial elms can still be found (Fig. 8.13), but the prospect is still for a fuzzy segregation between the few long-lived, slow-growing individuals on dry sites and potentially fast-growing, short-lived individuals elsewhere. And, our advice to elms is still to stay slow, low and out of reach of deer.

8.12 Limes and Elms in Natural Woodland

The pre-Neolithic, or 'Atlantic', forest of north-west Europe was first characterised as mixed oak forest, but it has long since been recognised as a mosaic of different compositions in which limes, elms and oaks were the most important trees (Iversen, 1960;

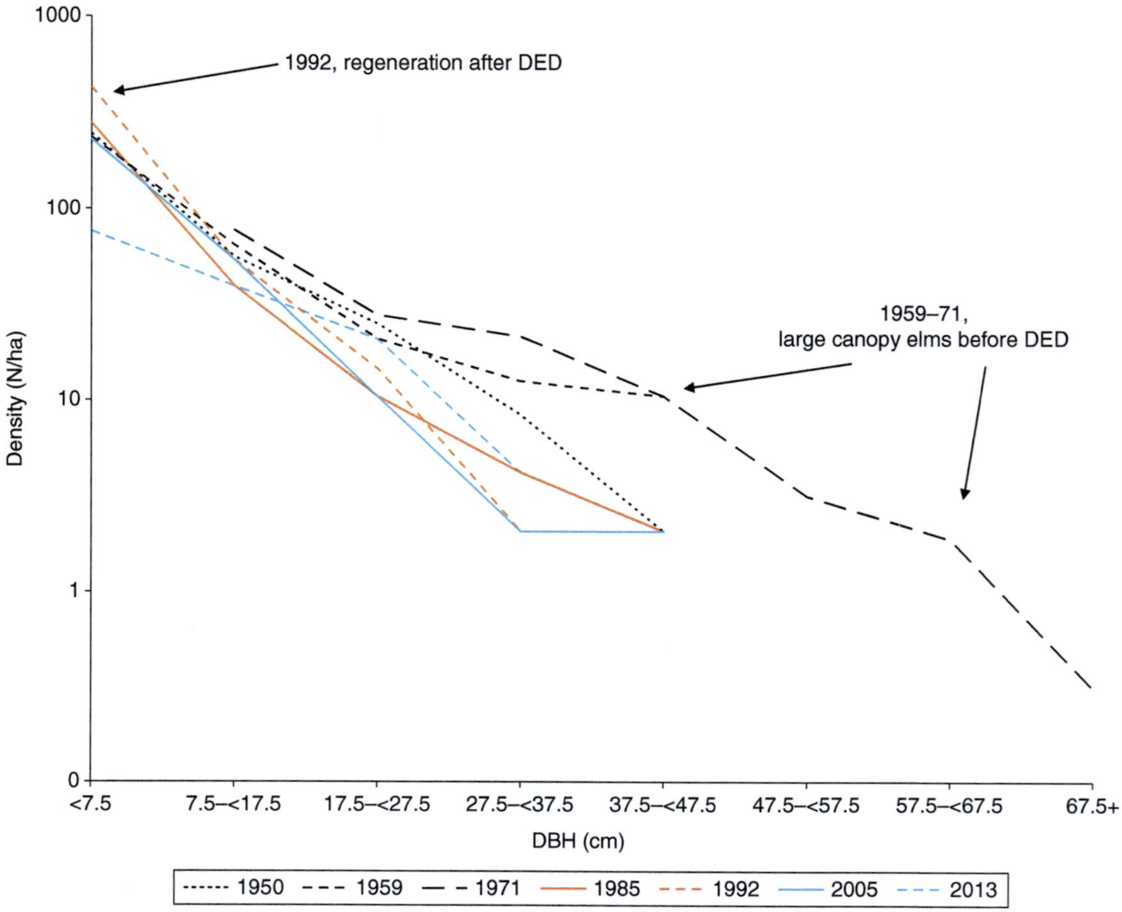

Fig. 8.12. Changes in the size distribution of wych elms below the cliff, 1950–2013. Note log scale of vertical axis. The lines combine the records from transects VII to IX, with the exception of 1971, which is taken from a complete enumeration of the whole stand below the cliff by a group led by J.F. Archibald. The largest trees in 1971 were not within the transects, so the impression of rapid growth of large trees up to 1971 and the retreat of the population to small sizes is slightly exaggerated.

Godwin, 1975). Limes reach their natural limits in northern England and do not reach Ireland, so the northern and western extremities were more a mixture of elm, oak, hazel and alder, while limes were dominant in much of the English lowlands. The outcome revealed in numerous pollen profiles summarised by Rackham (2006) was a lime province in the lowlands, an oak–hazel province to its west and north and a hazel–elm province further west, with Lady Park Wood on the boundary between the first two. Beech and hornbeam were still very restricted, and ash was not as prominent as it has since become.

Within this broad pattern lay considerable local variation (Iversen, 1960; Rackham, 2006), related partly to land form, but also to soils. On low-lying ground with a high water table, the growth of limes was inhibited, enabling ash, alder and pedunculate oak to play a greater role. The growth of elms was strongest on fertile soils, generally flushed sites and deeper soils on lower slopes, but on a wide range of less fertile ground, limes had the advantage, LLL particularly on calcareous ground, SLL elsewhere. Vegetative spread, limited dispersal of seed and other clustering mechanisms ensured that some trees tended to be gregarious (Rackham, 1992),

Fig. 8.13. A substantial wych elm in the thinned portion of the managed compartment. It escapes disease by leaning under a canopy tree, but retains sufficient strength to generate numerous basal sprouts. Photographed in July, these shoots will be grazed back before the next growing season.'

particularly limes. The outcome in eastern England was differentiation into several, more or less distinct woodland types that can be related to the various stand types recognised in ancient, semi-natural woodland today (Rackham, 2006). In at least one type, hazel must have been a canopy tree – had it been merely subordinate, it would not have dominated the pollen rain.

Pollen tells us little about forest structure, so we rely on inference and speculation. Taking the European Urwälder as his guide, Iversen (1960) believed that the pre-Neolithic forests of Denmark regenerated in small groups and that catastrophic disturbances, and thus birch and other pioneer trees, were rare. In most circumstances, elms, limes and hazel had many advantages over oaks. They are well adapted to shade and capable of vegetative reproduction in various forms, which prolongs life and, especially in the case of limes, promotes clustering. Indeed, limes are virtually immortal once established, since a large tree can replace itself with basal sprouts, even if it is blown down. In dense forest, oaks had only one advantage, their great longevity, but in a mosaic of openings and scrub, they would colonise and regenerate faster than limes and elm. The several species of limes, elms and oaks would each have their particular preferences. LLL the species of calcareous soils; wych elm the principal elm of richer soils, especially in the west; sessile oak on light, dry and shallow soils; pedunculate oak more on wet, waterlogged, heavy soils.

More recently, Vera (2000) has argued that forest comprised not just dense stands of trees but also parkland, open areas and scrub, kept open in a dynamic equilibrium maintained by large herbivores, and that this allowed oaks to remain a prominent component of pre-Neolithic forests and hazel to generate pollen in the open. This idea has generated heated debate ever since, and strong reasons have been advanced against British and Irish forests taking the form Vera proposed (Mitchell, 2005; Rackham, 2006; British Wildlife, 2009), but there must have been enough open ground for the components of wet meadows to be part of the scene (Peterken, 2013).

Since the start of the Neolithic, much has changed. Much woodland was cleared for agriculture, especially from the richer soils where elms and limes were prevalent. Elms were sharply reduced, probably by a combination of disease and lopping for fodder. Limes too were much reduced, but by clearance and lopping. Beech has spread widely through southern Britain, aided perhaps by the spread of woodland pasturage. This has also reduced limes, which were preferentially browsed. Limited pollen sources from near the lower Wye Valley confirmed that changes in the southern Welsh borderlands conformed to the general pattern: lime and elm decreased, while ash increased (Peterken, 2008a: p. 118). Charcoal evidence from the Wye gorge itself indicated that woodland containing hazel, ash, oak and yew had developed by the Bronze Age.

Viewed against this background, Lady Park is seen to retain the essence of the pre-Neolithic forests, in particular the large populations of the two limes, with LLL concentrated on limestone, SLL on a wider range of soils and wych elm on the richer soils below the cliff. We see also that it has changed, for beech and ash are co-dominants and elm is much reduced by disease.

Though the pre-Neolithic forests cannot be recovered, recent events can be regarded as a swing back towards the pre-Neolithic composition. Both LLL and SLL are not only potentially immortal but also untouched by grey squirrels and free of disease, and even wych elm has shown some resilience. Meanwhile, other forest trees have been impaired. Drought and grey squirrels damage beech, while oak regeneration is suppressed by leaf fungi and perhaps squirrels. Since 1976, ash has seemed a likely co-dominant with limes, but now its disease is on the doorstep. Until a new factor emerges, the prospect is lime dominance with wych elm subordinate on base-rich soils, and already the elite limes of both species are among the most vigorous and impressive trees in the wood. Both limes have demonstrated a capacity to regenerate from seed, which would presumably increase if the climate warms, and this might help them to spread within the wood. The main danger to all three species is browsing by high deer populations, which – as the managed compartment shows – can even prevent vegetative reproduction.

Fig. 8.15. A gap created by the fall of a large spreading beech near transect VI has filled with Bramble, which protects mixed seedling regeneration of Small-leaved lime, Sallow, Ash and Hazel, as well as a few Silver birch and Sessile oaks.

Fig. 8.14. Subordinate stems of Small-leaved lime coppice re-growth bent over by a late spring snowfall, which lay on the ground and trees for no more than 6 hours. Photographed in 1984, three years after the event.

9 Birch and other Short-lived Canopy Trees

These are species whose survival strategy is best summed up as quick-witted opportunism. They often dominate new growth after felling, especially where the ground has been left bare by timber extraction, but they cast only light shade and cannot therefore exclude the long-lasting shade species, such as beech, wych elm and lime. Fast growing in their youth, provided they are well lit, they generally outpace the most vigorous ash, oak and beech enough to dominate the canopy, but their growth rate declines and eventually their crowns are confined by more vigorous neighbours. Individual trees rarely last long and cannot replace themselves in a mature stand until the canopy is opened up by heavy thinning or a substantial natural disturbance. Even then, they succeed only by responding quickly. In theory, they will be favoured when the rate and intensity of disturbances – natural or silvicultural – increases, but long periods without substantial disturbance should reduce them to minor stand components.

Two subgroups can be recognised, the suckering species and the non-suckering. The main suckering species are gean and aspen, both of which maintain clones from which clusters of new shoots will grow when openings occur. Neither will regenerate from cut stumps (though gean does so half-heartedly), so they survive in coppices by suckering and occasional seeding. Both are widespread in ancient woods. Gean produces cherries that are readily eaten by thrushes, so it sometimes spreads into secondary woodland. Aspen, on the other hand, produces little seed, so is restricted more to ancient woods.

The non-suckering species, birches, alder and sallows, regenerate from seed, which they produce in prodigious quantities and distribute far and wide. They are the pioneers, not only colonising gaps created in established woodland but also invading unwooded ground with ease. They regenerate from stumps, so populations in woodland commonly consist of a mixture of coppice regrowth and maidens grown from seed. As a group, they tend to be particularly abundant in secondary woodland, but towards the north and on the outermost extremes of woodland on swampy or extremely dry ground they may dominate permanently. Thus, birch woods dominate the north of Scotland and the tundra fringes of mainland Europe, while alders, poplars and willows dominate on wet ground and disturbed riparian sites. In Lady Park, which is an ancient wood well within the ranges of most long-lived forest dominants, they prosper best after disturbance.

9.1 Birch

Birches are ephemeral dominants in southern Britain. They are perfectly capable of seeding in droves on to open ground and clear areas in woods and growing faster than other trees, but birch woods rarely last. They cast light shade, so other trees grow steadily beneath them and eventually take over the canopy when the birches die after 70–80 years of dominance. If perchance no other trees seed below them, the birch groves simply degenerate into open scrub. Much further north, birches grow slower, live longer and seem able to perpetuate their dominance.

Both birch species are widespread in Lady Park. Silver birch is characteristically a tree of dry sites on shallow soils, whereas downy birch is more a species of wet ground and peats, but the preferences are far from absolute and no particular pattern has ever been detected in Lady Park. One or other is capable of growing anywhere in the wood, but their distribution and abundance at any particular time is patchy, which reflects their opportunities over recent decades. They must have been part of the wood throughout the wood-pasture and coppice

Fig. 9.1. Birch was once dominant in the young-growth, but is fast dying out. Here, near transect IV, it remains a prominent component of the stand, but in the background a litter of dead birch testifies to the general high mortality.

phases, but probably declined to small numbers during the early 20th century under an increasingly beech-dominated canopy. Their greatest opportunity was, of course, the wartime fellings, which allowed them to regenerate in huge numbers.

9.2 Birch Regeneration

Birch seedlings were abundant in the felled areas in 1945 and formed dense thickets by 1955. In March 1949, EWJ noted (LPW EWJ A24) that:

> Birch (*B. pendula*) regeneration is very abundant. Much of it is 3–4′ tall, but in good places it is up to 10′ tall. It is patchy; some groups of strong saplings occur amongst Rubus, but some patches of dense Rubus have not even got any birch. It is often present throughout grassy patches in shelter of standards, but only 6″–1′ tall.

On New Year's Day 1958, EWJ noted that some of the birch thickets on transect VI had been cut, presumably by foresters 'cleaning' the new growth, but otherwise growth was natural. On the colluvial fan, a few birch of seedling origin had reached 23–25 cm GBH, but on the thinner soils of the steep slopes above, the best had reached only 14 cm. In both zones, most were much smaller and failing. Some had been crushed by clematis tangles and several had been girdled by grey squirrels. There were also a few birch of coppice origin, presumably 1902 and 1930 poles that had been felled, the best of which were taller and up to 36 cm in girth, but they had also attracted grey squirrels, which had girdled and broken some at 3–5 m.

Since the canopy closed in the early 1950s, birch regeneration has been very limited and transient. Thus, seven small recruits were recorded on the transects between 1985 and 1993, but they had all gone by 2005, some having been eaten by deer. Birch seedlings are frequently seen in the clearings, and saplings can always be found somewhere in the wood. At present, saplings are starting to prosper in tree-fall gaps near transect VI and on the top slopes, but most gaps contain no young birch.

9.3 Birch Growth and Survival

The later history of the birch in the young-growth can be illustrated by the two quarter sections recorded in 1948 (see Chapter 5, Section 5.11, this volume). In 1944, the density of birch seedlings was 10,400 and 20,700/ha, and many more must have been present but were too small to mark. By 1960, the young growth could be described legitimately as a birch wood, so dominant and conspicuous were these trees, but the population declined steadily to 130 and 200 birch/ha by 2013. This severe thinning of the post-war cohort was not recorded in detail until 1977. In 1958, some larger trees were stripped of bark by grey squirrels, but they survived and generated new leaders (Fig. 9.2).

We can reconstruct the 1976 population approximately by adding the standing dead birch of 1977 to the live population (Fig. 9.3). With girths of 6–79 cm GBH (median 28 cm), most were small, but there was a scatter of larger trees. Essentially, a few had won the race to stay in the canopy, but most of the population was struggling lower in the thicket. Thereafter (Table 9.1), there was no lasting recruitment, for the two recruits recorded in 1993 died before 2002, whereas the mortality rate remained unflinchingly high. The larger birch continued to grow strongly, so basal area remained reasonably high until 2002 (see Chapter 5, this volume), but it declined steeply thereafter. Seventy years after it was initiated, the post-war generation of birch was rapidly coming to its end (Fig. 9.1).

The 1977 record also shows the impact of the 1976 drought (Fig. 9.3). The trees recorded as 'standing dead' in 1977 must be almost all the trees that died in 1976, for any trees that died before 1976 would already have fallen and only the smallest of the trees that died in 1976 would have fallen that winter. A limited number of trees were listed as 'crown-dead or crown half-dying' – although over half were still alive in 1983, most had died by 1993. Birch of all sizes succumbed, but the smaller birch suffered most; that is, drought-administered euthanasia was inflicted on individuals that were already on the way out. The fate of the larger trees mattered most, and here there was great variation through the wood. Over 80% of larger trees (10 cm GBH or more in 1977) in transects V and VI survived, whereas less than 20% survived in most of transects III and IV (see Peterken and Jones, 1989: p. 418). This patchiness probably reflected the depth of soil – better survival on deeper soils – but the pattern was far from clear.

The subsequent fate of the birches that survived the drought depended on their size. The smallest had all died by 2015, together with some 75–80% of the medium and large birch. Above 8 cm DBH in 1977, mortality rate was much the same for all sizes, but larger trees grew slightly faster (Fig. 9.4). Most of the largest trees had died by 2015, and

Fig. 9.2. Photographed in the 1980s, when they were still in prime condition, the distorted crowns of these birch were the result of bark stripping by grey squirrels in 1958.

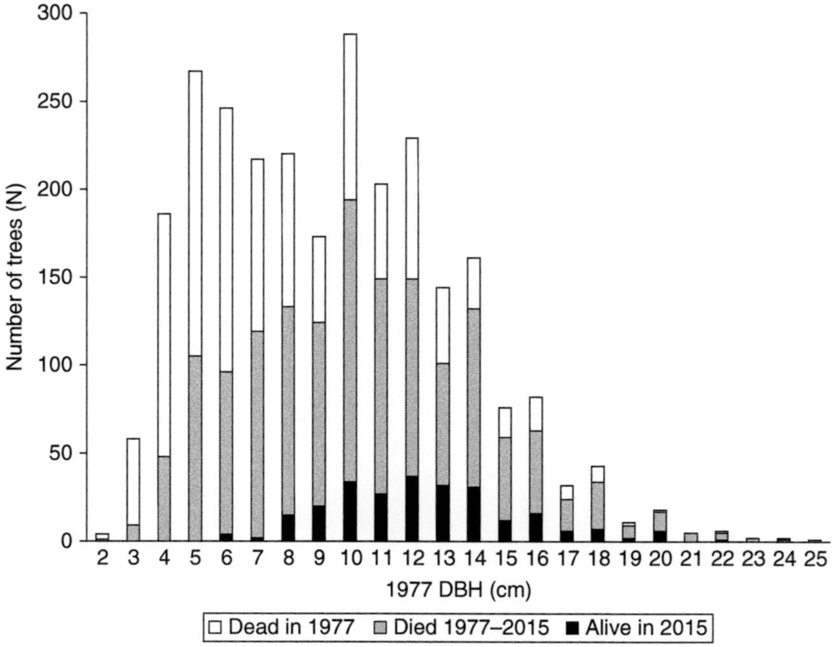

Fig. 9.3. Fate of the birch population in the young growth in relation to size in 1977. Most of the trees recorded as dead in 1977 had died in 1976. Thus, three populations are shown: 1976 (whole column), 1977 (black and grey parts) and 2015 (black part).

Table 9.1. Population change of birch in the young-growth stands.

	Seasons (year)	Initial density (N/ha)	Live at start	Losses	Gains	Mortality rate (%/year)
(1976)–1977	1–3[a]	1305[a]	2674[a]	–1085[a]	0	15.9–40.6[a]
1977–1983	6	775	1589	–408	0	4.83
1983–1993	9.5	576	1181	–368	+2	3.85
1993–2002	9	398	815	–283	0	4.63
2002–2013/15	13.5[b]	260	532	–245	0	4.47
2013/15		121	247			

Notes: [a]The 1085 trees recorded as standing dead in 1977 have been added to the live trees in 1977 to estimate the population in or slightly before 1976. The range of the mortality rate has been calculated on the basis that these represent the mortality of 1–3 years. [b]Roughly half the transects were recorded in spring 2013 and the rest in September 2015, so the records have been merged and deemed to be representative of late 2014.

those that survived grew slowly. Throughout these decades, their growth rates declined steadily: if we take only the trees that survived to 2015, their median radial growth rates declined from 2.22 mm/year in 1950–1977 to 0.96, 0.62 and 0.61 in 1977–1993, 1993–2002 and 2002–2015, respectively. (The equivalent medians for basal area increments were 4.2, 3.9, 3.0 and 3.7 cm^2/year.) Despite the very obvious and substantial impact of the drought on the 32-year-old birch population, the iron rule that larger trees survived better than smaller trees prevailed, but the trees with the fastest radial growth rate after 1977 were spread across the entire size range, excepting only the very smallest.

By 2015, the 71-year-old birch were nearing the end of their allotted span (Fig. 9.10). Most retained only small crowns, which were being crowded out by the continued height growth and crown expansion of ash, beech, lime and oak. As they near their end, some break in the wind; some blow down, leaving small

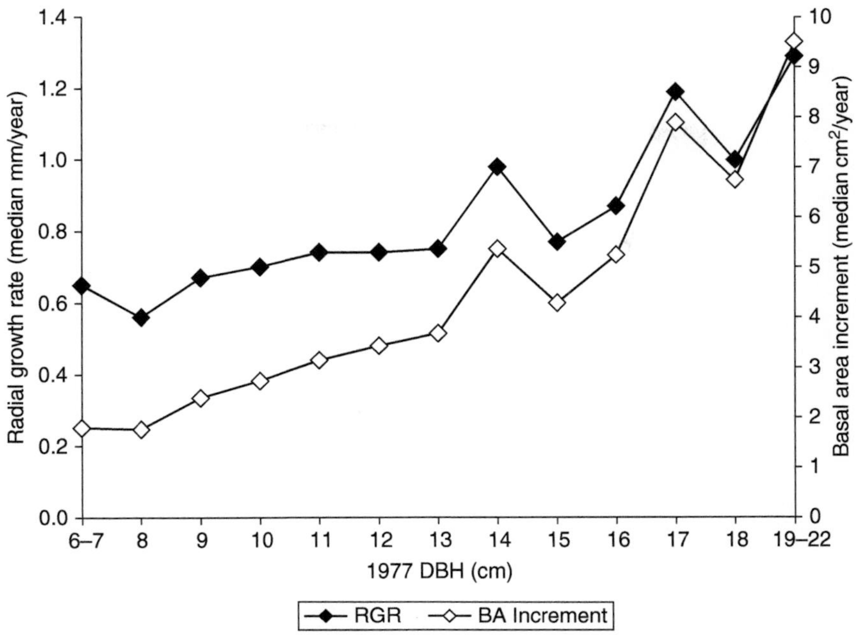

Fig. 9.4. Growth rates between 1977 and 2015 of surviving birch in the young-growth stands in relation to size in 1977. Growth is given as medians of radial growth rate in mm/year (filled symbols) and basal area increment in cm²/year (open symbols).

pits and root plates; and some, being lightweight trees, are easily crushed when larger neighbours fall. Many just die standing, sprout a crop of birch polypores and quickly fall, leaving rapidly decaying logs whose bark, oddly, lasts longer than the wood. The clusters of high mortality recorded in the drought (Peterken and Jones, 1989) have been repeated in a different form since. For example, we have witnessed the collapse, without obvious cause, of a substantial birch stand just above the re-entrant cliff between transects V and VI, and the crushing of a cluster of birches by the fall of a large standard beech near transect VI.

If birch generations last little more than 70 years, the old-growth would have included very few in 1945. In fact, there were 38 birches (density 42/ha), none above 21 cm DBH, all presumably the response to thinning in 1902, or more likely about 1930. Most had gone by 1993, having never grown large, and only one tree remained in 2016, growing with a full crown among ash near the top of the wood. Below the cliff, birch has always been infrequent. In 1950, there were nine in the transects, all but one of which had gone by 1985. The fastest grew at a cracking 2.44 mm/year in 1950–1959, but it was the slowest that survived to 2005, growing at only 0.22 mm/year.

Birch regeneration has failed almost completely since 1950. Despite the formation of many gaps, especially below the cliff and in the old-growth, birch saplings are rare and have not survived. In recent years, groups of young birch have developed in openings on the upper plateau (Fig. 9.5), but in the body of the wood, it appears to be en route to extinction.

9.4 Sallow

Sallow (*Salix caprea*) is widespread, if thinly scattered, through the ancient woods of the lowlands, but infrequent in the upland woods. Three factors seem to control its distribution: (i) disturbance (which allows it to regenerate from seed); (ii) well-lit situations in gaps and on margins; and (iii) moisture in the soil. Still frequent and widespread in lowland coppices on heavy and other poorly drained soils, it may once have been commoner when coppices were cut regularly, for it is capable of coppicing and will regenerate in large gaps associated with disturbed ground. Further west, it seems to be more associated with clay loams derived from limestone, but it will also occupy freely drained, base-rich sites in the oceanic west. Ever an opportunist, it will colonise places in the chalkland

Fig. 9.5. Dense cluster of birch saplings in a gap on the upper plateau. One of the old hawthorns is flowering in the background.

beechwoods where trees have been felled and extracted from heavy ground on the plateau. Mostly, it remains a small tree or large shrub, but the most vigorous individuals can grow surprisingly tall if they keep up with the leading trees. Eventually, the trunk rots at the base and the tree collapses.

In Lady Park, sallow has consistently behaved as a gap-phase pioneer capable of living for little more than 70 years. It colonised in patches after the 1942–1944 fellings, where some saplings even outpaced the most vigorous birch. By autumn 1948, saplings were thinly scattered over the upper plateau and upper steep slopes, but were frequent lower down and had reached 1–3 m on the mid- and lower slopes and the colluvial fan, though many growing by paths and in grassy glades below standards were browsed by deer (LPW EWJ A9). By 1977, 77 remained standing on the young-growth transects, and 28 of these were dead. Thereafter, they continued to thin out rapidly (Table 9.2; Fig. 9.6a and b),

Table 9.2. Population changes in sallows in the young growth.

	Initial density (N/ha)	Live at start	Losses	Gains	Mortality rate (%/year)
1977–1983	24	49	−20	0	8.37
1983–1993	14	29	−9	+1	3.84
1993–2002	10	21	−13	0	10.17
2002–2013	4	8	−7	0	17.22
2013	<1	1			

Fig. 9.6a and b. The last of the post-war sallows near transect I falls in September 2012, when the full crown proved to be too heavy for the hollowed trunk.

with only a slight respite in the decade after the 1976 drought, when the mortality rate decreased and one sapling was actually recruited – though it had died within a decade. By 2015, the only survivor on the transects was also doomed, having broken at the base of the trunk, and only a few large trees remained elsewhere.

Initial growth was rapid: radial growth rates from 1950 to 1977 of up to 5.37 mm/year (median 2.55) were faster than after 1977. Trees of all sizes survived each period, but the larger and faster growing proved to be marginally more durable. Thus, for example, the trees that survived beyond 1993 had grown at 1.30–5.37 mm/year (median 2.69) before 1977, whereas the trees that died between 1977 and 1993 had grown at 1.11–4.07 mm/year (median 2.41). Shade killed the smaller trees, while the larger trees seemed vulnerable to wind: in 1993–2002, for example, the four fast-growing trees that died in that period were all windthrown. Clearly, survival depended on fast growth, but even the most successful were becoming moribund and vulnerable to wind by the time they reached 50 years old. The largest girth achieved on the transects was 113 cm.

In the old-growth, a group of nine saplings entered on the margins and had reached 1.3–6 m by 1945, but they all vanished before 1983. Below the cliff, in contrast, sallow almost maintains a permanent presence, but the size of the trees present has fluctuated over the decades (Table 9.3). A cohort was admitted in the 1930s and 1940s, partly in response to wartime felling, which lasted perhaps 35–40 years and reached no more than 63.5 cm GBH: the last survivors were all dying in 1959 (LPW EWJ A29/6). A new generation was recruited in 1985–1992, when the canopy was still open in the wake of DED, but this lasted less than 30 years. This is the classic behaviour of an ephemeral gap-phase pioneer.

9.5 Gean

In early spring, wild cherries are a conspicuous feature of the lower Wye Valley. Their white flowers stand out so clearly that for a fortnight the species can be mapped in detail from across the valley, thereby revealing immediately that they are commonest on the steep valley sides and that they tend to grow in clonal groups. Such clones generally include not only mature trees but also suppressed saplings and poles.

In accordance with the general pattern, the geans in Lady Park Wood are almost confined to the sloping ground below the cliff. In 1950, 14 trees were recorded, all on transects VII and VIII, the largest being 35.6 cm GBH. Like the sallows, they were probably recruited in the 1930s and 1940s. Further individuals were recruited in each period until 1992–2005, but the population peaked at 18 in 1959, then declined steeply: all the original trees died before 1992 and the recruits had vanished by 2005, but one tree of 72 cm GBH survived just outside transect VIII in 2013. Thus, the generation inherited in 1950 continued to recruit until 1950–1959, but had virtually expired by 2013, roughly 80 years after it was initiated, and failed to replace itself, even when the canopy opened after the elms died (Fig. 9.7).

9.6 Aspen

Aspen is a clonal species (Fig. 9.8) that can grow on drought-prone sites, including both rocky outcrops in Highland pinewoods and poorly drained clays in lowland coppices and, in both kinds of site, clones generate almost pure stands. Aspen will also coexist with other species in much the same fashion as sallow, by growing rapidly when it gets the chance. It achieves this by means of sucker

Table 9.3. Performance of sallow population in the stand below the cliff.

	Initial density (N/ha)	Live at start	Losses	Gains	Mortality rate (%/year)	Recruitment rate (N/ha/year)	Largest (cm GBH)	Median (cm GBH)
1950–1959	27	13	−9	0	12.27	0	52.1	38.7
1959–1985	8	4	−4	+24	(100)	1.91	63.5	52.8
1985–1992	50	24	−13	+4	10.55	1.18	16.0	5.0
1992–2005	31	15	−11	0	9.67	0	34.0	11.0
2005–2013	8	4	−4	0	(100)	0	50.5	41.5
2013	0	0						

Note: The largest and median trees at the start of each period are indicated.

Fig. 9.7. The mortal remains of a large gean, one of the last to die out from the young-growth stands.

shoots that look weak and sickly under shade but which grow rapidly into trees if canopy space is vacated above them (Fig. 9.9). Most aspen clones are fairly limited in extent, but occasionally they take over large tracts of coppice/standards; for example, Brasenose Wood, part of Shotover Forest (Oxon). Being clonal, and apparently not seeding often, it persists where it is, but spreads only slowly.

At least five clones are known in Lady Park: two on the mid-slopes on transect V; two on the shallow soil over the limestone outcrops in the managed stand on transect X; and a large clone just above the limestone cliff lining the Whippington Brook. The clones in the reserve grew up after the 1942–1944 fellings and may initially have formed quite a thicket, but by 1977 just 19 individuals remained alive, with 12 snags. These thinned out rapidly (mortality rates 3.3–6.1%/year), leaving just three large individuals in 2013. The smallest and slowest-growing died first, but even the survivors were slowing down rapidly and hardly grew at all after 2002. Radial growth rates from 1950 to 1977 of the live trees in 1977 were 1.30–3.15 mm/year (median 2.04), well above the growth rates achieved in any period after 1977. Three suckers were recruited in 1983–1993 when the stand was still fairly open after the drought, but they had all gone by 2002, probably eaten by deer.

In the managed stand, thinning clearly benefited aspen, generating both appreciable regeneration and

Fig. 9.8. This aspen on the plateau of the managed stand grew from a clonal root system. Three trees on this clone have died and rotted, leaving holes in the exposed roots. Fallen leaves show that it was growing with beech and small-leaved lime.

much faster growth than in the unthinned young growth. On transect X, the five individuals retained were impressive trees by 2014, 69.5–109 cm GBH (median 93.5) and had grown at 1.30–2.94 mm/year (median 1.86). Numerous suckers grew up after thinning and had reached up to 39 cm GBH by 2014: no doubt browsed, their growth rates were not as fast as the retained trees. The most impressive clone was sampled by Williams plot 10. It stands on gentle slopes just above the long limestone outcrop overlooking the Whippington Brook, where it resembles a well-maintained plantation. Here, the four trees sampled grew at 0.86–3.14 mm/year (median 1.38) in 1979–2014.

9.7 Alder

Alder is a tree of riverbanks, swampy ground and other wet places. Dominant in the fen woodlands around the Norfolk Broads and in other carrs, it lines the rivers and streams of Wales and other hilly regions. Less obviously, it also grows on plateaux where drainage is slow, and on steep slopes with a consistent supply of water flowing down in the subsoil. Thus, in the lower Wye Valley, it not only lines the river but also grows sparingly on former marshes in the plateau farmland and on flushed slopes on the sides of the gorge. Prehistorically, it was abundant on lowland flood plains before rivers

Fig. 9.9. A small thicket of aspen suckers regenerating in the managed compartment after thinning in the 1980s.

were civilised and the bottomland was turned into meadows, and it can still be found as fairly natural flood plain woodland near Loch Ness, on the lower Spey and in the New Forest. Alder was also mixed into the lowland coppices in depressions and on shallow elevations, where the water table remained high but the water kept flowing. Thus, in Norfolk, it coexists with bird cherry in shallow valleys within Swanton Novers and other ancient woods. Both medieval land surveyors and modern phytosociologists know an alder wood as an *Alnetum*.

In Lady Park Wood, alder is a species of the flood plain and the very bottom of the slopes. Most of this is outside the reserve, but a small fragment of flood plain, which escaped clearance to meadow, falls within the reserve, where it is crossed by transect VII. In 1950, three alders were recorded there. One grew very slowly and was dead before 1985. The other two grew very fast (4.24 and 4.44 mm/year) until 1985, when the largest reached 128 cm GBH, but were static after 1985. After 1992, one toppled, the other leaned and both were felled as part of the safety fellings in 2000. Several stems grew from existing trees after the elms died and the canopy opened, but all save one were cut off in 2000. Three alders were recorded in 1977 on the colluvial fan in transect VI, but despite growing fairly fast (2.27–3.15 mm/year), they had all died by 1983, when one was still visible as a snag. Outside the reserve, on the flood plain and riverbank, alders grow rapidly, but many suffer from *Phytophthora alni* and now provide gaunt perches for resting cormorants.

This is a tiny and battered population, but it is enough to show that the natural flood plain woodland must have included alder along with ash, wych elm and hazel, and that it grows well and regenerates freely when it gets the chance. Curiously, a few alder regenerated *above* the cliff after 1942–1944. (A single bird cherry was also recorded, possibly in error for gean.) They did not survive, but, in passing, they demonstrated strikingly that species would grow well beyond their normal tolerances when competition was temporarily removed by felling.

9.8 Review

These species formed a consistent functional group. They all relied on quick responses to gap formation and rapid early growth, but mortality rates were high, the growth of survivors declined steeply, only the most vigorous survived much beyond 50 years, and when they died, they rotted away rapidly. No doubt they would all prosper after a catastrophic stand disturbance, but their role in relatively undisturbed mature natural stands is uncertain. Birches, sallows and alder all distribute seeds far and wide, but they regenerate better if the soil is disturbed, and this is only the case locally in tree-fall gaps of the scale found so far in Lady Park. Aspen and gean regenerate from suckers, and may do better if gaps appear close to clones. All suffer if deer numbers climb.

They should have regenerated well in the thinned stands of the managed compartment, but deer prevented seedling regeneration. One aspen clone did generate a cluster of new growth, of which 25 stems survived after 30 years, but their growth was poor (maximum radial growth rate 2.07 mm/year, median 1.17). Thinning allowed the small sample of retained trees to grow somewhat faster than their contemporaries in the unmanaged reserve. Thus, the 15 retained birch grew at 0.53–3.74 mm/year (median 1.49), compared with 0–3.10 mm/year (median 0.80) in the unthinned young growth.

Interestingly, birches, alder and aspen can behave differently further north, where they face little competition from other tree species. Thus, in the woods of the Boreal regions, birches grow slower and can live for over 200 years in Britain and 400 years in Scandinavia. They have also lived long enough to be treated as pollards. Alder, too, can live to great ages, forming giant stools and pollards in upland wood pastures. Aspen clones grow on rock outcrops in northern birch and pine woods, where they also appear to be much more persistent than the fugitive aspens of southern mixed deciduous woods.

Fig. 9.10. Part of the young-growth stands passes from its early dominance by birch towards dominance by ash and small-leaved lime.

10 Field Maple and Hazel, the other Coppice Species

In regularly cut coppices, field maple and hazel grow every bit as vigorously as the other trees – indeed, hazel and field maple are major contributors to the underwood – but as soon as coppices are neglected, their slower height growth and their ultimately smaller stature oblige them to grow in shade and become subordinate in high forest and most forms of natural woodland. Together with ash as coppice and pedunculate oak as standards, they form the characteristic lowland coppice type on neutral-alkaline clays and loams, and jointly dominate a few coppices in Essex and Dorset, but both become less frequent on light, acid soils, especially maple. In Lady Park, maple is never more than frequent, even on the most alkaline soils, whereas hazel is abundant and still dominates parts of the young-growth stands. Both could be characterised as either large shrubs or small trees.

10.1 Field Maple

Field maples were common components of the underwood in the ancient coppices of the lowlands, south-east lowlands and borderlands of England and Wales, especially on alkaline soils. They can grow into trees, but have rarely been allowed to do so, and they never attain the size of oaks and other canopy dominants. Widespread also in mixed hedges, where they have sometimes been pollarded, they can grow into sizeable boundary trees. They compete in the scrub that colonises open ground, but not vigorously. Their lives became more difficult in the 20th century with the spread of grey squirrels, which debark them almost as enthusiastically as they debark sycamore.

In Lady Park, maples were a common component of the coppice, especially on the colluvial fan and the shallow limestone soils of the upper plateau. They were probably cut back during the 1902 and *c.*1930 thinnings, which perpetuated their coppice form and ensured that new growth was weak in the shade. None was retained in the 1942–1944 fellings, but they were a common component in the new young growth. In 1977, 388 stems were recorded on the young-growth transects, of which 37 were dead (Table 10.1). Most originated from coppice, just 18% appearing to have originated as post-1942 seedlings. The population thinned out steadily at a slowly increasing mortality rate, countered by both limited recruitment after the drought and new growth from existing trees after the reserve was fenced. The pattern of losses was unusual, for at each period the stems that died were much the same size as the stems that lived. Why the unconventional thinning pattern? Seemingly, small trees are vulnerable to deer, larger ones to squirrels, falling trees and the weight of ivy, and both are reasonably relaxed about shade by birch and ash.

In the old-growth, maples were concentrated on the limestone plateau and the upper steep limestone slopes, with outliers on more acid soils lower down. All languished in the subcanopy and underwood. Most were of coppice origin, but 26% were judged to be maiden. The population was still recruiting in 1945 (Table 10.2), and a group of saplings established themselves at the top of transect II after the drought, but otherwise the population declined steadily. The mortality rate was generally faster than in the young growth, which is understandable given that the shade was cast mainly by beech, not birch. The pattern of mortality changed. Only the smallest individuals were lost in 1945–1955, but later, as beech became more dominant, larger trees died at much the same rate as smaller individuals. By 1992, all but one of the remaining maples were in the underwood, and by 2010 just seven maples remained, one a recruit protected within the safety exclosure, the rest dating from before 1945. They

Fig. 10.1. A hazel on the upper steep slopes of transect V, with the original post-1942 growth still healthy and a new generation of stems that have grown up since the wood was fenced.

Table 10.1. Population changes in maple in the young-growth transects.

	Initial density (N/ha)	Live at start	Losses	Gains	Mortality rate (%/year)	Recruitment rate (N/ha/year)	Size at start (cm GBH), range (median)
1977–1983	171	351	–43	0	2.15	0	3–50 (19)
1983–1993	150	308	–61	+2	2.30	0.10	Not recorded
1993–2002	121	249	–59	0	2.67	0	6–63 (23)
2002–2013/15	93	190	–56	+9	2.99	0.38	6.5–68 (25)
2013/15	70	143					10–72 (30)

Table 10.2. Population changes in maple in the old-growth transects.

	Initial density (N/ha)	Live at start	Losses	Gains	Mortality rate (%/year)	Recruitment rate (N/ha/year)	Largest at start (cm GBH)
1945–1955	68	61	–16	+4	3.00	0.44	48
1955–1977	54	49	–35	0	5.30	0	48
1977–1992	16	14	–7	+4	4.83	0.32	47
1992–2000	12	12	–1	0	1.05	0	61
2000–2010	11	11	–3	0	3.50	0	71
2010	8	7					77.5

had been weak and deformed since at least 1992, and several had lost their leaders, although the strongest were generating new shoots (Fig. 10.2). One might easily conclude that maple is incapable of living long and forming trees when competing against beech and lime, were it not for a single maple in the subcanopy that remains healthy, albeit with a narrow crown and growing slowly.

Maple performed much better below the cliff (Table 10.3). The 1950 population, which ranged up to 43 cm GBH, included many derived from coppice. Most of the losses in 1950–1959 were small individuals, i.e. this was conventional thinning under heavy shade. Thereafter, mortality was spread across all sizes except the largest, and was faster than above the cliff, though the open canopy after elm disease appears to have afforded some relief. Recruitment, mainly in the form of new stems on old trees, picked up in the wake of DED, but fell away thereafter. By 2013, the population was much reduced: just one tree survived from before 1950 and the rest post-dated the canopy opening brought about by elm disease. Many survivors had arched stems and trunk sprouts, indicating poor growth. Some were broken, one flattened, and several had been debarked by squirrels.

Growth was clearly slow in all stands and became slower, but it was difficult to measure accurately, for stems were often irregular, patchily debarked and clothed in weak twigs and adhesive ivy. In the young growth, the 137 trees that survived from 1977 to 2015 slowed from 0.37–2.59 mm/year (median 1.11) before 1977 to 0–1.74 mm/year (median 0.41) thereafter. Very little can be said about the small population in the old-growth: eking out an existence under beech and other canopy trees, the six survivors from 1945 all grew faster after 1977 than they did from 1945 to 1977, presumably because the canopy was more open, but the rate of 0–1.92 mm/year (median 0.96) was still slow. Below the cliff, the only tree that survived from 1950 to 2013 grew at 0.53 mm/year. Even in the managed stand, the 17 stems retained from 1984 to 2014 grew at only 0.03–0.69 mm/year (median 0.37).

Maple seems barely able to sustain itself in mature high forest, especially when browsed by deer, damaged by grey squirrels and shaded by beech and lime. Even when relatively well lit, its growth is extraordinarily slow, and in the shade many are almost static, though persistent. Below the cliff, where the canopy is always open and the soil strongly alkaline, maple still grows slowly and remains particularly vulnerable to damage when larger trees fall. Seedling maples are infrequent, but they have prospered in gaps near the plateau, several are growing into gaps on the colluvial fan and more would have recruited but for deer. Under natural conditions, maple would probably remain a moderate component of woodland on the thin soils of the plateau and slopes and under the fragmented canopy below the cliff, regenerating more in gaps

Fig. 10.2. A once-large maple in the old-growth stands with the strength to generate new stems, even though it is shaded by beech and its bark has been stripped by grey squirrels.

Table 10.3. Population change in maple below the cliff.

	Initial density (N/ha)	Live at start	Losses	Gains	Mortality rate (%/year)	Recruitment rate (N/ha/year)	Largest at start (cm GBH)
1950–1959	62	30	−10	+7	4.41	1.61	43
1959–1985	56	27	−24	+21	8.41	1.74	48
1985–1992	50	24	−4	+8	2.57	2.37	50
1992–2005	58	28	−12	+1	4.21	0.16	54
2005–2013	35	17	−7	+2	6.42	0.52	58.5
2013	25	12					64

than in shade. Elsewhere it would be rare, and everywhere it would lead an apologetic existence in the shade. Its best chance would come after a storm has generated large canopy openings.

10.2 Hazel in General

Hazel was *the* coppice species of Britain. Except on very wet and very poor soils, or where other species had been planted, it was the matrix within which other coppice components were scattered. Sometimes reaching over 10 m, it is usually seen as a shrub, streaming with catkins in early spring, but it will grow tall enough to qualify as a tree and can maintain itself in the canopy, where the growth of all trees is restricted by poor soil. It dominates the coppices of the Dorset and Hampshire chalk lands, the centre of wattle hurdle-making and the stamping ground of Thomas Hardy's *Woodlanders*, but it also thrives in the extreme north-west Highlands, where it grows as the sole dominant of lichen-festooned hazel woods (Coppins and Coppins, 2012).

Capable of rapid early growth after coppicing, hazel brakes quickly as it ages and rarely grows above 60 cm GBH. Old hazels usually remain multi-trunked and wide spreading, but eventually old trunks break or start to rot and, either way, they fall. Their tendency to fall is reinforced on slopes, where old stems usually arch downhill, before either breaking or gently unbalancing and uprooting the whole edifice. As an underwood tree, hazel is often crushed when larger trees or their main branches fall, but it fights back. It will root near the tips of fallen branches (a feature that the woodmen of old employed to propagate hazels in coppices) and, more importantly, it is strongly self-coppicing: arched and fallen stems reiterate new shoots from the upside, which can take over the trunk completely, and tipped-over stools usually re-sprout vigorously, the new shoots becoming dominant and the old stems slowly dying as the shade increases. Even on undamaged stools, new shoots regenerate from the base as old trunks lose their vitality, and arching trunks sprout new shoots when they lean 'too far' (Fig. 10.1: see p. 166). The outcome is that old hazels are usually untidy, straggling, complex structures with stems and branches of various ages. Once dead, old trunks rot quickly on the stool, leaving the younger stems unaffected. Thus, when one finds a small hazel with young stems, it may indeed be the young hazel it seems to be, but it could equally be an old individual from which all the old stems have rotted.

In Lady Park, hazel was once a major coppice species. Some of the present plants may be survivors from stools that were coppiced before 1870, but it is impossible to be sure, for hazels regrowing after the 1902 thinning could now either be monsters or look like young stools. Somewhat less abundant on the acid soils and scarce in the zone above the cliff where beech and sessile oak are common, it is particularly abundant on the moist soils of the colluvial fan.

10.3 Recording Hazel

Surveying hazel is a nightmare. The surveyor's heart falls when he confronts a thicket of straggly old hazels, especially when they also have numerous strong new shoots: large hazels can have well over 10 old stems and 30 new shoots reaching above 1.3 m. Given this, it is perhaps excusable that hazel recording has suffered under pressure. Before 1977, EWJ measured individual stems of 20 cm GBH and above, but otherwise simply recorded plants with smaller stems as having 2–10 stems or more than 10. In the young growth, only the 'principal' coppice stools and 'most conspicuous' seedlings were recorded, and in both 1948/9 and 1955–1958, the only measurement taken was the height of stools and seedlings. In 1977 and 1992–2010, the girths of all stems above a certain size threshold were recorded, together with some smaller stems (see Chapter 4 for details). In 1983, the stems recorded in 1977 were recorded as dead or alive, new stems were counted and the few recruits were measured, but in 1985/6 small subsamples were measured. In 2013, only the largest stem was measured, but in 2015 (young-growth transects I, II, III, V: 100–300 only) all stems from 5 cm GBH upwards were measured and smaller stems reaching 1.3 m were counted. Collapsed and arching trunks often generate new stems (Fig. 10.3), but these reiterations have not been recorded separately, except those that emerged close to the stool.

Hazel performance can be analysed at the level of individual plants or stems. The former is straightforward, but the latter has been difficult because we have sometimes been unsure which earlier recorded stem has survived. Determination of whether a stem has survived from a previous recording was based on field appearance, combined with assumptions that trunks did not shrink and that the size order of several trunks on one plant remained unchanged between recording dates. The detailed stool diagrams prepared by EPM meant that most stems could be matched with certainty during 1992–2010; to a lesser extent, this applied to the stool diagrams shown on the 1977 transect charts.

Fig. 10.3. An old, multi-stemmed hazel stool that tipped over, leaving the prostrate trunks alive and generating a thicket of new growth along the entire length of the old trunks. At this stage, the girths of the prostrate trunks would be measured and the number of new shoots reaching 1.3 m counted.

10.4 Hazel in the Young-growth Stands

Coppice regrowth and seedling regeneration after 1944 quickly restored the hazel underwood, but the process could not be quantified completely due to the limitations of the record before 1977. Most coppice stools had sprouted to 1.2 m by 1948/9 and the tallest had reached 4.9 m, the strongest and densest growth being on the colluvial fan and the lower steep slopes. More was recorded in 1955/58 and even in 1977 (Table 10.4), but these were probably not new plants but small plants that were ignored earlier. Seedlings were recorded patchily in 1944 and 1948/9, but more thoroughly in 1955/58. In the latter recording, they were thin on the ground in the lower slopes and lower plateau, and none was recorded on the plateau above the cliff. Some that were recorded in sections where none was recorded in 1944 and 1948 must have been genuine late recruits. Recorded increases were greatest on the upper plateau and upper steep slopes. From the outset, the growth of seedlings was far weaker than coppice growth. Thus, in 1955/58, 14% of the seedlings were browsed, 35% were overtopped and only 33% remained vigorous, so only the stronger and luckier could thrive in the developing thicket.

After 1955/58, the mortality rate of hazel individuals greatly exceeded recruitment, and density decreased steadily. The mortality rate dipped in 1983–1993 at the same time as recruitment increased, a belated response to the opening of the canopy while deer browsing was not as severe as it later became. The advantages of starting early and vigorously were clear: individuals first recorded in 1948/9 had a lower mortality rate to 2013/15 (0.78%/year) than the individuals that were too small to record then or which were recruited later. Judging which 1977 individuals had originated as seedlings is a hazardous business, but the best guess is that only 38% grew from post-felling seedlings. Most were small: the largest stems were 6 cm DBH and below. Density was highest on the colluvial fan, the lower steep slopes and the upper plateau (Table 10.5) – all highly calcareous zones – and was lower on the base-poor zones, where competition from lime, beech and oak coppice had long been severe, but the differentials were much reduced by 2013/15 because mortality rates were generally higher in the denser populations. The population on the colluvial fan stood out, having a high density of coppice stools and high mortality rates associated with intense competition. The later increase in mortality rate was associated with the flattening of many stools by the fall of a large beech, but some may yet recover if they can grow through the branch debris and vigorous bramble.

The analysis of changes in the population of hazel stems gives a different perspective. Stem density must have peaked about 1950, and by 1977 it had been much reduced. The evidence for this was that many dead stems were recorded in 1977, and a substantial fraction of the stools recorded in 1945 and 1955 could not be matched with individuals in 1977.

By 1977, the hazel populations consisted mainly of stems generated after the 1942–1944 felling. New

Table 10.4. Population changes in hazel individuals in the young-growth.

	Initial density (N/ha)	Live at start	Losses	Gains	Mortality rate (%/year)	Recruitment rate (N/ha/year)
1948/9–1955/58	141	273	–4	+367	0.20	25.2
1955/58–1977	369	612	–148	+56	1.28	1.6
1977–1983	293	600	–63	+2	1.83	0.2
1983–1993	263	539	–45	+25	0.91	1.3
1993–2002	253	519	–81	+1	1.87	0.1
2002–2013/15	214	439	–122	+2	2.67	0.1
2013/15	156	319				

Note: Based on recording areas of 1.9416 ha (1948/9), 1.6569 ha (1955/58), 2.0496 ha (1977–2013/15); only the principal coppice stools and a few conspicuous seedlings were included in 1948/9.

Table 10.5. Mortality and recruitment in hazel individuals in the young-growth stands, 1977–2013/15.

	Individuals in 1977		Mortality rate (%/year)					
	N	Density (N/ha)	1977–1983	1983–1993	1993–2002	2002–2013/15	Recruits, 1977–2013/15	Density in 2013/15 (N/ha)
Upper plateau	50	545	1.74	1.08	2.01	4.19	+1	239
Upper steep slopes	51	277	3.97	1.40	0.60	0.99	+4	190
Mid-slopes	115	294	1.04	0.19	1.67	1.59	+7	238
Lower slopes	126	155	0.95	0.73	2.78	2.62	+4	85
Lower plateau	31	190	1.68	0.38	1.70	1.44	+1	135
Plateau above cliff	0	0					0	0
Lower steep slopes	61	498	1.12	1.79	1.21	2.84	+2	294
Colluvial fan	166	677	2.80	1.19	1.89	5.05	+12	281

shoots were being produced – roughly 900 were noted in 1983 – but after 1993, many were browsed down before they reached 1.3 m. This state of affairs continued through to 2007, by which time the old stems were thinning out and new shoots were mostly still being browsed down. Thereafter, deer were excluded and the recruitment rate of new stems vastly accelerated. We can quantify these changes in the lower slopes of transects I–III (Table 10.6) and the mid-slopes of V (Table 10.7). Mortality on the lower slopes accelerated as debarking by grey squirrels increased, falling birch struck them down and as the older stems became moribund under a closing canopy of increasingly dominant small-leaved lime (Fig. 10.4), but recruitment has gone through the roof since the deer were excluded (though the rate is exaggerated by incomplete recording of small stems before 2015 (see Fig. 10.1).

By 2013/15, just 36% of the 1977 stems survived. The middle-sized stems survived best, for the smallest (1–2 cm DBH in 1977) were presumably excluded by competition or destroyed by deer, and the larger stems (12–14 cm DBH in 1977) tended to break. In the middle range, mortality was marginally lower in the 6–9 cm DBH stems of 1977, except in transect VI where a substantial area was flattened indiscriminately by a falling beech. Mortality rates increased substantially after 1993 (see Tables 10.6 and 10.7). Clear differences emerged between different parts of the wood. Thus, the mortality of larger stems was much greater in the southern parts (transects I–III).

Changes in growth rates broadly reinforce the picture that emerges from the analysis of population changes. Hazel stems are frequently irregular and difficult to girth accurately. Moreover, it can be difficult to match records of the stems on large stools from one date to the next, though generally we have assumed least change. Restricting attention to the 125 surviving stems whose identification through the record is reasonably reliable (mostly the larger stems that originated after the 1942–1944 felling), these grew at an estimated 0.66 mm/year before 1977, 1.07 mm/year from 1977 to 1993, then 0.30 and 0.31 mm/year in 1993–2002 and 2002–2015, respectively; that is, they accelerated after 1977 (which must be a response to reduced competition after the drought) but slowed down hugely after 1993. Since 2002, the fast-growing stems could be found right across the size range, with only a slight tendency for larger stems to grow faster.

The emerging pattern is reasonably coherent. Hazels must sprint at the outset to survive, which gives a strong advantage to established coppice stools, but even the survivors find themselves in the underwood and slow down severely. If the canopy

Table 10.6. Population change of hazel stems in the young-growth stands of transects I, II and III.

	Initial density (N/ha)	Live at start	Stems per individual (mean)	Losses	Gains	Mortality rate (%/year)	Recruitment rate (N/ha/year)
1977–1983	573	429	3.3	−48	+41	1.96	9.66
1983–1993	576	422	3.6	−46	+19	1.21	2.76
1993–2002	541	395	3.4	−107	+13	3.45	1.88
2002–2015	414	301	2.9	−140	+732	4.53	77.50
2015	1228	893	11.4				

Table 10.7. Population change in hazel stems in transect V: 100–300.

	Initial density (N/ha)	Live at start	Stems per individual (mean)	Losses	Gains	Mortality rate (%/year)	Recruitment rate (N/ha/year)
1977–1983	1697	208	3.20	−2	+7	0.16	9.5
1983–1993	1737	213	3.28	−14	+0	0.71	0
1993–2002	1623	199	3.16	−39	+12	2.39	10.9
2002–2015	1403	172	3.02	−66	+544	3.52	341.3
2015	5294	650	12.98				

Note: 2015 figures include 17 fallen stems.

Fig. 10.4. A large, old hazel stool that has fallen on hard times. After the 1942–1944 felling, new growth had reached 12 feet by the winter of 1948/9. It reached its zenith about 1993, when it had 20 live stems up to 32 cm GBH, but by then it was growing slowly – the girth of the largest stem had only grown by 4 cm since 1977 – and by 2002, most of the largest stems were dead and just ten remained alive. By 2015, it had five stems (largest 23 cm GBH) and a few new sprouts. Why the decline? It grew too close to a beech that shaded it heavily from the 1990s onwards. Descriptions in 2002 showed that some of the old stems had broken, so mechanical weakness may have played a part, and before 2007 deer prevented new sprouts from rejuvenating the stool.

opens in the first 40 years, growth rates can still accelerate, survival rates improve and new stems on established plants can prosper, but after 50 years or so, trunks become moribund and the largest become more vulnerable. Why? Large trunks spread, especially on slopes; they tend to split; they present ever-larger targets for falling canopy trees; and they may grow heavy and one-sided enough to topple stools. Even though they age and old stems often rot, large and medium individuals retain enough vitality to generate new stems from the base if deer are kept in check. Thus, by 2015, many hazels possessed a cluster of post-2007 stems growing up around the remaining older stems. Stools inherited from before 1942 have proved to be stronger than new individuals, and a few have effectively coppiced themselves twice since 1942. Almost all regeneration has been vegetative, growing either from stool bases or from trunks, especially those that have fallen. Seedling regeneration has been very limited, but may now be increasing as the stand continues to break up and diversify vertically and horizontally.

10.5 Hazel in the Old-growth Stands

Many hazels that thrived in the pre-1870 coppice must have been shaded out as the stand was promoted towards beech high forest, but 38 stems of 6–10 cm DBH were recorded in 1945 on 23 well-scattered individuals, so some survived the shading as well as the cutting and crushing when the old-growth was thinned in 1902 and *c.*1930. The population was still recruiting after 1945, as new stems on established stools (Tables 10.8 and 10.9), but mortality rates were very high until the drought brought some relief from beech shade. After the drought, recruitment resumed and new stems grew again on established individuals, but by 1992 the canopy had closed, deer proliferated and hazels reverted to the status quo ante. Weak though they were, they re-sprouted continuously, especially after the drought and again after deer were excluded in 2007.

Compared with hazels in the young growth, the population density was much lower and mortality rates much higher, which must reflect the depth of shade more than the age of the stand. The survivors were limited almost entirely to the upper plateau, where ash was co-dominant. Changes in growth rates from 1945 to 2010 cannot be examined because only one stem survived the 65 years. Radial growth rates (RGRs) in 2000–2010 were related loosely to stem age: medians for stems arising before 1992, 1992–2000 and after 2000 were 2.71, 1.11 and a minimum of 0.64 mm/year, respectively. Despite the high turnover of individuals and the shift from many small plants in 1945, most of which were multi-stemmed, the overall size distribution changed little in 65 years (Table 10.10).

10.6 Hazel below the Cliff

This hazel population clearly followed a different course from those above the cliff, but it could not be followed properly because the stand was not recorded in 1971 when elm disease struck. It was still recruiting rapidly in 1950–1959, following the disturbance of the wartime thinning (Table 10.11), but recruitment evidently declined as the stand closed. After 1971, population changes must have gone into reverse, with mortality rate declining as recruitment

Table 10.8. Changes in population of hazel individuals in the old-growth stands.

	Initial density (N/ha)	Live at start	Losses	Gains	Mortality rate (%/year)	Recruitment rate (N/ha/year)
1945–1955	130	117	−40	+14	4.10	1.56
1955–1977	101	91	−65	+2	5.30	0.10
1977–1983	31	28	−9	+2	6.26	0.80
1983–1986	23	21	−4	+4	8.11	1.78
1986–1992	23	21	−3	+9	2.76	1.82
1977–1992		*28*	*−13*	*+12*	*4.36*	*0.95*
1992–2000	30	27	−3	+1	1.30	0.12
2000–2010	28	25	−9	+2	4.84	0.25
2010	20	18				

Table 10.9. Changes in population of hazel stems in the old-growth stands.

	Initial density (N/ha)	Live at start	Stems per individual (mean)	Losses	Gains	Mortality rate (%/year)	Recruitment rate (N/ha/year)
1945–1955	999[a]	899[a]	7.68[a]	−358[a]	+381[a]	4.95[b]	42.35[b]
1955–1977	1025[a]	922[a]	10.13[a]	−858[a]	+7[a]	10.95[b]	0.34[b]
1977–1992	79	71	2.54	−22	+165	2.61	13.10
1992–2000	238	214	7.93	−78	+28	4.91	3.46
2000–2010	182	164	6.56	−68	+57	5.78	7.04
2010	170	153	9.56				

Notes: [a]Estimated on the basis that hazels recorded with 2–10 small stems actually had 6 stems; while those with more than 10 stems had 15 stems; [b]minimum values based on changes in the number of stems counted on each individual.

Table 10.10. Size distribution of living hazel stems in the old-growth stands measured in 1945 and 2010.

Stem size (cm GBH)	<5	5–<10	10–<15	15–<20	20–<25	25–<30	30–<35	35–<40	40+
1945			1396[a]		25	8	1		
2010	41	51	20	20	13	5	3	1	1

Note: [a]Small stems were not measured. Estimated on the basis that hazels recorded with 2–10 and more than 10 stems had 6 and 15 stems, respectively.

Table 10.11. Change in hazel individuals below the cliff.

	Initial density (N/ha)	Live at start	Losses	Gains	Mortality rate (%/year)	Recruitment rate (N/ha/year)
1950–1959	356	172	−35	+89	2.50	20.5
1959–1985[a]	468	226	−105	+51	2.47	4.2
1985–1992	356	172	−9	+44	0.76	13.0
1992–2005	428	207	−71	+2	3.18	0.3
2005–2013	286	138	−51	+5	5.60	1.3
2013	190	92				

Note: [a]The long interval between recordings will have lowered the rates of gains and losses, because many individuals could have come and gone within the 26 years.

resumed, and this continued until 1992. Since 1992, mortality has increased sharply and recruitment has been limited: but why?

Shading increased as the canopy recovered from elm disease; hazel stems competed with neighbours on the same stool and with nearby lime, yew, beech and elms; browsing by deer and debarking by squirrels increased; some hazels were cut during the 2000 safety fellings; others were struck by falling trees and branches. Nevertheless, after the stand was fenced in 2007, stem recruitment substantially increased (Table 10.12). As an example of the changes, a large hazel (enthusiastically described as a 'world record' at the time!), with seven stems over 40 cm GBH and nine smaller stems in 1985, had been reduced to one stem of 23 cm and ten post-2005 shoots.

Closer examination of the changes and how individual plants grow provides some pointers. Some 52 of the original 170 hazels survived in 2013, but only 40 of the 197 post-1950 recruits (19 date from 1950 to 1959, 12 from 1959 to 1985), so the problem is less survival of old individuals but more failure of recruits to establish themselves. The 52 individuals that survived from 1950 to 2013 are at least 63 years old, but there is no convincing sign of loss of vigour, though most are currently much smaller than they once were. The number of live stems remaining from 2005 or earlier is 0–11 (median 3), i.e. some plants only have young sprouts. The largest stem each hazel has generated ranges from 11 to 68 cm GBH (median 26), but the largest live stems in 2013 were smaller, 4–43 cm GBH (median 19.5). Only 20 of the 52 hazels still possessed their largest ever stems in 2013, and the majority (27 ex 52) failed to generate any new stems in 2005–2013, despite freedom from browsing after 2007: the remainder generated 1–45 stems (median 3). There was no correlation between the present or historic size of hazel and the recent vigour of sprouting.

Since 1950, many hazels have generated two and sometimes three crops of mature stems. Of the 875 stems alive (and measured) in 1985, just 18.5% (162) remained alive in 2013. Remarkably, survival rates were above average in medium-size stems (6–15 cm GBH): indeed, stems above 25 cm GBH had the lowest survival rate – just one of the 13 survived. Smaller stems have been vulnerable to deer; large stems to physical damage. Fourteen of the hazels first recorded in 1950 perished in 2005–2013: they were generally smaller and less vigorous than those that survived to 2013. Despite the open structure of the stand, stems rarely last more than 50 years, and many individual plants eventually fail to generate new sprouts.

The stand below the cliff ought to be ideal for hazels – the soil is base rich and underwood stems are not heavily shaded – but in practice stems suffer constant bombardment from the fall of canopy trees;

Table 10.12. Change in hazel stems below the cliff.

	Initial density (N/ha)	Live at start	Stems per individual (mean)	Losses	Gains	Mortality rate (%/year)	Recruitment rate (N/ha/year)
1950–1959[a]	2320[a]	1121[a]	6.52	–301[a]	+716[a]	3.41[b]	165[b]
1959–1985[a]	3179[a]	1536[a]	6.83	–992[a]	+323[a]	4.07[b]	27[b]
1985–1992	1795	867	5.04	–138	+694	2.48	205
1992–2005	2946	1423	6.87	–903	+262	7.45	42
2005–2013	1619	782	5.67	–511	+358	12.42	133
2013	1302	629	6.84				

Notes: [a]Estimated on the basis that hazels recorded with 2–10 small stems actually had 6 stems; while those with more than 10 stems had 15 stems; [b]minimum values based on changes in the number of stems counted on each individual.

stools often topple under their own weight and sometimes tear themselves out of the ground; and individual stems lean and are often smashed. Vegetative regeneration since 2007 has been patchy and generally limited, and regeneration from seed is slow in the carpet of mosses and ferns and cannot have been helped by unrestricted grey squirrel populations. Hazel plants last over 65 years, but most look as if they will have difficulty making it to 100 years.

Hazel stems grow rapidly at first, but 'top out' early and grow slowly until they collapse or die. That is the expected pattern, but it cannot be verified easily because of the high turnover of stems and the uncertainties about tracing individual stems through the record. Three tests produced ambiguous results. In the most recent periods, when identifying stems across periods was most reliable, the median (and maximum) RGRs of all live stems were 0.91 (2.96), 0.37 (2.69) and 0.40 (3.08) mm/year for 1985–1992, 1992–2005 and 2005–2013, respectively. If, instead, we look at the median RGRs of the 67 stems that were measured in 1950 and survived, we find 0.34, 0.57, 0.68 and 0.65 for 1950–1959, 1959–1985, 1985–1992 and 1992–2005, respectively. (The last six died after 2005.) A third perspective comes from the median RGRs in 2005–2013 for stems originating in 1959–1985, 1985–1992, 1992–2005 and 2005–2013, which were 0.20, 0.40, 0.40 and at least 0.60 mm/year, respectively. (The 'at least' for the youngest stems recognises that some reached 1.3 m long after 2005.)

On this basis, the recent decline in hazel below the cliff appears to reflect an inability to regenerate from seed, combined with the loss of vigour of established individuals. However, ash and wych elm have suffered similar declines in recruitment and increases in mortality. Perhaps there is a particular explanation for each species, but a collective explanation seems more likely. Deer might be suspected, but fallow have been excluded since 2007; muntjac have only just arrived, and low growth does not look browsed.

10.7 Hazel Summary

Hazel was a prominent species in pre-Neolithic woodland (Godwin, 1975), but its role in natural woodland must have varied with site and the trees it had to compete with. In Lady Park, its performance has been fairly strong on thin soils over dry limestone, the disturbed and chaotic conditions below the cliff and the calcareous loams of the colluvial fan, but weak on the drier and more acid soils, where it has persisted below birch and ash but struggled under oak, beech and small-leaved lime. Growth, self-coppicing and fruiting would naturally be weak under closed old-growth stands, but it sprouted vigorously in gaps and irregular canopies and must have relied on nuts produced in gaps to generate the new seedlings necessary to replace the old individuals that eventually lost their vigour in shade. Layering and regrowth from collapsed trunks must have been common, especially where they were crushed by falling canopy trees. Unrestrained grey squirrel populations must consume many nuts, but hazel still regularly produces a scatter of seedlings. Most are browsed by deer or swamped by bramble and other low growth, but a few survive. Formerly, it remained abundant in coppices because it was cut before it lost vigour; sprouted vigorously; produced abundant nuts that were not consumed by grey squirrels; seedlings stood a better chance of establishing themselves on the disturbed soils left after coppicing; and competing bramble would have been limited by dense coppice cut in rotation.

11 Minor Trees and Shrubs

These subordinate tree species form a mixed bunch. Several can grow into the canopy, but they never achieve the stature of oak and other canopy dominants. Mostly, they remain in the underwood, where they influence the growth of the canopy dominants by restricting space for regeneration. The evergreens – holly and yew – can dominate underwood, but they and the other species are generally found as a thin scatter. In Lady Park, the performance of each is distinctive enough to encourage the idea that tree species have their own personalities – that each contributes something unique to the woodland community.

11.1 Hawthorn

All the individuals in Lady Park are common hawthorn, *Crataegus monogyna*, a ubiquitous species, common in many kinds of woodland, which fruits abundantly, employs birds to distribute these fruits widely and thereby becomes one of the characteristic colonists of secondary scrub and a characteristic underwood in secondary woodlands on former arable. Populations thin out on base-poor soils, but hawthorn seems well suited to most other soil types. The quickset hedges with which many open fields and commons were enclosed, and the scattered, now elderly, bushes that colonised upland pastures during periods when grazing pressures were low, ensure that hawthorn seed can reach almost anywhere.

True to form, hawthorn was one of the species that regenerated in the stands felled in 1942–1944. By 1977, these regenerants were reduced to 141 individuals (Table 11.1) and 12 smaller saplings. Many grew on the lower plateau, colluvial fan and the lower steep slopes in the northern parts of the reserve. Most were small (median 9.5 cm GBH), but two at the base of transect VI, which had reached 50 cm, must have survived the 1942–1944 felling as small bushes. Losses thereafter were heavy, especially after 1993 when browsing became severe, though some survived as bitten-down stumps for a decade or more. A steady trickle of new stems was recruited: between 1993 and 2002, all were new stems on existing individuals. The growth of the survivors slowed sharply after 1993, but by 2015 the largest had grown to 77 cm GBH. The 2015 population remained wholly in the underwood and comprised 10 new stems recruited since 2002, 5 recruited in 1993–2002 and 43 that had survived from 1977, one of which had been recorded as dead in 2002, but regrew. Today, they survive in the underwood, where the smaller bushes had been vulnerable to deer browsing until the reserve was fenced. In the managed stand, where deer have access, they are widely scattered but understandably small, despite receiving more light.

In the old-growth (Table 11.2), hawthorns were well scattered in 1945. The size of the individuals implied that they had regenerated after the 1902 and *c.*1930 thinnings but had never grown well, though a few had reached the middle stratum. They continued to recruit after 1945, and again after the 1976 drought, albeit mostly as new stems on existing individuals. The population thinned out steadily, however, and by 2010 it had been reduced to just five individuals, three from 1945 and a two-stemmed sapling that entered after 1983. Curiously, the largest hawthorn remained roughly the same size (32–40 cm GBH) at each recording, but it was not the same individual: the largest one in 1945 died between 1983 and 1986, and the largest in 2010 was tiny (6.3 cm GBH) in 1977. By 2010, some were at least 65 years old.

Below the cliff (Table 11.3), the story was much the same, the main difference being that losses were slower and recruitment rates higher, no doubt a consequence of the more open canopy and lower intensity of browsing. The hawthorns of 1950 were

Fig. 11.1. One of three large yews on a limestone outcrop at the north end of the reserve, where they have remained unshaded. Yews have not grown this large in shade.

Table 11.1. Population change and growth of hawthorn in the young-growth.

	Initial density (N/ha)	Live at start	Losses	Gains	Mortality rate (%/year)	Recruitment rate (N/ha/year)	Radial growth rates (mm/year) Range (median)
1950–1977							0.19–3.00 (0.74)
1977–1983	69	141	−15	+4	1.86	0.32	0–2.29 (0.68)
1983–1993	63	130	−28	+8	2.52	0.41	
1993–2002	54	110	−43	+7	5.36	0.38	0–1.41 (0.18)
2002–2013/15	36	74	−27	+11	3.43	0.44	0–1.82 (0.23)
2013/15	28	58					

Table 11.2. Population changes in hawthorn in the old-growth.

	Initial density (N/ha)	Live at start	Losses	Gains	Mortality rate (%/year)	Recruitment rate (N/ha/year)	Largest stems (cm GBH)
1945–1955	32	29	−6	+7	2.29	0.78	35, lower and mid strata
1955–1977	33	30	−19	0	4.27	0	35, mostly lower stratum
1977–1992	12	11	−4	+5	3.18	0.40	34.6, all lower stratum
1992–2000	13	12	−6	+1	7.41	0.12	32, all lower stratum
2000–2010	8	7	−2	0	3.31	0	38
2010	6	5					40

Table 11.3. Changes in the population of hawthorn below the cliff.

	Initial density (N/ha)	Live at start	Losses	Gains	Mortality rate (%/year)	Recruitment rate (N/ha/year)
1950–1959	50	24	−6	+15	3.15	3.45
1959–1985	68	33	−26	+11	6.01	0.91
1985–1992	37	18	−4	+6	3.53	1.77
1992–2005	41	20	−10	+1	5.19	0.16
2005–2013	23	11	−6	+2	9.39	0.52
2013	14	7				

mostly small and scattered throughout the stand, with just one of 22 cm GBH that must have originated before 1940. Significant losses and gains occurred at all periods, mostly as new stems on old individuals. The large tree of 1950 perished between 1959 and 1985, but we do not know its fate. The larger surviving trees grew steadily, if slowly, from 26 cm GBH in 1985 to 36 cm GBH in 2013.

Within woodland, hawthorn grows slowly, regenerates sparsely and its spines afford only limited protection from deer, though it is resilient enough to generate shoots repeatedly from apparently dead stumps and will recruit new stems on existing individuals in the presence of deer. Its main advantage ought to be its ability to disperse far and seed like a pioneer into gaps, but there is little sign of it taking advantage of naturally formed gaps in Lady Park. Its key characteristic is persistence, just ticking over as a population of individuals that can grow strongly, flower and fruit if a gap opens above them.

11.2 Wild Service

This is an intriguing species. Known also as the chequer tree, the inspiration for several pub names, it is widely distributed through the woods of southern Britain, but is hardly ever abundant in any wood. It is strongly associated with ancient woods, where it is often confined to margins, a feature that has been interpreted as a slow colonising ability combined with little practical value and elimination by woodmen from the coppices. Most new trees are

Fig. 11.2. A cluster of wild service saplings concentrated on the lip of the gorge overlooking the Biblins campsite.

suckers, so the tree is usually found in loose clusters, but it does generate viable seed too. It coppices well enough and can, but rarely, achieve co-dominance in patches of underwood. Given the chance, it forms substantial, shapely trees, which just occasionally grow as large as standard oaks. There was once a scatter of service standards in the woods north-east of Oxford and GFP well remembers searching hard for the source of service tree leaves on the ground in Tortoiseshell Wood, Lincolnshire, before finding it disguised as a standard oak. It is principally associated with dry soils across the whole pH range, but these can include thin soils over outcrops, well-drained light soils and drought-prone, poorly drained clays.

In Lady Park, service is concentrated around the cliff, where it grows both on the limestone itself and the acid soils of the lower slopes and plateau just above the cliff (Fig. 11.2). It is also scattered across the upper steep limestone slopes and the thin soils over limestone outcrops in the managed stand. In both young growth and old-growth, its population turnover has been astonishingly low. Thus, in the young growth, only 3 of the 23 stems present in 1977 had been lost by 2002, while 2 had recruited, both before 1993. In the old-growth (Table 11.4), where the population has always been confined to the plateau above the cliff and adjacent parts of the mid-slopes, the 31 individuals present in 1945 were evenly distributed between the canopy, subcanopy and underwood, but they all quickly slipped back into the subcanopy and underwood and by 1977 several small individuals had died. Thereafter, save for the post-1976 drought period, a steady, slow mortality rate was balanced by an equally steady, slow recruitment rate until recently, when losses

Table 11.4. Changes in the population of wild service in the old-growth.

	Initial density (N/ha)	Live at start	Losses	Gains	Mortality rate (%/year)	Recruitment rate (N/ha/year)	Largest (cm GBH)	Median (cm GBH)
1945–1955	34	31	−1	+4	0.33	0.44	51	Small
1955–1977	38	34	−9	+2	1.33	0.09	51	Small
1977–1992	30	27	0	+4	0	0.32	94	22
1992–2000	34	31	−2	+3	0.74	0.37	107	16
2000–2010	36	32	−8	0	2.84	0	113	16
2010	27	24					119.5	26

Note: The largest and median trees at the start of each period are indicated.

accelerated and recruitment failed. The 2010 population comprised 18 trees that dated from before 1945, supplemented by 6 others recruited at all periods since and clusters of seedlings (Fig. 11.3) and small suckers.

Growth has, like the population, been remarkably slow and steady. The growth rates of the 18 survivors from 1977 to 2013/15 were from 0.02 to 2.05 mm/year (median 0.18): the largest in 1945 grew from 47 cm GBH to 50 cm GBH in 65 years. They were slowing down from 0.37 to 3.70 (median 1.11) from 1950 to 1977, and were particularly slow after 1993. In the old-growth, the growth rate from 1945 to 2010 was 0–1.89 (median 0.16). In effect, most of the population was static, adding negligible increment. The largest trees had reached 86 cm GBH (young growth) and 119 cm old-growth. There is one tree in the managed stand that has reached 161 cm and is shaped like a standard tree, so moderate growth rates can be sustained, but most of the population just ticks over, hardly growing at all over decades.

The performance of service in Lady Park is consistent with its conservative reputation. As a clonal species able to bear deep shade, it is well designed to hang on where it is, but there has been little sign that it can colonise at a distance. The established trees in the old-growth are surrounded by a scatter of seedlings and suckers that have grown slowly but steadily behind fences since 1993 in the safety exclosure and more widely since the reserve was fenced in 2007. There are also some small 'saplings' that show it can spread a few metres by seed.

11.3 Whitebeam

Whitebeams are either very easy to identify or very difficult. Their rounded, grey-green leaves with white, felty undersides make them distinctive, but they comprise several microspecies separated on their leaf shapes and fruit characteristics. Commonly associated with limestone outcrops and highly calcareous soils, some microspecies also thrive on strongly acid soils in the south-east lowlands, Inverness hinterland and oceanic districts in the south-west and the western Highlands. In the New Forest, whitebeams form scrub with holly, yew, hawthorn and rowan on strongly acid gravels. The unifying factor seems to be an ability to bear drought. Like the related wild service, it is capable of coppicing and of forming moderately large trees, but it tends to occur more on woodland edges and in scrub, which suggests that it is a better colonist.

The Lady Park whitebeams form part of a swarm of different microspecies and hybrids in the upper Wye Gorge (Rich *et al.*, 2010), but they will be treated here as one species. They conform to their calcicole character, being strongly associated with the upper steep slopes, the limestone outcrops of the managed stand and the cliff and the slopes below the cliff. Throughout, the population has proved to be moderately resilient. In the young growth, the 13 stems present on 4 individuals in 1977 all survived to 2015, growing slowly (RGR 0.39–0.95 mm/year, median 0.78) to a maximum of 53 cm GBH. In the old-growth, the 16 stems present in 1945 comprised three coppice stools and a scatter of small stems, located mainly in the canopy and subcanopy. By 1955, they had been joined by two others, but all had died by 1977, except five stems on one stool on the upper limestone and another on the upper steep slopes. The two smaller stems had died by 2010. The four survivors grew at 0.72–1.08 mm/year (median 0.90) to a maximum of 85.5 cm GBH.

Below the cliff (Table 11.5), where whitebeam is associated with the steeper ground and limestone outcrops on transects VII and IX, the constant instability provides it with more opportunities, but

Fig. 11.3. Young wild service tree, probably a seedling.

Table 11.5. Changes in the population of whitebeams below the cliff.

	Initial density (N/ha)	Live at start	Losses	Gains	Mortality rate (%/year)	Recruitment rate (N/ha/year)	GBH (cm) of original mature trees Range and median	
1950–1959	35	17	0	+4	0	0.92	20–61	40
1959–1985	43	21	−13	+2	3.79	0.16	27.3–62.9	44
1985–1992	21	10	−5	+24	9.43	7.10	33–69	61.5
1992–2005	60	29	−26	0	16.01	0	54–71	62.5
2005–2013	6	3	−1	0	4.94	0	71.5–77	74.5
2013	4	2					71–80	75

at the cost of too much excitement. In 1950, 17 stems were recorded on eight well-established trees and three saplings, so whitebeam must have been regenerating regularly before 1940. All survived to 1959, and four more saplings were recruited on a mid-slope outcrop. However, by 1985, the saplings and smaller individuals had died, and several of the largest trees had also gone. In late 1985, two large whitebeams were crushed when a large beech fell, but by 1992, over 20 new stems had grown up from the fallen trees and a single sapling had regenerated from seed in the gap. By 2005, all that regrowth had died, together with one earlier sapling and one larger tree, which left just two large trees dating from 1950 and a trunk shoot of 10.5 cm. The mortality rate was largely determined by the special events that took place at the base of transect IX. The slow growth and hazardous existence of whitebeams on the slopes can be illustrated by the growth of the three largest trees present in 1992. The strongest grew steadily from 49.5 cm in 1950 to 80 cm in 2013, despite a severe lean and no clear leader. Another grew slowly from 61 cm in 1950 to 71.5 cm in 2005, by which time it had been uprooted by other falling trees and survived, hardly growing, only by leaning on a neighbour. The weakest grew from 40 cm in 1950 to 56 cm in 1985, did not grow at all up to 1992, and then died, excluded by shade.

Whitebeams evidently lead a marginal and hazardous existence. They have shown only a limited ability to bear shade. Rather, they sustain themselves in the canopy, where growth is limited on thin soils over limestone and where the canopy is irregular and open round the cliff. Growth of strong trees is slow and steady, but they compete only moderately in young stands and fall behind within a few decades. Survival in closed high forest thus depends on finding a niche where canopy growth is poor or broken – outcrops, thin soils and cliffs. On steep slopes, they are clearly liable to being crushed by larger trees. On the evidence of transect VII, they do not establish on stable scree below a closed canopy. Nor, on the evidence of the 1942–1944 felled areas, do they colonise gaps easily. Despite the deer fence, they have not recruited since 2007.

11.4 Crab Apple

Widespread in Britain outside the Highlands, crab apple forms small underwood trees and just occasionally stretches into a high woodland canopy. Within woodland, it is a loner, invariably found as scattered individuals. The carpets of fallen crabs below mature trees never seem to generate new individuals, but, at a distance, and presumably after passing through the gut of a blackbird, seeds must germinate and occasionally become established. It also grows in hedges, scrub and wood pastures, where, again, it is normally found as scattered individuals. Hedge trees produce much heavier crops of small apples.

There is little to be learned from its presence in Lady Park, where it is scattered very thinly. Single, small individuals were recorded from old-growth and below the cliff, but they did not last long. The trees in the young growth, including the four on the transects, were widely scattered in different zones. The largest is on the lower steep slopes near transect VI. Growth was slow at 0.02–1.39 mm/year (median 0.46) from 1977 to 2013/15: the largest recorded tree grew from 31 cm GBH in 1977 to 44 cm GBH by 2015.

11.5 Yew

Yew is the only native conifer in English and Welsh woods and, with holly and box, one of the few evergreens in temperate woods. Famous for living to great ages – probably several thousand years in unshaded conditions in churchyards – it forms dense woodland on chalk uplands at Kingley Vale (West Sussex) and Hambledon Hill (Dorset), where stands develop slowly and may well also live to great age. It is characteristic of cliffs, such as Yew Cogar Scar in upper Littondale (Yorkshire). Like whitebeam, it is usually associated with lime-rich sites, but also colonises acid heaths, the New Forest for example, along with holly, rowan and hawthorn. Within woods, it forms dense underwoods in high forest, principally on chalk scarps, Welsh borderland limestones and Magnesian Limestone at Castle Eden Dene. In shade, it generally remains small and is probably not long-lived. Despite a reputation for poisoning stock, it is grazed by deer.

In the Wye Valley, yew is found on cliffs with whitebeam and beech, and as an underwood (Pierce Wood). In both locations, it has been abundant enough to incite the Ordnance Survey to map the woodland as coniferous. In Lady Park, it is almost confined to the limestone cliffs and outcrops, extending sparingly on to steep limestone slopes (Fig. 11.4). Even in such locations, it behaves more as an underwood tree in the sense that it never becomes large. The shade of trees growing up from below and out from above may be one limiting factor, the other being the precarious rootholds, from which it regularly falls off completely or

Fig. 11.4. This view from a recent tree-fall gap near the top of the cliff shows yew concentrated on the limestone crags. On the opposite bank of the Wye, the former meadows have been transformed into the Biblins campsite.

leans down, hanging on by roots that become the highest point of the tree. Within Lady Park, just to the north of transect VI, however, there is a trio of large yews with secure rootholds growing almost unshaded on the cliff, which have grown to 297, 167 and 338 cm GBH (2015), respectively in order from the north (Fig. 11.1: see p. 178). There is also a curious even-aged grove of yews located within what may be a long-abandoned enclosure (see Chapter 2, this volume).

In the young growth, it failed completely to regenerate in the 1942–1944 felled areas, even on dry alkaline soils, but one large bush was spared the axe. It fell in 1983, but lives on, growing steadily and generating a new upright tree from the trunk. In the old-growth, the yews have always been confined to limestone crags at the top of transect V and the base of IV and V. However, they were only recorded irregularly, so no population changes can be calculated. The most complete recording was in 2000, when 24 live and 6 dead stems ranged from 6 to 132 cm GBH (median 53.5): judging by size and growth rates, most date back to well before 1945. Incomplete information notwithstanding, population turnover is clearly low. At least seven trees survived from 1945 to 2010, and the five that could be measured safely in 2010 grew at 0.26–1.21 mm/year (median 0.42). Even the dead yews hang around: one that died before 1977 was still present as a snag in 2010.

Below the cliff (Table 11.6), yews grow mainly in the steeper transects, IX and VIII, where they root into outcrops and shallow soils. Unlike the yews above the cliff, there was a constant turnover of population – literally, since many grew on crags and fell off. Rates of loss and recruitment were both slow and steady until 2005, save for the increase in recruitment after DED, but like most other species in the stand, mortality then increased and recruitment ceased. The dead included trees of all sizes, but causes of death differed across size range: small yews were sometimes browsed, but large ones often fell off their perches, regenerating only if enough roots hung on. Deer were mainly responsible for the recent lack of regeneration: in 2005, the lower trunks and exposed roots of 12 yews of all sizes were severely debarked. Size distribution in 1950 (Table 11.7) indicates that many yews survive from before 1940, and perhaps 14 survive from before 1900, judging by growth rates.

Growth rates have been surprisingly high in the managed stand and on and below the cliff. Three trees on the upper cliff in the managed stand grew at 0.69–2.39 mm/year from 1984 to 2014. Below the cliff, growth was steady and, in the case of the largest yew, even faster than those in the managed stand. Thus, the 16 trees for which information is available grew at 0–2.58 mm/year (median 0.84) between 1950 and 1985 and 0–2.33 mm/year (0.81) between 1985 and 2013, and the faster growers before 1985 remained the faster growers after 1985. Near the top of the cliff in the old-growth, the two yews were slower at 0.26 and 0.98 mm/year from 1945 to 2010. The one large tree in the young growth growing on the upper steep slopes of transect IV was reasonably fast, even though it toppled over between 1977 and 1983. Still thriving in a gap, it measured 113 cm in 1977 and by 2013 the main trunk had grown to 142 cm (RGR 1.30 mm/year), and three new shoots had developed from the trunk, all somewhat gnawed, seemingly by both deer and squirrels.

Like service, yew is inherently a conservative tree, capable of long, slow, steady growth, but its predilection for crags and shallow soils on steep slopes turns it into an unstable species, which repeatedly falls. Provided some of its roots cling on, yew generates new shoots, growing into the gaps formerly occupied by its previous crown, yet its old crown generally survives in the shade until what was once a single yew becomes an extremely straggly collection of trunks growing at all angles from vertical to pendant. Despite this literal turnover, the population below the cliff has remained effectively constant: its median size has hardly shifted, but its size range has become much wider. Growth rates can be high, but only in the open, and it is only there that it seems capable of growing to great ages and sizes. In natural woodland, it may be divided between two populations. One, on cliffs and outcrops, can grow to a large size and age with a generally low turnover, but it can grow fast. Elsewhere, it suffers from shade and browsing, and appears unable to grow fast or large, or to live long.

11.6 Holly

Holly is sensitive to both frost and waterlogged soil, so it tends to be more abundant in the west, near the coast and on well-drained soils, mostly base-poor sands and loams, but also limestone rubble. Like whitebeam, yew, sessile oak and to some degree beech, holly has a bimodal distribution in relation to soil pH, but it is closest to sessile oak in being biased to acid soils. Despite its prickles, it is readily browsed

Table 11.6. Changes in the population of yew below the cliff.

	Initial density (N/ha)	Live at start	Losses	Gains	Mortality rate (%/year)	Recruitment rate (N/ha/year)	Largest (cm GBH)	Median (cm GBH)
1950–1959	89	43	−2	+4	0.53	0.92	109	37.5
1959–1985	93	45	−10	+3	0.96	0.24	122	40.6
1985–1992	79	38	−2	+10	0.77	2.96	151	48
1992–2005	95	46	−4	+4	0.70	0.64	158	46.5
2005–2013	95	46	−15	0	4.81	0	167	46.5
2013	64	31					175	69

Note: Population counts include fallen trees still alive, which declined from six in 1950 to two in 2013. The largest and median trees at the start of each period are indicated.

Table 11.7. Girth (cm) distribution of yew below the cliff in 1950 and 2005.

	0	10	20	30	40	50	60	70	80	90	100	110	120	130	140	150	160
1950	5		8	11	5	6	3	2	1	1	1						
2005	9	4	4	3	3	6	3	2	5	1	1	1	2	1	0	0	1

Note: Population counts include fallen trees still alive, which declined from six in 1950 to two in 2013.

and was traditionally pollarded for late-winter fodder in hollins. Being evergreen, it grows well in shade, so in many respects its role is close to that of yew, but growth is much stronger in the open. In exceptional circumstances, it will grow into large trees, even overtopping pedunculate oaks in Staverton Park (Suffolk). Its bright red, bird-distributed berries allow it to develop dense thickets in the underwoods of some wood pastures, apparently as a response to release from browsing, this development being helped by its ability to survive as a dense, knee-high, thick hedge in moderately browsed woods. Heavy browsing will, however, reverse this, for example in Denny Wood (New Forest) (Mountford *et al.*, 1999), where a dense holly underwood has been almost eliminated by a combination of browsing and bark gnawing by deer and ponies. Given the chance to grow to maturity in woodlands without being browsed, its long, trailing branches will root into deep litter and generate clonal clusters of new trees. It is uncommon in coppices, even though it will sprout vigorously when coppiced, and this must be because other species are more useful and grow faster, but it seems to be gaining ground in the shade now that coppices are neglected. Outside woods, it forms dense scrubs in the New Forest and is common in hedges in the south and west, where it was often preserved as winter browse.

In Lady Park, it has never been abundant, but it is scattered through the whole wood. Its performance in the old-growth (Table 11.8) has been chequered. The few hollies of 1945 mostly originated as coppice, presumably cut when the stand was thinned. Many were recruited in 1945–1955, but these were largely eliminated in 1955–1977, mostly by deer: low-growing, well-browsed old holly bushes were quite widespread in later records. This coming and going was then repeated, steady recruitment in 1977–2000 being reversed by substantial mortality in 2000–2010. After 1992, deer browsed and debarked all accessible hollies, inhibited regeneration and generated a browse line. The residue population of 2010 comprised a four-stemmed stool from 1945 and small individuals on rocky ground dating from all periods after 1983. The largest tree throughout was one of the original 1945 trees, growing very slowly indeed (RGR 0.07 mm/year). There was also a small population on the lip of the cliff that was relatively free of browsing: it was excluded from the population count because it was not recorded consistently.

Holly was not one of the regenerants after the 1942–1944 fellings, but there were still 17 small, widely scattered stems in the young growth by 1977 (Table 11.9), presumably brought in by birds.

Table 11.8. Changes in the population of holly in the old-growth.

	Initial density (N/ha)	Live at start	Losses	Gains	Mortality rate (%/year)	Recruitment rate (N/ha/year)	Largest (cm GBH)	Median (cm GBH)
1945–1955	7	6	−1	+20	1.81	2.22	63.5	20.4
1955–1977	28	25	−21	0	7.66	0	63.5	Small
1977–1992	4	4	0	+7	0	0.56	64	8
1992–2000	12	11	−1	+9	1.05	1.11	65	5
2000–2010	21	19	−11	+4	8.29	0.44	65	6
2010	14	12					66.5	10

Note: The largest and median trees at the start of each period are indicated.

Table 11.9. Changes in the population of holly in the young growth.

	Initial density (N/ha)	Live at start	Losses	Gains	Mortality rate (%/year)	Recruitment rate (N/ha/year)	Radial growth rates (mm/year) Range (median)
1950–1977							0.56–1.48 (1.21)
1977–1983	8	17	−5	0	5.64	0	0.60–1.45 (1.13)
1983–1993	6	12	0	+3	0	0.15	
1993–2002	7	15	−7	+1	6.75	0.05	0–0.71 (0.49)
2002–2013/15	4	9	−3	+27	3.46	1.15	0.51–1.21 (0.64)
2013/15	16	33					

Fig. 11.5. An old holly rejuvenated. Its two trunks were so severely debarked by deer that they rotted and died, and eventually fell. Base sprouts, which had previously been bitten to the stump, can now grow strongly.

By 1983, four had been knocked down by falling canopy trees: they survived for a decade, but were eventually gnawed to death by deer. Others were ringbarked by deer, their regrowth browsed, and eventually trunks rotted enough for the trees to collapse. Just six of the original trees remained in 2015, their growth rate much slower than it had been before 1983. After 2007, however, the exclusion of deer allowed these and other old trees to regenerate prolifically from the base of the trunks (Fig. 11.5).

Until 2005, the hollies below the cliff (Table 11.10) increased rather more than those above, mostly as additional shoots on established trees. They have only been recorded on the steeper slopes (transects VIII and IX), where the seven individuals in 1950 included one of 28 cm GBH that must have pre-dated 1940. Steady gains thereafter as new individuals grew on outcrops and other ground were partially negated by losses in 1959–1985 and 1992–2005, some of which survived as flattened individuals. One of the flattened hollies was the large individual of 1950, which fell downslope between 1959 and 1985 and survived almost to 2005, even though it was shaded by yew: it was eventually browsed out by deer and its roots became detached from the soil. The peak 2005 population comprised mostly small (largest 32 cm) individuals, plus several new, short saplings and a few flattened older individuals. Deer browsed and debarked most accessible stems. Growth was very slow: the 17 trees that survived from 1992 to 2005 grew at 0–1.35 mm/year (median 0.24). Like most other species, mortality increased steeply after 2005, while recruitment lapsed.

Despite their prickly, slow-growing, unchanging image, holly populations in woods change remarkably quickly in response to changes in deer populations. Given no more than moderate browsing, they establish themselves in shade and can respond quickly if given more light, but they are vulnerable to a surge in browsing and falling trees and branches. Individuals persist for a decade even under the severest pressure, because: sprouts regrow even when repeatedly bitten tight; it does not give up when it is flattened below a fallen branch; and basal sprouts grow vigorously when given the chance.

11.7 Shrubs

Several shrubs species were recorded in the regeneration after the 1942–1944 fellings, and the larger stems were mapped and measured in the transects. They were never a major component of the developing stands, nor did they ever monopolise the underwood. By far the commonest was dogwood, which developed a substantial population in the young growth in the northern half of Lady Park. By 1977, 158 individuals were recorded, of which 122 remained both alive and standing (Table 11.11). Clearly, they were dying off fast, and this high mortality rate was maintained unabated until just one remained in 2015, by then probably 65 years old. Deer had debarked at least two-thirds of those that died between 1993 and 2002. A few regenerated when the stand was opened by the drought, mostly layers from a leaning bush, but they perished within a decade or so. Dogwood was also scattered through the stand below the cliff, but the 19 stems recorded on the transects by 1959 had all gone by 1985, and a single recruit in 1992 had gone by 2005. By far the largest stem, at 46 cm GBH, grows in the managed stand, where it was large enough to be confused with a maple.

Small blackthorn populations have survived on the upper plateau, upper steep slopes and the moister parts of the lower slopes in the young growth (Table 11.12). They grow as clones, constantly generating new stems, which persist only under light shade and where deer allow. Mortality rates are high, but three stems survived from 1977 to 2015, having grown at 0.34–1.03 mm/year to reach 17–34 cm GBH.

Table 11.10. Changes in the population of holly below the cliff.

	Initial density (N/ha)	Live at start	Losses	Gains	Mortality rate (%/year)	Recruitment rate (N/ha/year)
1950–1959	14	7	0	+7	0	1.61
1959–1985	29	14	−9	+16	4.03	1.32
1985–1992	43	21	0	+6	0	1.77
1992–2005	56	27	−2	+10	0.59	1.59
2005–2013	72	35	−31	+1	23.75	0.08
2013	10	5				

Table 11.11. Dogwood population change in the young growth.

	Initial density (N/ha)	Live at start	Losses	Gains	Mortality rate (%/year)	Recruitment rate (N/ha/year)	Girth (cm GBH) at start of period Range (median)
1977–1983	60	122	−68	0	12.70	0	3–21 (12)
1983–1993	26	54	−33	+10	9.46	0.51	
1993–2002	15	31	−24	0	15.24	0	Survivors, 8–29 (19); recruits 1–9 (3)
2002–2013/15	3	7	−6	0	15.57	0	
2013/15	<1	1					23

Table 11.12. Blackthorn population change in the young-growth.

	Initial density (N/ha)	Live at start	Losses	Gains	Mortality rate (%/year)	Recruitment rate (N/ha/year)
1977–1983	5	10	–2	0	3.65	0
1983–1993	4	8	–3	+16	4.83	0.82
1993–2002	10	21	–12	+2	8.98	0.11
2002–2013/15	5	11	–7	+7	8.42	0.30
2013/15	5	11				

Spindle has been recorded mostly on the exposed limestone of the upper plateau, upper steep slopes and below the cliff, but it is scattered elsewhere. Ten stems, all small, were recorded in the young growth in 1977. Three lived to 2002 and one survived in 2015 (38 years old). Of the two saplings and three sprouts recruited between 1986 and 2002, one remained alive in 2015. The largest grew to 30 cm GBH in 1993. Three small plants were recorded in the old-growth in 1945, but they died before 1955, or shortly after, and were later replaced by two browsed saplings. The two saplings recorded below the cliff in 1959 had gone by 1985.

Nine rose plants were mapped in the young growth in 1977, of which two *Rosa canina* were still there producing new canes in 2015, at least 38 years old. One comprised eight small, live stems growing from a stock 20 cm in diameter on which the stumps of larger stems persisted. The other was the complete opposite, a large forking stem that had grown up with the canopy of the adjacent hazel, reaching 6.2 cm GBH in 1977, 9.5 cm in 2002 and 13 cm in 2015 (Fig. 11.6). Both produced new canes in 2015. *Rosa arvensis* is rare, but may be able to survive just as long: one plant on the edge of an exclosure has thrived since at least 1988.

Other shrubs were rare or fleeting. Many brooms regenerated after the 1942 fellings, but had mostly gone by 1958, though new plants still spring up occasionally in gaps. Guelder rose regenerated on transect I after the 1942–1944 felling, but had gone by 1977. Elsewhere, it was recorded on transects VIII and IX, but the seven saplings in 1959 had all gone by 1985 and the five recruits of 1985–1992 had gone by 2013. A few short-lived wayfaring tree bushes were recorded on alkaline sites at the top of transect VI and the base of transect IX. Five elder were recorded below the cliff: a single bush from 1950 to 1959 had disappeared by 1977; and four recruits in 1985 that grew until 1992 but had more or less died off by 2005. Privet was occasionally recorded, but did not last long. Two rowan seedlings were recorded below the cliff in 1959. A larch in the young growth has been allowed to grow from 57 cm GBH in 1977 to 112 cm in 2013. There was also a sycamore in the old-growth, but we felled it.

11.8 Small Trees and Shrubs Generally

Small woodland trees must be able to bear shade, but their performance under shade is much less impressive than in full light. Yew is the obvious case in point. It forms a dense understorey in several woodland types, but grows slowly, generally remains small, and rarely, if ever, achieves great age; but outside woodland – notably in churchyards – it can reach great ages and immense girths. To take the most notable example, the Fortingall yew is several thousand years old and achieved a girth of nearly 16 m in its heyday. Large yews can be found within woodland, but they grow in long-standing openings (such as the largest yews in Lady Park) or they were once, like the King Yew at Tidenham Chase (Gloucestershire), in open woodland that has since become closed.

Similar contrasts are shown by other small trees. If we take the girths of the largest specimens known to Elwes and Henry (1906–13): crab apple can reach 213 cm; service 246 cm; whitebeam 251 cm; the Hethel (Norfolk) hawthorn reached a scarcely credible 452 cm in 1895 and is still protected as a nature reserve; and holly reached 457 cm on Innisfallen Island, Killarney. As experience at Lady Park shows, they all remain small in woodland, except in the case of service, which can grow to the size of standard oaks, and holly, where it forms a major component of wood pasture, whether it is allowed to grow to full height (Staverton Park) or survives for over 200 years as a pollard (New Forest). Rowan, which is surprisingly rare in Lady Park, likewise grows large in the open.

Fig. 11.6. This dog rose growing on the mid-slopes of transect V is at least 38 years old and is still vigorous. The main cane, which ascends into the canopy of the adjacent hazel, grew from 6.2 cm GBH in 1977 to 13 cm in 2015.

Collectively, small trees and shrubs come into their own on the margins, which in natural woodland means the cliffs and any other natural openings. In Lady Park, yew and whitebeam are prominent on the cliffs and outcrops, often with roses. In the body of the wood, one would expect them to regenerate in gaps and be more prominent in stands with a high gap creation rate (i.e. the stand below the cliff), but experience so far hardly supports this, possibly because deer have masked regeneration patterns. Rather, they survive by sheer persistence, as clones (dogwood, blackthorn, service) and individual trees, by repeated sprouting (holly) and a marked ability to bear shade (yew, holly, hawthorn and service).

12 Habitats

Woods contain a host of habitats that enable particular species to thrive within their bounds. Within Lady Park, we have, for example, tracks, rock faces and tufa-forming seepages, but not streams, save for the not inconsiderable bulk of the Wye at its foot. This chapter gives more detail on the habitats most directly related to stand dynamics, dead wood and open spaces.

12.1 Dead Wood

If there was one feature that foresters sought to minimise, it was dead wood. Managers who grow utilisable timber see dead wood as waste, timber defects, a source of disease and, in some forest types, an invitation to fire. And yet, in natural forests, dead wood is not just inevitable but an important habitat, and a state through which the nutrients are recycled into new growth. Moreover, just as bald heads and gammy legs are part of the human condition, so dead branches and hollow trunks form a natural stage in the life of trees.

In Lady Park Wood, dead wood takes many forms. Stumps of trees felled in 1942 are still prominent (see Fig. 7.19). Coppice stools remain, some completely dead, most the foundation of living trees (Fig. 12.1: see p. 192). Fallen wood of all sizes, from twigs to the trunks of large trees, litters the ground (see Fig. 9.7); some remains detached but caught up in the canopy ('widow-makers'); and dead trees of all sizes stand as snags. Living trees have dead twigs and branches. Some of the larger trees have rotting trunks, rot holes at major branches, and rotting crown branches debarked by grey squirrels. Out of sight, below ground, there are dead roots, not just of dead trees, but also on living trees.

We estimated dead wood volumes in 1995 (Green and Peterken, 1997), both within the fenced reserve and in various kinds of managed stand elsewhere in the lower Wye Valley. In each stand sampled, we placed ten short transects randomly. Within each, we measured the length of the line that intersected with dead wood on the ground, and estimated the girth and length of dead branches in trees rooted within 5 m of the transect. The former yields an estimate of the fallen wood volume (Kirby *et al.*, 1998), while the latter yields a rough estimate of the volume of dead wood elsewhere. We also knew the girth of standing dead trees from which a third component of the overall volume could be estimated.

The results from Lady Park showed how the dead wood built up as the years passed (Table 12.1). The volume in the pre-1870 coppice would have been well below 10 m^3/ha, but the young growth must have contained larger volumes after 1944, if only because stacks of small wood were left that were still detectable into the 1980s. Volumes in all stands must have increased substantially in the 1970s following DED and the drought. The much greater volume in the old-growth was principally due to the snags (Fig. 12.2) and fallen logs (Fig. 12.3) generated from canopy beech. On a long timescale, and in so far as the young and old-growth can be regarded as a chronosequence, we see that volumes increase with age because larger trees are involved.

No further estimates of total dead wood have been made, but changes in the basal area of larger snags (Table 12.2) show that dead wood volumes have not been constant. While snags in the old-growth have increased and plateaued since 1977, those in the young growth have declined since their post-drought peak, and much the same must have happened below the cliff since the early 1970s. Since 1995, extensive tree falls have increased the volume of fallen timber in the old-growth, but much of this is still alive. The pulse of dead wood that entered

Fig. 12.1. Beech stub on the upper plateau, draped by an old clematis. A relic of pre-1870 coppicing, this picturesque structure provides dead wood in a living tree, which will stand until the remaining shell can no longer support the increasingly heavy post-1942 growth.

Table 12.1. Dead wood volume and distribution in Lady Park Wood and nearby stands.

	Volume (m^3/ha)	Main components of dead wood
Young-growth at 50 years	28–45	Fallen wood, snags, dead stools
Old-growth at 125 years	47–129	Fallen wood, snags, dead branches on living trees
Semi-natural managed woodland	18–31	Fallen wood, stumps
Continuous cover, mixed broadleaves	31–39	Fallen wood

with DED and drought had largely passed through the system 30 years later. The other general feature apparent in Lady Park is the increasing small-scale *range* of volumes with time. Whereas the dead wood in the young growth, arising mainly from competitive thinning, was spread fairly evenly, the dead wood in the old-growth was patchy, due to localised tree falls and post-drought group mortality of beech.

How does this compare with the most natural stands containing beech that are available elsewhere? Estimates from 86 minimum-intervention forest reserves in Europe (Christensen *et al.*, 2005) found greater volumes in montane forests (62–552 m^3/ha, median 161) than in submontane (10–285 m^3/ha, median 72) and lowland (43–220 m^3/ha, median 63) forests, due to the greater conifer component in montane reserves and the greater length of time that had elapsed since timber had been extracted. In almost all cases, the volume of fallen logs was much larger than the volume of snags. Estimates from six British woods that lay in the path of the great storms of 1987 and 1990 were much higher at 113–486 m^3/ha (median 269). Such values could easily be exceeded in mainland Europe if woods containing beech were levelled like Noar Hill Hanger (Hampshire), for their live volume reached as high as 876 m^3/ha. Against this background, the average volume of 81 m^3/ha from Lady Park was low: when it was measured in 1992, 65% had fallen, whereas in the early 1980s most took the form of snags.

In the La Tillaie reserve (Fontainebleau, France), where no felling has taken place for centuries, two sample plots had 256 and 142 m^3/ha in 2000, compared with 92 and 145, respectively, in 1982 (Mountford, 2002). Most of this volume was derived from trees felled or broken in storms. Here, and elsewhere, fallen trees may obstruct progress in places, but in general visitors do not have to clamber over piles of rotting timber to move around the reserve. Dead wood is unevenly distributed, concentrated in canopy gaps and towards the base of slopes. It also enters in pulses; storms in the case of La Tilliae, drought in the case of Lady Park.

Managed woodland also contains dead wood, but much less than in near-natural old-growth. Two stands in Highmeadow Woods on much the same ground as Lady Park, from which some timber had been extracted, contained slightly less dead wood than the young growth in the reserve (see Table 12.1). Elsewhere in the lower Wye Valley, volumes were much lower in both coppice stands and conifer plantations, except where marginal blowdowns had been too small to warrant salvaging: these values are comparable with those found by Kirby *et al.* (1998) in British managed woodland. Traditionally, the dead wood was concentrated in the wood pastures, where it stood in the decay columns of old pollards and other veteran trees. In long-unmanaged stands, an exceptional 201 m^3/ha was found in Denny Wood (New Forest) (Mountford *et al.*, 1999), and the 1987 blowdown of beech–oak wood pasture at Toys Hill (Kent) generated over 400 m^3/ha.

Fallen trees may be orientated in all directions, but frequently they are biased to certain directions, either downslope due to gravity (Fig. 12.4) or downwind of the storm that felled them: examples from the 1987 storm are given in Peterken (1996). Snags and leaning trees are present in most old-growth. If a stand is completely levelled in a storm, the dead wood volumes will temporarily be very much larger, as at Toys Hill, but decades later there will be a complementary dip as the new growth generates only small amounts of new dead wood each year. If a stand is consumed by fire, dead wood is also consumed, leaving a relatively low volume as snags of larger trees.

Interpreted historically, dead wood volumes were clearly very low in ancient woods managed as coppices. Every stick could be used as firewood or brushwood for reinforcing roads, even where no specialised market existed. Even the roots of felled trees were dug out for firewood. In the medieval landscape, dead wood was concentrated in wood pastures and boundary trees (some of which were on the boundaries of coppices). Dead wood volumes would have crept up as coppices became neglected,

Fig. 12.2. A victim of the 1976 drought, this beech snag is in the last stages of decay 30 years later.

Fig. 12.3. Fallen beech, now a substrate for porcelain fungus.

Table 12.2. Basal area of larger snags,[a] 1977–2010.

	Old-growth	Young-growth	Stand below the cliff
1977	0.75	3.57	
1983/5	1.65		1.61
1992/3	2.01		0.83
2000/2002	2.09	2.72	
2005			0.97
2010	2.07		

Note: [a]Snags attaining 5 cm DBH or 15 cm GBH.

but the plantations with which they were replaced were equally bereft. Latterly, however, lop and top has been left in piles after felling, and 'habitat piles' have been left deliberately, but they generally lack large pieces of dead wood and the volumes generated are still no more than a quarter of those in mature, natural stands.

12.2 Open Spaces in General

Woods appear on maps as wall-to-wall trees, but most have open spaces where trees do not shade the herbaceous vegetation. They are usually small in relation to the tree-covered ground, but they are crucial for habitat and species diversity, for they provide interfaces between woodland and non-woodland which are shielded from the farming environment that influences woodland margins. Depending on the condition of the adjacent stand and how the opening is managed, these margins may be abrupt walls between woodland and grassland or a stepped transition through scrub. The habitats they provide are distinct enough for German ecologists to give them a separate name – the Zaum – where there is semi-shade, light traffic and tall herbage. They are the richest habitats for invertebrates, for they provide not only a range of aspects from

Fig. 12.4. Gaps are inevitably places where dead wood is concentrated. This large gap formed in the winter of 2014/15 when an oak on the lip of the cliff lost its footing and brought down a stand of lime, beech and yew as it fell.

cold north-facing to hot south-facing but also the mixture of shelter and nectar sources that some species need.

Open spaces in woodlands and plantations come in several forms, each with its distinctive characteristics. The main distinction is between permanent and temporary openings. Permanent openings comprise both artificial habitats created to facilitate woodland management and natural habitats within the woodland. The commonest artificial habitats are the forest roads and rides required for access, which make up 5–10% of the ground in larger woods and plantations, and these are supplemented by loadings bays, sheds and other means of handling timber. Less common, but widespread, are strips kept open below power lines and even grassy openings over pipelines. The natural habitats include mires, watercourses, pools, lakes, cliffs and landslides. All these habitats will retain or develop distinctive 'non-woodland' vegetation, often with a scatter of trees and scrub. In addition, there are margins that admit side light and nutrients drifting from the surroundings, and whose effects may be 50 m wide.

Rides are a form of semi-natural grassland that is also shielded from the nutrient enrichment and drainage associated with farmland. Many are lightly grazed by deer, and most will be kept in trim by mowing and light traffic. In effect, they constitute meadow-like grassland, where tall herbs get a chance to flower and maintain themselves in the sward. Indeed, in arable lowland parishes, they, the churchyard and road verges will often be the only places where semi-natural grassland survives, and the least eutrophicated of these are woodland rides. Depending on the topography and soil type, rides can support a form of heathland and, more often, patches of marsh.

Forest roads add open, gravelly surfaces that provide sunspots for invertebrates. Woods, in short, do double duty as grassland reserves and more.

Temporary openings are mostly coupes where trees have been felled and gaps created by the death and/or fall of canopy trees. Such openings last for up to a decade and supposedly fill with a distinct group of plants known as gap-phase species, mostly species that have excellent colonising ability and/or a capacity to survive for decades as dormant seed, buried in the soil. There is little chance of a heath or grassland community developing, except in secondary woodland planted on to heathland or grassland, respectively (whose species may survive on site as dormant seed or nearby in permanent openings).

Some measure of the influence of open spaces on the woodland flora came from central Lincolnshire (Peterken and Francis, 1999). The open-space species (defined as those that would probably not survive under a closed woodland canopy) increased with the total area of open spaces, which in turn increased with the area of the wood. However, the proportion of the total woodland flora increased in steps according to wood area. Below about 3 ha, open-space species made up no more than 30% of the total flora; above 30 ha, they made up 50% or more; and between 3 ha and 30 ha, they made up an increasing proportion. The explanation seemed fairly straightforward. Below 3 ha, most woods were too small for rides; above 30 ha, they were too large to be neglected, so the rides were kept open for and by forestry operations and for shooting pheasants. Between 3 ha and 30 ha, individual woods were increasingly likely to include rides and be 'actively managed'.

12.3 Open Spaces in Lady Park Wood

Now, and in the past, several kinds of temporary opening have been associated with Lady Park and its surroundings. Felled areas were regularly generated in coppice, and also presumably in the preceding wood pasture, but now they no longer occur within the fenced reserve. Canopy gaps develop naturally within the fenced reserve and formerly (one supposes) in coppice and wood pasture after storms. Stands of all kinds have margins that allow side light to penetrate. Today, the upper margin of the reserve lies against a woodland road, and the lower margin, once bordered by meadow and then the railway, is now open woodland with a cycle track.

Likewise, there are and have been several forms of permanent opening. The main tracks were formerly part of the local road system. Up to 3 m wide, they would have been shaded intermittently by marginal coppice growth, but probably maintained narrow, tall-herb verges. Today, they are shaded. Charcoal hearths were in use until some time in the 19th century. They must have fluctuated between open but scorched patches and fairly shaded clearings, but, like the tracks, they are now all shaded. Cliff faces remained open while the woods were managed (Fig. 12.5), but most were too small to remain unshaded when the trees rooted below were allowed to grow to full height. They form the centrepiece of a range of specialised habitats, from dry outcrops around the lip to stabilised screes below. The main cliff in Lady Park is just tall enough to remain unshaded by trees, but is so covered in great ivy that very little limestone is exposed. The rock towers of the Seven Sisters opposite Lady Park have fine natural rockeries with a specialised light-demanding flora that survives on the summits because trees are reduced to scrub by droughts. The top of the main cliff in Lady Park approaches the conditions of rock towers, but is now almost wholly shaded. Finally, bracken glades can remain open in both managed and natural woodland, but those in Lady Park are too small.

The suite of openings has changed considerably over the centuries. Under wood pasture, one assumes that the Hadnock Wood was a patchy mosaic of trees, scrub and open ground, bordered by a meadow between the wood and the Wye. Under coppicing, permanent openings would have been largely replaced by a regular supply of temporary openings: at a rotation of 20 years, 5% of the wood would always have been a newly created gap in addition to perhaps 5% of the wood as rides and other permanent openings. When coppicing was replaced by high forest management, the permanent openings would have been more shaded and the supply of temporary openings would have diminished. In 1942–1944, nearly 60% of what is now the reserve was felled, but once the canopy of the new growth had closed in the 1950s, permanent openings disappeared, except near the cliff face, and few, if any, gaps appeared above the cliff until both old- and young-growth stands were opened by drought. Since then, gap creation has been slow but steady, but may now be accelerating as the rate of tree falls increases. The long-term gap creation rate in natural woodland would probably be about

Fig. 12.5. The main cliff face. Far taller than the trees growing at its foot, one might expect it to be a refuge for light-demanding herbs, but it is actually covered in old ivy.

1% of the area per annum (Runkle, 1982), but it was briefly much higher in Lady Park's upslope old-growth after the drought.

12.4 Cliffs as Open Spaces

Natural cliffs within woodland are largely confined to the Highlands, western uplands and borderlands, and even there most are not high enough to escape shading by mature trees. Lady Park, like the rest of the lower Wye, is thus exceptional in containing spectacular tall cliffs whose tops remain open permanently, as well as smaller cliffs and outcrops that generally remain shaded. Collectively, they add exposed rock faces, thin soils, sheltered crevices, rock shelters and caves to the list of habitats.

The main cliff overlooks the Wye just upstream of the Seven Sisters on the Herefordshire bank. Formed in Crease Limestone, it is over 30 m high. It attenuates north into an intermittent small cliff interspersed with steep slopes that runs up the northern boundary. To the south, it attenuates above the Wye as a long steep slope with outcrops, rising to small cliffs opposite the Slaughter. It runs on up the Whippington Brook valley, then bends into a dry gorge high up in the managed compartment, where it faces another low cliff. The tallest cliff faces north and east.

The contrast between permanently exposed high cliff and the smaller, sheltered cliffs is not as large or as stark as one might suppose. The tops of the tallest cliffs may seem open, but they are covered in ivy (see Fig. 12.5), shaded from above by trees leaning over the lip and obscured by shrubs growing from crevices on the face. Small cliffs, which are shaded by trees rooted at their foot, are often well lit because the trees growing near them lean out and frequently fall. Collectively, the cliffs form a concentration of habitat diversity, from parched, unvegetated rock faces, through rock overhangs, shallow caves and outcrops shaded to every degree, grading below into screes and rubble, bordered above by a narrow zone of dry soils.

12.5 Canopy Gaps in General

Gaps form when a tall tree falls or dies and thereby allows more light through to the underwood and ground. In theory, this should stimulate regeneration and provide an opportunity for both light-demanding herbs and the fauna that prospers in the open and on the edges.

Gaps vary in several important characteristics. Some are large, clearly defined holes in a tall stand, whose boundaries are unambiguous and whose area can be measured easily, but most are not. When a small-crowned tree expires, its gap may be hardly larger than the spaces normally found between crowns in a 'closed' stand. When a group of birch or ash departs only to reveal an understorey of beech, lime or other shade bearers, a canopy gap is formed but the ground remains shaded. Many substantial gaps have complex shapes with ill-defined boundaries and individual trees standing isolated within the gap, as well as a smattering of underwood shrubs and saplings of various heights. The manner of gap formation – death standing, trunk breaks, or uprooted – determines the amount of soil disturbance and the degree to which the underwood is crushed and broken. In general, the gaps that form in an even-aged closed stand are well defined, but those in a stand that already possesses a complex vertical and horizontal structure are irregular and diffuse.

Gaps start to close as soon as they form. Crowns of marginal trees expand, undamaged underwood trees and shrubs grow vigorously, while crushed underwood and fallen trees may generate new trees from prostrate trunks. Seedlings, especially those already established, may fill the remaining spaces. On the other hand, once initiated, gaps may enlarge, because trees on their margins are more exposed and themselves become more likely to fall. Thus, new gaps will either decrease or increase as the seasons pass, or, more likely, migrate amoeba-like through the stand. Eventually, gaps lose their individual identity and the stand becomes increasingly complex, vertically and horizontally, until there is a stand-destroying disturbance that initiates a new, even-aged stand.

12.6 Gap Formation in Lady Park Wood

Few gaps formed in both the old- and young-growth stands until the 1976 drought opened both, principally by killing beech in the old-growth (see Fig. 7.10) and birch in the young growth. Gaps continued to form and enlarge thereafter as weakened trees died, and by 1983 the total gap area in the upslope part of the old-growth was estimated at 26% (see Fig. 5.11). Gaps formed in the drought left the soil largely undisturbed because snags either snapped, leaving a stump, or tipped over after the main roots rotted. Crown expansion reduced the gap area, which by 1992 was estimated at 9% in the upslope old-growth.

Gaps were still forming in this way in 2016 as drought-weakened trees finally succumbed, but as the stands grew older, gap formation was increasingly due to tree falls. Large and seemingly healthy canopy trees were either torn up by the roots in storms or simply toppled when the increasing weight of unbalanced crowns could not be resisted by trees rooted on steep slopes. Gaps thus formed create a pit where the tree once grew (see Fig. 6.10), a corresponding mound of displaced earth, rocks and roots, a patch of debris-covered ground where the tree fell, and ripped soil where roots were dragged out by the fall. Mounds slowly diminish, but remain visible long after the tree has decayed to nothing.

The rate of gap formation changed and varied between stands. From the outset, gaps have appeared regularly on the steep slopes below the cliff. Indeed, a major gap formed just after transect VIII was established when a large beech fell into the ash stand below. Above the cliff on the mid- and upper steep slopes, the gap formation rate was high for a few years immediately after the drought, but later settled to a slow and steady rate as more beech succumbed to drought-induced injuries and other trees on the gap margins fell. On the upper plateau, windthrow increased when the plantations above the wood were felled. After 2000, gap creation remained steady in the young growth as the canopy birches reached their natural span, but in the old-growth it accelerated and became more violent as the increasingly decrepit beech collapsed or were blown over.

Gap formation can be simple when a single tree falls or dies standing, or a large branch breaks, but in some cases it involves many trees and several predisposing factors. The illustration (Box 12.1) involved not just the skittles effect often encountered on slopes, but its origins can be traced to a large, forked canopy beech (F2) that eventually split, a weakened beech (F4) that had struggled on for 38 years after the 1976 drought, and an oak that had the misfortune to grow from a small bank. The most spectacular tree-fall gap formed when an old oak on the lip of the cliff fell into a stand of yew, large-leaved lime and ash below the cliff, thereby exposing the cliff, scraping the soil of the upper slopes bare, removing the shade inhibiting several yews and limes on the margins, and leaving a mound of fallen and smashed trees lower down the slope (Fig 12.4).

In so far as they are discrete and can be delimited, the individual gaps generated recently have been fairly small (Figs 12.7 and 12.8). The largest, at 1063 m^2, formed when a group of birch was blown down on the lower slopes, and the runner-up, at 902 m^2, formed when a huge beech standard growing on the colluvial fan fell on to a thicket of birch and hazels. Most are smaller, even those formed by dramatic falls. For example, when a drought-damaged canopy beech on transect I finally broke, it knocked over an ash and another beech, but still produced a gap of only 290 m^2. Gap shape can be expressed as an aspect ratio by measuring the longest diameter (M) and the largest diameter at right angles to it (L) and dividing L/M. A circular gap would score 1, but seven Lady Park gaps ranging from 290 to 1063 m^2 scored 0.36–0.80 (mean 0.56), nowhere near circular. (Gap areas were estimated as $M \times L \times \pi$.)

12.7 Stand Responses to Gap Formation

The responses of each tree species to gap formation are described in earlier chapters, but summarising:

- The crowns of trees on gap margins expanded, both on major crown branches and pendant growths, especially on beech, limes and the few wych elms. Epicormic shoots grew from the exposed trunks of ash, lime and especially oak.
- Sprouts grew from rootstocks, notably lime and beech, but also weakly on ash. Since 2007, new shoots have grown strongly from the base of limes and beech, less so on ash. Both old and newer basal sprouts have grown faster where gaps have opened above them.
- Some established, but moribund, underwood shrubs and trees have been rejuvenated, particularly some hazels. Most hawthorns and hollies have not responded.
- Fallen trees have sprouted from head to toe, especially limes (see Fig 8.7) and beech. Fallen ash and oak sprout weakly, but many trees nevertheless eventually expire.
- The few saplings established before gap formation grow faster and often re-sprout if they have been flattened.

Box 12.1

Early in 2012, a large tree fall in old-growth blocked the main path between transects II and III (Fig. 12.6, 12.10). This was the culmination of a gap-forming process that started many years earlier when a slender tall beech (F1) finally died of drought-induced injuries and, later, a large, forked beech (F2) did the splits, broke an underwood lime (T1) and injured another tall beech (F3), itself previously weakened by drought. Beech F3 eventually died and stood as a tall snag until it snapped near the base and fell against another beech (F4) and a large oak that had once been a standard growing among coppice. Normally, such an oak would have survived the blow, but it was rooted in the small bank above the main track, so it also fell, exposing a huge root plate, knocked over a tall coppice-origin lime with 16 trunks (T2), broke the top of an underwood beech, leaving a 4 m high snag, and crushed several lesser trees. An oak near the top of the cliff was whipped by falling crown branches, but remained upright. The outcome was an irregular, elongated gap. The broken lime regenerated as an informal pollard, and in 2016 the fallen trees were still generating new stems from the prostrate trunks, though shoots from beech F4 were wilting. The eventual beneficiaries may be the small ash, beech and holly saplings established before 2012 and sprouts from the fallen lime.

Continued

Box 12.1 Continued.

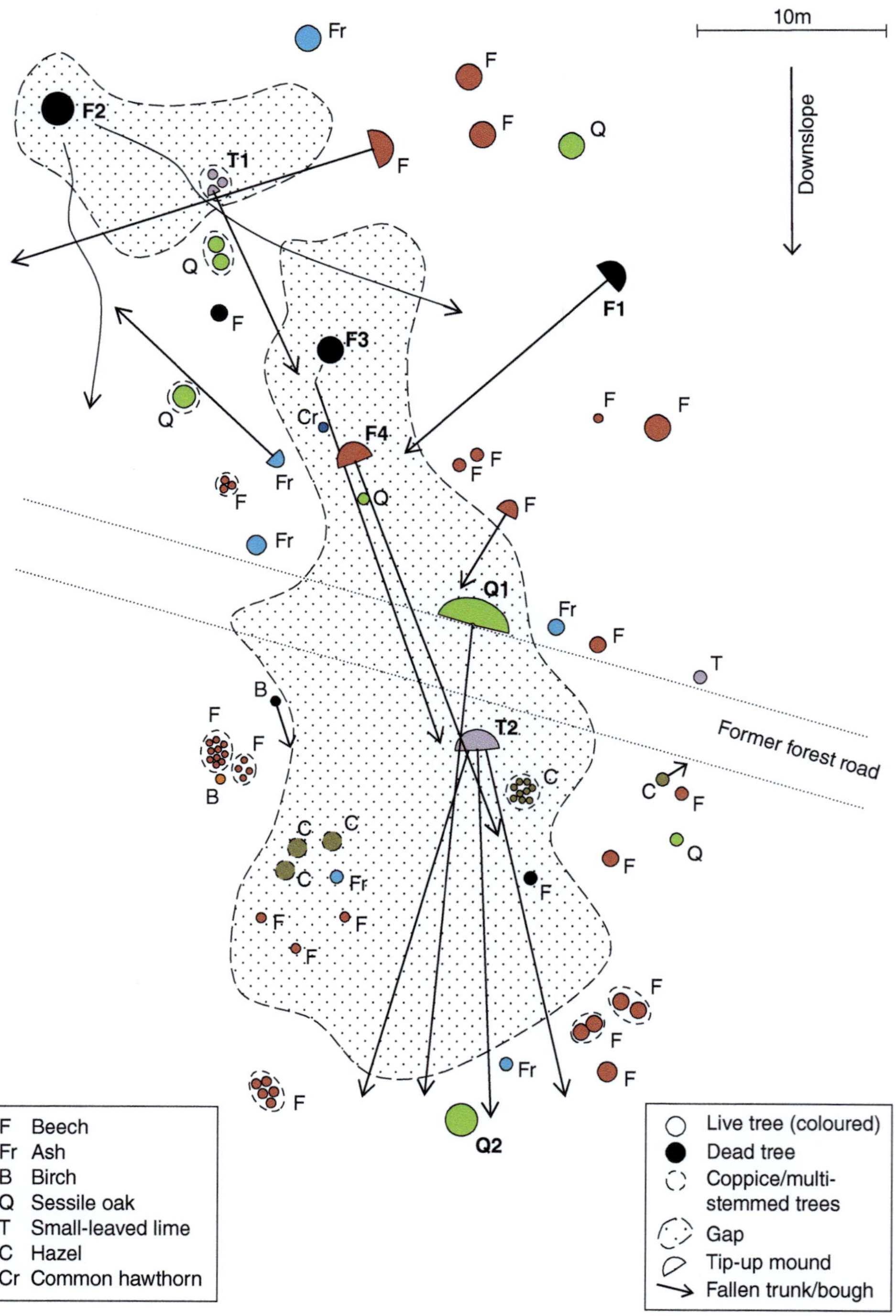

Fig. 12.6. Large tree-fall and canopy gap lying across the main path between transects II and III. Its formation is described in Box 12.1.

- Seedlings germinate and grow, especially ash (see Fig 6.5) and some beech, but also occasionally lime and hazel, rarely oak, elm, maple, holly, birch or hawthorn. We would also expect aspen and service suckers to grow well, but the few that emerged have not happened to grow under gaps.
- Gaps expand as more beech die on the margins and marginal ash, oak and limes are blown over.

Small gaps have been closed after a few years by crown expansion, but larger gaps have generally remained at least partly open (Fig. 12.9). Before 1992, bramble smothered saplings and deer browsed almost all regrowth to death or gnawed barely sprouting stumps. After 2007, basal growth prospered and saplings survived, but after 10 years the vegetative sprouts have reached only 1–5 m and the saplings are still mostly below 1 m, although a few have reached 4 m. Recruitment has, however, been limited by the recovering bramble thickets, which became almost impenetrable to new recruits in less than a decade.

12.8 Overview of Chapters 5–12

Even after more than 70 years of minimum intervention, the Lady Park stands are still growing out from their managed state. No informed visitor has any difficulty recognising standard oaks and coppice stools inherited from traditional coppice management. This is particularly true of the young-growth stands, which are typical of long-neglected coppices, but less true of the old-growth and the stand below the cliff, which grew out from coppicing 146 years ago and were merely thinned occasionally until 70–85 years ago. Nevertheless, natural processes have dominated all stands since 1944, and increasingly we see how these processes operate, the structures they develop and how individual tree species respond to circumstances and each other. Despite this, Lady Park can never be completely free of the indirect impacts of people and, just occasionally, direct impacts.

Much of what has happened has been eminently predictable, especially the progressive thinning as stands have aged and the build-up of dead wood. Less predictable have been the disturbances and their variety. The most important have been the 1976 drought and the impacts of deer. One was a discrete event with long-lasting consequences for structure, dynamics and the relationships between tree species. The other has been more a chronic condition in which the fluctuations are the events, brought about by changes in deer populations in the district and the presence and maintenance of fences. As the wood has become more natural, so the variety and impacts of disturbances have increased. This is due partly to the growth and ageing of trees, but is also due to globalisation, which has successively visited grey squirrels, elm disease, muntjac deer, feral pigs and, in the very near future, ash disease on the wood. Stand development over decades integrates steady, predictable growth and unpredictable disturbances. Far from developing towards a stable 'climax' community, the prospect is uncertain, varied and ever changing.

At the species level, we have been enlightened and surprised. *Inter alia*, we have:

- witnessed how self-thinning occasionally allows some small trees to get lucky and survive against the odds;
- found that external events temporarily change relationships, notably between beech and oak after 1976;
- realised how important vegetative growth is for generating new potential trees;
- discovered the extreme contrast between the high turnover species (e.g. aspen, dogwood) and the low turnover species (e.g. service, hawthorn, yew) in the underwood;
- learned the meaning of dominance, which aptly describes the way beech has 'made the weather' for other species;
- found that vegetative responses to natural gaps differ from responses to felling coupes;
- found that apparently weak and insubstantial elements can be long lasting, such as the dog rose and spindle bushes over 40 years old and the beech and service 'saplings' now past 70 years;
- discovered that mortality rates fluctuate, but are still relatively steady compared with regeneration rates;
- accepted that the understanding of species behaviour reached over decades may still have to be modified by unexpected turns of event (e.g. elm and ash regeneration below the cliff);
- observed how the performance of particular species varies between stands, depending on their competitors;
- developed respect for the persistence of understorey hazels in testing circumstances;

Fig. 12.7. A large tree-fall gap formed in the old-growth on transect II involving the fall of several slender limes, and ash, a large oak and a beech to form an irregular gap elongated down the slope. Unusually, the trees at the top of the fall were smaller than the larger trees they apparently knocked down. Perhaps the large trees fell first and the smaller limes fell on them after they became exposed.

- increasingly understood the wood as a community of species, each with a distinct character;
- and confirmed that unmanaged stands will become progressively less diverse in many respects as the years pass.

What was once a clear progression to ever-greater beech dominance has been replaced by a swing towards ash–lime, then ash–lime–beech and soon lime–beech dominance. Essentially, while we can extrapolate present trends, we have no idea what will actually happen next, and not just because we cannot know whether the wood will be blown flat by a storm before this book is published. These matters and their implications are explored in the following chapters.

Fig. 12.8. The canopy gap formed by the fall shown in Fig. 12.7. Despite the fall of several canopy trees together, the gap so formed is narrow, irregular and likely to close by crown expansion.

Fig. 12.9. A large, irregular gap formed after the 1976 drought killed many beech in the upslope, old-growth stand and was enlarged by later tree falls. Since formation, the gap has largely closed, almost wholly by crown expansion. Deer, Bank voles and now bramble have prevented any saplings becoming established.

Fig. 12.10. Fallen beech and oak in the gap charted in Fig. 12.6. In March 2016, the oak was still sprouting sparsely along the length of its trunk. Crossing heaps of fallen timber is only a minor challenge for the younger generation.

13 Species

The wild flora, fauna and fungi of Lady Park have changed in concert with changes in stand age, deadwood volume and the amount and form of open space, but recording has so concentrated on the trees that the records available are insufficient to examine the changes thoroughly. Only the ground flora has been recorded in any detail, so analysis concentrates on the plants, however much we might wish for a more balanced appraisal.

13.1 Ground Flora

Woodland ground vegetation reflects the texture, drainage and reaction of the soil, but the amount of light reaching the ground determines its expression at any one time. Many species grow and flower in spring, when the amount of light reaching the ground rises to its annual maximum before the tree leaves develop, but then they die back, leaving dry stalks (e.g. bluebell) or nothing at all (e.g. ramsons). Summer-growing plants remain weak under thickets, but compete strongly in the higher light levels afforded under mature stands and may reduce the performance of spring-growing species. Some, like the evergreen brambles, continue to grow in winter. Perennials predominate in the shade, but when light floods in after felling, they can be joined by annuals and biennials, which colonise from elsewhere or awake from dormant seed stored in the soil since the last felling. Wood anemone, dog's mercury and other perennials produce only limited seed, but expand vegetatively into dense patches. Many, like primrose (Fig. 13.2), drop most of their seed at their feet. Either way, ground vegetation tends to be patchy, a mosaic of overlapping clones spreading out from colonisation points and clusters of progeny from one pioneer colonist. With their limited capacity for colonisation, many woodland plants fail to spread to new woodland and recover only slowly from major disturbances within woodland (Peterken and Game, 1984; Rackham, 2006).

Not all plants within a woodland bear shade. Rides and other open spaces of managed woods are occupied by de facto meadows comprised of light-demanding species, albeit mixed with shade bearers. Many species find optimum conditions on ride margins, where they receive more light than plants in the woodland interior but less competition from grassland species weakened by shade. In large woods, especially those with a steady programme of felling and restocking, more species will be associated with open spaces than with shade (Peterken and Francis, 1999).

As an ancient coppice wood in a species-rich district with a variety of soil types and land forms, the flora of Lady Park must once have been rich, and it still contains species of note. Until 1870, its ground flora must have combined both herb-rich grassland on the rides and the boom-and-bust cycles within the coppice compartments, but we have no explicit record of the species involved. In the wood-pasture phase before 1600, grassland would have been more extensive, but some shade bearers (which are often sensitive to grazing) would have been restricted to thickets and rock outcrops.

Today, the ground vegetation covers a very wide range. The extent of the upper limestone outcrops is immediately visible in spring as a bright green carpet of dog's mercury. Lower down on the acid sandstone drift, patches of bramble and great wood-rush are set off by clusters of wood anemone and thin sweeps of bluebell. Close to the cliff, the strongly acid soils are marked by hard-fern and bilberry. Below the cliff, dog's mercury is again conspicuous, but so is hart's tongue. On the colluvial fan in the north corner, ramsons (Fig. 13.3), dog's

Fig. 13.1. Wood sedge, *Carex sylvatica*, yellow pimpernel, *Lysimachia nemorum*, and germander speedwell, *Veronica chamaedrys*, thriving in a large tree-fall gap.

Fig. 13.2. Primroses, *Primula vulgaris*, remain scattered through the southern part of the fenced reserve and in the managed compartment, despite the presence of deer.

mercury, bluebell and wood anemone vie for dominance in spring. Until the 1990s, much of the wood was dominated by brambles, but the underlying herbaceous pattern was still clear.

The records of individual vascular plant species come from the usual mix of systematic survey, casual findings and deliberate searches for rarities. The earliest recorder, S.G. Charles, who lived on the Kymin overlooking Monmouth, noted 45 species in and around Lady Park. EWJ surveyed the flora of young-growth transects: together with casual sightings, he recorded 85 species before he lost interest in 1960. In 1970, J.F. Archibald recorded species as he directed a group of tree surveyors. In 1977/8, Alan Orange listed 113 species for the reserve as a whole. In 1977, GFP recorded 63 species in nine 900 m^2 plots, and Vanessa Williams recorded 100 species in 288 8 m^2 plots scattered in groups of four throughout the reserve and managed compartment. In 1984, Tim Barfield noted 43 species while establishing transect X. In 1997, Alastair Monument resurveyed many of Vanessa Williams's plots, and in 2009 a team from Leuven and Ghent in Belgium resurveyed all those in the unmanaged stand without adding new species. GFP had noted a few species after 1983, but from 2012 until July 2016 he compiled a new whole-site list. Leaving aside some ambiguities and counting dandelions, hawkweeds and brambles as single species, 258 species of herbs, ferns and small shrubs have been recorded in or near the reserve since the early 20th century. The *Flora of Monmouthshire* (Evans, 2007) records an additional 83 species in the tetrad of which Lady Park is part.

Fig. 13.3. Ramsons, *Allium ursinum*, have spread within the fenced reserve and now dominate tracts of woodland on the upper steep slopes and the colluvial fan.

13.2 Floristic Changes at Small Scales

After what is now the managed compartment was added to the reserve in 1984, we fully expected that resumed silvicultural operations would diversify the ground flora, or at least retain species already on site, but by 1990 we suspected that floral diversity was declining throughout the reserve. This belief was reinforced by informal checks on some of Vanessa Williams's plots, and later by Alastair Monument's resurvey, which found that floral diversity had halved since 1979 at the 8 m^2 scale.

Naturally, we blamed the deer, which by then were preventing all tree and shrub regeneration and had almost eliminated brambles, but the 2009 resurvey, which found that the number of plant species in each 32 m^2 plot had declined from 21.3 in 1979 to 15.5 in 2009 (Baeten *et al.*, 2012; Vanhuyse *et al.*, 2012), reached a more nuanced conclusion. Analysing the changes in terms of Ellenberg numbers (Hill *et al.*, 1999), the Belgian team concluded that floristic impoverishment was due to deer, increased shade arising from increasing dominance of beech and limes, and lack of disturbance (what elsewhere would be called lack of management) acting in concert. Deer graze ground vegetation, especially in gaps and clearings, curtail growth and largely prevent flowering, while prolonged shade further reduces flowering and seed production and thereby attenuates the seed bank. Lack of disturbance and a probable increase in the amount of falling leaves as the stands mature allow litter and nutrients to accumulate, while wide-scale pollution reinforces the nitrate build-up. The leaf litter acts as blanket that discourages seed germination and the establishment of herbs, when small disturbances might have given them a chance.

The floristic impoverishment between 1979 and 2009 had reduced the total number of species in the 35 plots resurveyed (total area 1120 m^2) from 93 to 61, with 36 being lost from the plots and just four species gained. The few beneficiaries included ferns (e.g. male-fern) and nitrophiles (ramsons). The spring geophytes (wood anemone, bluebell, ramsons) had scarcely been affected; but many species had decreased, including graminoids (wood sedge, tufted hair-grass, great wood-rush), small shrubs (brambles, dog rose) and forbs (wood spurge, slender St John's-wort, common dog-violet, yellow pimpernel, yellow archangel, rosebay willowherb).

13.3 Floristic Changes at the Whole-wood Scale

The 8 m^2 plots show only how diversity has changed at a small scale, but had species merely thinned out or had some been eliminated completely? To answer this question, the new whole-site resurvey was started in 2012. Ground flora species were noted at every visit, and particular searches were mounted around the cliffs and margins. The resulting list is no doubt incomplete, but it forms a basis for comparison with all the earlier records. By July 2016, 125 species had been found within the fence, of which six were confined to fragments of the open ride above the wood that had been enclosed within the fence.

Within the reserve, the ground vegetation is conspicuously species poor. Most of the ground is either bare or populated by tree seedlings that will not survive. Patches are covered by dominant dog's mercury, bluebell, brambles, great wood-rush, wood anemone, tufted hair-grass, bracken, bilberry and glaucous sedge (Fig. 13.4), to the near exclusion of all others. A few other species are widely scattered at low density, notably wood melick, male-fern, wood spurge, lords-and-ladies, false brome, ivy and hart's tongue, but otherwise we have a very long tail of almost 100 species that remain rare, or at best occasional (Figs 13.5 and 13.6). Most cling to specialised niches, such as:

- The cliff. The face itself, the dry and moderately well-lit zone just above it, and the steepest scree below, e.g. wood fescue, fingered sedge, wall lettuce.
- Small patches of strongly acid soils just above the cliff, e.g. common cow-wheat, pill sedge, goldenrod, heather, wavy hair-grass, wood sage. Most of these cannot stand the shade of a closed canopy: since 1979, the main remaining patch has been greatly reduced as the gap to which these species are confined has almost closed.
- Very dry, lime-rich soils on steep slopes near the top of the wood, e.g. early purple orchid, toothwort.
- The moist, base-rich soils of the colluvial fan, e.g. herb Paris, enchanter's nightshade.
- Seepages, mostly at the base of the slope, e.g. opposite-leaved golden saxifrage.
- Disturbed ground, notably the sunkway to the south and the margins of the main track, both of which leave soil free of litter, e.g. yellow pimpernel, wood speedwell, wood sedge.
- Canopy gaps (see below).

The present 119 species includes 13 that had not previously been recorded within the reserve. Much more significantly, 52 species have not been found recently, yet were present at one or several dates between 1944 and 1979, and a further 22 species were recorded by S.C. Charles, whose earlier presence in the reserve is less certain. Eight overlapping groups can be recognised.

1. Gap-phase species. Many woodland herbs thrive only in temporary gaps, having colonised from another gap or having survived the shade as dormant seed in the soil. Lady Park has lost columbine, small teasel, common hemp-nettle, heath cudweed and raspberry. In addition, rosebay willowherb, foxglove, nipplewort and heath groundsel were also missing, but Lady Park has lost numerous grassland associates, including strong calcicoles, such as yellow-wort, wild basil, ploughman's spikenard, wild marjoram, hoary plantain, centaury; and neutral grassland species, such as common bent, sweet vernal grass, common sedge, meadow buttercup, lesser trefoil and bush vetch; and the species of neglected meadows, cock's-foot and Yorkshire fog.

2. Ruderals. The forestry operations that disturb open spaces in managed woods enable ruderals to flower in woodland. Lady Park has lost tall herbs such as garlic mustard, hemlock, great mullein and hedge mustard; grassland 'weeds' such as spear thistle; short species of scuffed ground, such as annual meadow-grass and procumbent pearlwort; garden weeds, such as white dead-nettle and common

Fig. 13.4. Glaucous sedge, *Carex flacca*, a surprising species to find below closed-canopy woodland on limestones near the cliff.

chickweed; and waste ground associates, such as colt's-foot. All must formerly have been dependent on disturbance during forestry operations. In addition, creeping thistle and groundsel have recently been refound.

3. Small shrubs. The margins of open spaces provide opportunities for small shrubs. Gorse has been lost from Lady Park, but dog rose has recently been rediscovered. Broom, once common in the felled areas, occasionally appears in gaps.

4. Wet soil associates. Wavy bittercress, marsh thistle, toad rush, soft rush, water figwort, tansy and common valerian have been lost. The only damp soils in the reserve now are in shade at the foot of the slope. Perhaps soils were wetter in the felled areas until the canopy redeveloped and the trees pumped them dry.

5. Epiphytes. Surprisingly, neither mistletoe nor intermediate polypody have been seen in the reserve since the 1970s.

6. Rock associates. Pellitory-of-the-wall, bird's-foot sedge and perhaps rustback, brittle bladder-fern and hawkweed microspecies have been lost from Lady Park, and so have the naturalised species wallflower and yellow corydalis. All are most likely to have been on open cliffs and rock outcrops, which implies that these have become more shaded.

7. Shade species. Several shade species have been lost, which is surprising, given that Lady Park is not short of shade. They include the grasses false brome and bearded couch, dewberry and a variety of herbs (giant bellflower, wavy bittercress, sanicle, hairy violet). If we count S.G. Charles's records, lily-of-the-valley,

Fig. 13.6. Toothwort, *Lathraea squamaria*, an infrequent species associated with hazel on the upper plateau.

mountain melick, oak-fern, pale St John's-wort, Goldilocks buttercup, fly orchid and thin-spiked wood sedge have also been lost, but some may not have been within what is now the fenced reserve. Lily-of-the-valley and mountain melick were still present in 1979. Several other species have been found only with great difficulty: bitter vetch, for example, has recently been rediscovered as a single plant.

What do these departures tell us about habitat change? First, context: all the lost gap phase and most of the lost ruderals survive nearby, and several others are widespread and common in the upper Wye gorge. In fact, it is quite possible that *all* the lost species survive in the vicinity, now that fly orchid has been rediscovered, so the losses represent retreat rather than extinction. Some may still be present as dormant seed, as proves to be the case when broom appears in tree-fall gaps where seed was last deposited 60 years ago.

This floral retreat is no surprise. Under traditional coppicing, rides provided permanent havens for grassland and ruderal species, and repeatedly cut coupes allowed herbs to flower regularly and replenish seed banks. After coppicing ceased in 1870, ground flora diversity may have been maintained by the thinning in 1902/3 and *c.*1930, and certainly received a further boost when two-thirds of the wood was felled in 1942–1944.

The floristic benefits of felling can be judged from the records of ground vegetation in the young growth of transect IV. The number of recorded species rose from 26 in November 1944 to 46 in March 1949, but fell to 38 in the spring and summer of 2013, a season when one might have expected more. The shade bearers, tufted hair-grass, wood spurge,

Fig. 13.5. Broadleaved helleborine, *Epipactis helleborine*, a shade-bearing orchid, one of many species that survives as a very small population.

woodruff, perforate St John's-wort, great wood-rush, bramble, yellow archangel and bluebell had been most frequent in 1944, and these have persisted. Early colonists, such as lesser burdock, nettle-leaved bellflower, foxglove, great mullein and heath speedwell, vanished by 1948/9, but by then several others (deadly nightshade, creeping thistle, glaucous sedge, broom and hemlock) had colonised and some shade bearers had probably increased (ivy, male-fern, hairy St John's-wort, wood sedge, dog's mercury, bracken and spurge laurel). By 2013, male-fern, bluebell, dog's mercury, wood sedge and false brome were most frequent; false brome and several ferns (male-fern, scaly male-fern, broad buckler-fern, soft shield-fern, lady-fern and hard-fern) had increased; but lords-and-ladies, brambles and great wood-rush had long since retreated in the face of deer. The open-space flora probably vanished in the 1950s, leaving only weak dandelions as an occasional reminder.

Species associated with cliffs have declined throughout the lower Wye Valley. Until the 20th century, the limestone cliffs were adorned with the plants we normally associated with limestone grassland, such as small scabious, but most have been shaded out by the growth of trees on, above and below the rock faces. The best survival is opposite Lady Park on the Seven Sisters, where the rock towers face south and are tall and dry enough to maintain wild rockeries with horseshoe vetch, bloody crane's-bill, dropwort, soft-leaved sedge, dwarf sedge and many others.

Lady Park never supported limestone grassland – its cliffs are smaller, less prominent and they face north and east. Today, its cliffs are shaded or covered in ivy, leaving only fingered sedge, wood fescue, wild madder, some hawkweed microspecies and the *pachyrachis* form of maidenhair spleenwort to represent the rock flora. Immediately above the cliff, far from having the shallow lime-rich soils of the Seven Sisters, the soil is acid enough to support heather, wavy hair-grass, common cow-wheat and slender St John's-wort, as well as the shade-bearing bilberry and hard-fern. Most of these near-heaths have been eliminated by shade, leaving just tiny fragments lit by side light.

Explanations for the losses of the shade bearers are less obvious. Most of the species in any wood survive as small populations, so a few losses due to random population fluctuations are to be expected, and the heavy shade has reduced many to very small populations indeed. Further, the very limited response to both fencing and the recently increased rate of gap creation has gone far to confirm the Belgian analysis – that the seed bank is so attenuated, populations of woodland herbs are so thinly scattered and the litter is now so thick and nitrate rich that few plants respond when gaps appear.

In contrast to this generally bleak story, a mixed bag of 16 species has recently been found within the reserve, but had not previously been recorded. The naturalised species, butterfly bush, American willowherb and Indian balsam, must be genuine additions. So, too, is the slender rush that has become established on the flood plain. The rest are native species that were probably present but were too infrequent in the wood, or just too ordinary in the district, to be worth noting. They include three shade bearers (wild angelica, opposite-leaved golden saxifrage, early purple orchid) and several species of grassland and field margins, such as meadow vetchling and red campion.

13.4 The Role of Canopy Gaps

In theory, gaps provide an opportunity for light-demanding species to reoccupy mature stands, but few species have taken their chance. In 24 gaps examined in 2015, only betony, narrow-leaved bittercress, rosebay willowherb, creeping thistle, wayfaring tree (four gaps), wood avens, hairy St John's-wort, perforate St John's-wort, slender St John's-wort (two gaps), Indian balsam, nipplewort, wild madder, broom, common figwort (two gaps), common ragwort (four gaps), groundsel, red campion, smooth sow-thistle, dandelion and germander speedwell (Fig. 13.1: see p. 208) could be construed as incomers benefiting from gap formation, and several of these bear moderate shade. Others may have been present earlier, unobserved in the older gaps immediately after gap formation, but they would have been evicted after, say, 5–7 years, by which time gaps fill with bramble and shrub regrowth.

The single exception has been the large gap that formed at the top of the cliff and exposed rock faces, rubble slides and pockets of thin soil (Fig. 13.7, 13.12). In the second season after formation, it

Fig. 13.7. A relatively rich flora has developed quickly in a large tree-fall gap on the cliff, here including both species that were not apparent before the gap formed (e.g. hedge woundwort, *Stachys sylvatica*, deadly nightshade, *Atropa belladonna*) and others that were apparent before (e.g. wood fescue, *Festuca altissima*, bramble, *Rubus fruticosus*).

contained numerous ruderals (e.g. welted thistle, hedge woundwort), gap species (deadly nightshade, lesser burdock, hemp agrimony, heath groundsel) and grassland species (cat's-ear, heath speedwell), as well as characteristic shade and rock-outcrop species. Ten species were added to the reserve list in this gap alone. Its success must be due to its large size, extensive exposure of mineral soil and the long history of canopy disturbance in the vicinity.

The principal beneficiaries of gaps are shade plants that are presented with a chance to grow vigorously, set seed and spread, especially the various microspecies of bramble, which, within a few years, occupy most of the ground. The growth of brambles is somewhat limited by shade on gap margins, so the rest of the flora eventually comes to occupy a rough halo round the fringes. In effect, the rest of the flora has a few years to prosper, and then it is largely excluded.

Ferns also prosper in gaps, especially male-fern, which can dominate in small patches, often accompanied by lady-fern, broad buckler-fern, scaly male-fern, hart's tongue, hard-fern and soft shield-fern, and less often by hard shield-fern, bracken, hard-fern and narrow buckler-fern. The spring herbs, ramsons, wood anemone, bluebell and dog's mercury, do not seem to increase, but they persist in gaps and quickly resume their position when the gap closes. A few other herbs spread and flower vigorously, notably false brome, glaucous sedge, wood sedge, tufted hair-grass, wood spurge, tutsan, great wood-rush, yellow pimpernel, wood melick and common dog-violet. In addition, ivy regenerates sparsely, traveller's-joy prospers on the fringes and both black bryony and spurge laurel are found sparingly.

The vaunted high diversity of the gap-phase flora is a feature of managed woods, where gaps appear regularly and the soil surface is disturbed as the wood and timber is extracted. With particular exceptions, natural gaps, which form irregularly, where most of the ground is not disturbed, and which remain partially shaded by the fallen and regenerating trees, appear on the evidence of Lady Park to attract only a slight and transient gap-phase flora.

13.5 Floristic Changes in the Managed Compartment

What of Tony Joslin's bet that the Forestry Commission would make the managed compartment more diverse? By 2015, after at least 25 years of heavy grazing, we were not surprised to find that, far from becoming more diverse, the flora in nine of the Vanessa Williams plots had declined from 14.4 species at the 32 m^2 scale to 9.6. In frequency terms, bramble, common dog-violet, ivy, hairy wood-rush, false brome, wood melick, wood spurge and wood sedge had declined the most, and several other hitherto infrequent species had been lost entirely from the plots. In compensation, only lords-and-ladies and wood anemone recorded substantial gains, and the latter may be an artefact consequent upon late-season recording in 1979. Herbaceous cover decreased sharply in most plots, but increased where the stand had been almost clear-felled. A small exclosure demonstrated that grazing by deer was a major influence: not only were bramble and other declining species growing well but also cover increased, and wood anemone was actually flowering.

Some 75 species have been found recently in the whole compartment, which exceeds the 60 recorded in 1979–1984 by Vanessa Willliams and Tim Barfield, so Tony would have claimed victory. However, the earlier list was confined to the 12 plots and the line of transect X, whereas the later list comes from the whole compartment. More to the point, it does not look like a win. The vegetation is grazed flat; wood anemone hardly flowers outside the exclosure; and we have been unable to find several shade and marginal species recently (wood barley, hairy wood-rush, lily-of-the-valley, mountain melick, herb Paris, columbine). Most of the apparent gains were shade species that happened to grow outside the plots and transect: only selfheal and creeping cinquefoil might be recent immigrants.

One of Vanessa Williams's plots was located in the unmanaged old-growth on the outcrops above the Slaughter. In 1979, its ground flora contained 18 species that collectively covered 55% of the ground, but by 2015 it had been reduced to three, very thinly scattered, and among the losses was the main colony of lily-of-the-valley. With a closed canopy and deep litter accumulations, combined with constant access by fallow deer, this is a striking demonstration of how several factors have combined to impoverish the flora.

13.6 Floristic Changes in Context

First, the local context. Since 2012, 203 species have been recorded in the 'greater Lady Park'; that is, the reserve, the managed compartment and the immediate surroundings on the plateau above and the flood plain below. If one adds the species recorded previously, the total reaches 258 species, implying that 55 species may no longer be in the reserve or the immediate vicinity. The apparently lost species include lily-of-the-valley, oak fern, mountain melick and sanicle, which are all shade species. If we combine all past and present sources for Lady Park with the tetrad records shown in the *Flora of Monmouthshire*, we reach a total of 326 species, and there were and are more on the Herefordshire side of the Wye. Thus, Lady Park Wood contains only a small selection of the local flora: at a landscape scale, it is part of a rich and diverse district.

Within Lady Park, the non-intervention policy has combined with deer and pollution to impoverish the ground flora, and there has been only a very limited counter-flow of colonising species in response to fencing and accelerating gap creation. Outside the reserve, however, the rides provide the openings in which many of the evicted species still flourish (Fig. 13.8), though they remain exposed to fallow deer and now muntjac deer. Losses may just be more apparent than real, for many species have been reduced to very small populations, some of which have only been found again after years during which they were thought to be locally extinct.

On a world scale, the floristic impoverishment of Lady Park is one example of wider changes throughout the temperate forests of Europe and North America. Combined analysis of long-term changes in the ground flora of 23 woods throughout temperate Europe (one of which was Lady Park) indicated that shade and nutrient availability had increased, though not in all sites (Verheyen *et al.*, 2012). This was due at least partly to increased nitrogen deposition in rain and the retreat of coppicing, leading to general lack of management and increased litter input. Broad buckler-fern, narrow buckler-fern, ivy, common hemp-nettle and bramble were the species that increased the most, whereas the greatest losers were bugle, wood meadow-grass, dog's mercury, lily-of-the-valley and herb Paris. However, increased shade appeared to be masking the expected decrease in diversity associated with increased nutrients, and this might be setting up a potential 'nitrogen bomb', i.e. any reductions in shade (= resumption of management) might allow nitrophiles to increase at the expense of forest specialists. A further analysis, this time adding long-term plots in temperate North America, suggested that increased shade was cooling forest floors of unmanaged forests enough to suppress the 'thermophilisation' of woodland ground floras, i.e. the increase in warm-adapted species observed in managed forests (De Frenne *et al.*, 2013).

The invasion of wild boar early in 2015 and the continued presence of a small population has introduced a new factor that could influence the ground vegetation. This is a widespread phenomenon in Europe. Three years after colonising Dalby Söderskog, a long-established and long-recorded natural reserve in southern Sweden, pigs had reduced the quantities of spring species, such as wood anemone, but increased the diversity of summer-growing species (Brunet *et al.*, 2016). In Lady Park, they have disrupted patches of the less stony ground, concentrating particularly around well-rotted tree stumps.

13.7 Bryophytes (text by Sam Bosanquet)

Eustace Jones picked Highmeadow and then Lady Park for his teaching and recording, partly because the neighbourhood was famous for uncommon bryophytes. Thanks to his studies, building on pioneering searches by bryologists such as Eleanora Armitage and H.H. Knight, the site had become a *locus classicus* for bryophytes by the 1950s and 1960s. It was visited twice by large groups from the British Bryological Society, and since 2000 I have mounted detailed searches. The tall, east-facing limestone crag holds many of the reserve's most notable mosses and liverworts, although others occur on limestone boulders and smaller rocks on the slopes below the crag. At the east end of the site, alongside the Whippington Brook, is a concentration of particularly rare bryophytes on east-facing limestone, where water seeps out through crevices.

The bryophytes of Lady Park Wood are typical of the limestone of the lower Wye Valley. Characteristically abundant species on the cliff and the boulders below it include *Anomodon viticulosus*, *Ctenidium molluscum*, *Porella platyphylla*, *Marchesinia mackaii*, *Neckera complanata*, *Neckera crispa* and *Rhynchostegiella tenella*. Water seeping out of the limestone encourages the tufa-forming moss *Eucladium verticillatum*, sometimes accompanied by the liverworts *Jungermannia atrovirens* and *Leiocolea turbinata*. These species all indicate calcareous conditions, and many of the mosses that grow on the woodland floor are similarly calcicolous, particularly the bulky and abundant *Thamnobryum alopecurum*. However, leached patches, especially above the limestone cliff, support calcifuge mosses like *Mnium hornum*, *Polytrichastrum formosum* and *Pseudotaxiphyllum elegans*. Rotting logs are colonised by other acid-loving species, including the liverworts *Nowellia curvifolia* and *Riccardia palmata*, which are characteristic of western Britain and are rare in south-east Wales. Epiphytes on trees include *Frullania dilatata*, *Metzgeria furcata*, *Radula complanata*, five

species of *Orthotrichum* and the largest population of *Platygyrium repens* in Wales.

Among the larger mosses on the limestone grow two tiny liverworts – *Cololejeunea calcarea* and *Cololejeunea rossettiana* – and seeping limestone may hold equally small *Seligeria acutifolia* and *Gymnostomum calcareum*. These small bryophytes are scarce, though, both on the site and in Britain as a whole. As well as these scarce mosses, there are three particularly rare bryophytes that grow at Lady Park Wood. *Anomodon longifolius* has been known here for over a century, and suffered from unscrupulous collecting in the early 20th century that left huge patches of this rare moss in British herbaria and almost none on the limestone outcrops where it formerly grew. The last remaining patch on the main cliff was lost in the early 2000s when a tree that sheltered it collapsed, but three small colonies are still present near the Whippington Brook, and the population extends across the border into Gloucestershire, where there are several larger colonies. The Lady Park Wood population of *Seligeria campylopoda* also spans the border, wherever in the Wye Valley its habitat of small limestone rocks, half embedded in the woodland floor occurs. Despite its widespread occurrence here, its only other British colonies are at the eastern end of the Brecon Beacons massif. Similarly rare is the only known Welsh population of the tiny leafy liverwort *Lejeunea mandonii*, which grows on a limestone cliff near the Whippington Brook and is otherwise known in Britain only from the Lizard Peninsula in Cornwall and three sites in western Scotland.

Vanessa Williams recorded bryophytes in her quadrats: her records are summarised in Peterken (2008a: p. 353). After examining some of Vanessa's quadrats 25 years later, GFP was convinced that bryophyte cover had diminished and that some hitherto common species, such as *Atrichum undulatum*, *M. hornum* and *P. formosum*, had almost disappeared. Almost all of the bryophytes that had declined grow on soil rather than limestone. The only significant bryophyte cover in the plots in 2004 was on limestone blocks, tree roots or fallen wood: the ground flora had almost vanished. The only exceptions, where typically moss-rich woodland floor occurred in 2004, were on steep banks by the path that ran along the northern edge of the reserve – on ground that was too steep to be trampled regularly by deer – and in exclosures below the cliffs. It seems unlikely that pollution, a possible factor in woodland bryophyte decline elsewhere in Britain, would affect the ground flora everywhere except on steep banks, and genuine change appears most likely to result from deer numbers.

13.8 Lichens

A total of 74 species of lichen, including 65 on bark or wood, and 9 on rock, have been recorded by Alan Orange, but the figure would be higher if a more intensive survey were to be carried out. The diversity is limited by two main factors: shade and a low degree of ecological continuity. So-called old-forest lichens rely on the continuous presence of at least some mature trees at a site over the preceding several hundred years and are slow to colonise sites where tree cover has been lost. Lady Park Wood scores poorly on the New Index of Ecological Continuity (Coppins and Coppins, 2002), which is designed to estimate the likelihood of ancient origin and ecological continuity of a woodland site. The low score reflects the management history of the woodland and adjacent woodlands. Only four species on the main index occur: *Agonimia allobata* (on lime), *Cladonia parasitica* (on oak stumps), *Cresponea premnea* (on dead wood of beech and yew) and *Thelotrema lepadinum* (various phorophytes). *Porina rosei* is a fifth old-forest lichen that is treated as a 'bonus species' in the index, found on a yew and on an old ash stool. Nevertheless, these species provide interest that is unlikely to be found in the average coppiced woodland in the region, and four of the species were graded by Woods (2010) as Near Threatened in Wales. It is likely that a small degree of ecological continuity has been provided by yew trees (either deliberately not felled or on inaccessible cliffs) and perhaps the bases of old coppice stools, on which some lichens may have survived periods of felling. A similar situation is seen in the lower Wye Valley where *P. rosei* occurs on similar substrates, in the near-absence of other old-forest species.

The hard wood of the stumps of long-felled oaks provides a habitat for several species not recorded on other substrates, including *Cladonia digitata*, *C. parasitica* and *Lepraria umbricola*. Dead wood on a large standing dead beech supported *C. premnea*. These records illustrate the importance

Fig. 13.8. Weld, *Reseda luteola*, and foxglove, *Digitalis purpurea*, growing in the open above the reserve, along with many of the species that grew within the reserve after the wartime fellings.

of the dead wood habitat. Shade currently limits the diversity of lichens on tree trunks, and would limit most old-forest lichens even if they were present, although there is the possibility of elevated light levels on the cliff edge or where trees fall. There are no historic records for the site. Since regularly coppiced woodland typically has a poor lichen flora, with no old-forest species, the flora of Lady Park is likely to improve with continued non-intervention, despite the shady conditions. Diversity is also likely to be enhanced by an increase in dead wood.

13.9 Fungi

By November 2015, the Dean Fungus Group, led by Keith Davies, Val Davies and Dave Shorten, had recorded 419 species in 11 forays between 2012 and 2016 (Fig. 13.9), not far behind the 473 listed for Symonds Yat by 2012. The key question in advance of their visits was: has the increase in dead wood and depth of litter within the reserve allowed the mycoflora to become more diverse? Unfortunately, very few early records have come to light, so change cannot be demonstrated directly, and any proven change could be explained more by changes in the production of fruiting bodies than by actual extinction and immigration of species.

Some 313 species were associated with one or more tree species, the most catholic being *Kuehneromyces mutabilis* (sheathed woodtuft), *Schizopora paradoxa* (split porecrust) and *Trametes versicolor* (turkey tail) on six tree species. The tree with the most associates was beech (181 species), followed at a considerable distance by oak (73), ash (64), birch (51), hazel (36) and lime (30), the remainder having 11 or fewer. The number of associates is broadly in line with the quantity of each tree species in the stand, as measured by basal area, with birch and perhaps beech associates slightly over-represented, and lime, ash and maple associates under-represented.

Among the species recorded were *Eutypa spinosa* (spiral tarcrust), *Pholiota aurivella* (golden scalycap) and six species of *Pluteus* (*cervinus*, *ephebus*, *nanus*, *phlebophorus*, *plautus* and *salicinus*) (shield species), all species listed by Ainsworth (2004) as likely indicators of rich dead-wood habitats in beech woods. Species such as these presumably increased with the pulse of dead wood delivered by the 1976 drought, but they could have been present on the rotting cores of old stools before 1870.

The species that particularly interested the Dean Fungus Group were *Stephanospora caroticolor*, *Lactarius mammosus* and *Cortinarius rufo-olivaceus*. Of these, *S. caroticolor* (carroty false truffle) excited the most comment. Found in December 2012 under a yew, this is a declining and critically endangered Red Data Book species in the UK.

It has also been interesting to find *Hygrocybe coccinea* on the main track and other waxcaps elsewhere. Given that we associate them with semi-natural grassland, one is tempted to see them as survivors from the more open phases in Lady Park's history, but Dave Shorten assures us that in mainland Europe (and sometimes in Britain) they are at home in mature woodland.

13.10 Vertebrates

Fish in Lady Park would be more remarkable than salmon in Yemen, but the full set of Wye species passes along its lower margin and anglers can indeed be seen trying for salmon in its shade. We have yet to see a reptile or an amphibian in the reserve.

13.10.1 Mammals

Bats in the various caves, mines, scowles and clefts have been monitored by Peter Smith since 1989. Lesser (Fig. 13.10) and greater horseshoe bats have been found on almost all visits, but the former has always been far more numerous – more than 60 on one occasion. In addition, Brandt's/whiskered bat, Natterer's bat and brown long-eared bat have been found occasionally. A few of the greater horseshoes had previously been ringed in other local caves at Symonds Yat and on the Doward.

Have their feeding and roosting habits changed with the increasing availability of large trees and snags? An ongoing echo-location study by Andrew Carr of Bristol University has recently identified nine species that use Lady Park for feeding, of which four – whiskered/Brandt's, Leisler's, barbastelle and brown long-eared bats – were detected only in the unmanaged stands. By far the most significant of these is barbastelle. One of the rarest British species, whose range is restricted to south Wales and south-central

Fig. 13.9a,b,c etc. A diversity of fungi in Lady Park Wood. Species from top left to bottom right [*Amanita citrina, Lycoperdon perlatum, Pholiota flammans, Clavariadelphus pistillaris, Trametes gobbosa, Fistulina hepatica, Laccaria amathystina, Ramaria stricta, Oudmansiella mucida, Coprinus picacius, Krombholziella variicolor*]

Fig. 13.10. Lesser horseshoe bat, *Rhinolopus hipposideros*, which uses Lady Park cave and several other holes in the limestone.

England, it prefers to roost in woodland, especially in crevices and behind bark, and to forage over riparian and broadleaved woodland. It also lives as small populations that frequently move to new roost sites, so it prospers in large tracts of mature woodland. Indeed, the old-growth stands in Lady Park must have been ideal ever since the 1976 drought. The other species detected were greater horseshoe, lesser horseshoe, noctule, pipistrelle and soprano pipistrelle, of which the last was far more frequent than all other species together. These species also fed in the managed stand, but in smaller numbers. Overall, the sound registrations of all species were almost five times more frequent in the unmanaged stands.

Fallow deer were in Lady Park before 1944 and have remained there ever since. They are regularly culled by Forest Enterprise, but population densities must have fluctuated widely. After 1944, the area felled during the war regenerated thickets of bramble and saplings, but deer tracks pervaded the whole area, deer grazed hard under retained standards and they seemed to concentrate on a zone just above the cliff, where perhaps they liked the view. Certainly, judged by impacts, a substantial increase was allowed after 1993, for bramble was virtually eliminated from the reserve and the managed stand, which regenerated well after the 1983 thinning, hardly regenerated at all after 1994. Since 2007, they have been fenced out of the reserve, but still graze the managed compartment flat. Muntjac and roe deer have spread to the district recently, and the former now enters the reserve.

The most noticeable animals are grey squirrels, which are no longer controlled. They can be seen throughout the wood, but their impacts (see Chapter 7, this volume) sometimes appear patchy. Feral pigs arrived recently and managed to penetrate the fence early in 2015, and a small population remains inside. They explored the entire reserve above the cliff, turning over patches of litter and soil, especially on the edges of gaps and around the rotting bases of large trees. Badgers maintain setts within the reserve and move freely in and out of the reserve: lacking gratitude, they ignore the badger gates and scramble under the fence where the land dips.

The principal small rodents trapped by Graham Berry in 1994 were bank voles and wood mice, with smaller numbers of yellow-necked mice. The first two were most abundant where bramble formed a thicket in a gap in the old-growth and least abundant in the closed old-growth stand, while yellow-necked mice were trapped only in 'coppice'. Evidently, small rodents can be numerous in just those places – gaps – where tree regeneration might be possible. They are certainly influential, for, when bank vole populations peaked in the summer of 1985, all the small beech saplings were killed by ringbarking, and at other times well-established and vigorous small ash saplings, apparently protected by brambles, were killed when small rodents ringbarked them at ground level.

Mammal populations were, of course, very different in the past (summarised in Peterken, 2008a). The switch from red to grey squirrels took place in about 1943. Fallow deer were naturalised long ago, and their populations have fluctuated enormously over the centuries. The native deer would be red and roe, and the latter are returning. Prehistorically, the nearby caves and rock shelters have yielded the remains of an utterly different fauna from early Pleistocene deposits.

13.10.2 Birds

Birds operate on a larger scale than the reserve. Mute swans float past on the Wye; cormorants and kingfishers fly past; cuckoos can still be heard each spring; hawfinches move around Highmeadow Woods; peregrines nest in the vicinity; and buzzards are so common that they featured in the representation of the gorge on the cover of the *Wye Valley* New Naturalist book (Peterken, 2008a). These species are not entirely separate from the wood: buzzards nest in the wood, and we have even seen a mallard shepherding ducklings down through the reserve.

Within the reserve, we are indebted to three point-count surveys carried out in the 1980s, which collectively recorded 36 species and established that Lady Park was one of the richer woods along the lower Wye. Ian Bullock's 1985 survey found that the bird fauna was rich, and that the unmanaged areas were richer than the managed areas. Blue tits were three times more numerous in the unmanaged area; marsh tits, nuthatches (Fig. 13.11) and wood warblers were more numerous there; while tree creepers and chiffchaffs were recorded only in managed areas. The RSPB's 1986 survey found all three woodpeckers, with redstart, siskin, bullfinch, wood warbler, willow tit, tawny owl, pied flycatcher, nuthatch and tree creeper within the reserve. The most frequently recorded species was blackbird. Today, the reserve sometimes resounds to the hammering of woodpeckers.

13.11 Invertebrates

The Natural England reserve files contain several unpublished reports from the 1980s. Keith Alexander collected in both the reserve and the managed compartment in July and October 1984. He reported that 'the unmanaged area was clearly better for molluscs and for dead-wood invertebrates generally, (whereas) the lighter, managed area was better for flying insects, and especially butterflies'. He found two rare beetles associated with dead wood, *Diplocoelus fagi* and *Tillus elongatus*. These and two other beetle species, one fly, the brown snail and the ash-black slug were all regarded as

Box 13.1 Local status of species reported from Lady Park Wood by Keith Alexander

Coleoptera

Erotides cosnardi, Cosnard's net-winged beetle (Lycidae) – **Section 41 species**. This very rare relict old-growth beech species was first reported in Britain in 1944 from 'the Staunton Road', then found recently by Peter Kirby on Little Doward and further down the Wye Valley in Cadora Woods. Elsewhere in Britain, it has been found only in the ancient Forest of Arundel area of West Sussex.

Tillus elongatus (Cleridae, chequered beetles) – **NS** and **CS**. Found by Keith Alexander on a dead beech, 1984. A predator of other beetles on old broadleaved trees, especially larvae of Anobiidae (woodworm family), particularly *Ptilinus pectinicornis*, and usually in hard dead heartwood of beech; also reported from traditional orchards. The larvae hunt nocturnally under bark and on the outside of the tree.

Diplocoelus fagi – **NS**. Found in 1984 under dead beech bark by Keith Alexander, this is known also from the ancient beech woods and wood pastures of the Cotswolds. Larvae and adults feed on fungi; until recently, exclusively associated with beech, the adults occurring under bark on dead wood, particularly the loose outer layer. Adults are moderately mobile and are attracted to bracket fungi for feeding and overwinter in dead wood, including oak.

Cis festivus (Panzer) – **CR**. Found in 1984 by Keith Alexander. Generally associated with wood-decay fungi on aerial dead branches, such as standing dead hazel poles.

Leptura quadrifasciata Linnaeus – **CS**. Found in 1985 by D.J. Gibbs. Develops in dead and decaying stumps, especially birch, but also a wide range of other broadleaved trees, and once reported in spruce; may also develop in fallen branches and logs; larval sites permanently moist; adults visit flowers, sun-loving and fly on hot, sultry days. A relict old-forest species in the county, and largely confined to the north-west.

Diptera

Limoniidae

Achyrolimonia decemmaculata (Loew) – **CS**. Found in 1984 by Keith Alexander. A very local species that is usually associated with wood-rotting fungi; found in high-quality sites.

Mycetophilidae

Allodiopsis maculosa (Meigen). Found in 1986 by Ian McLean. An uncommon fungus gnat developing in various mushrooms and toadstools on the woodland floor.

Hybotidae – assassin flies

Oropezella sphenoptera (Loew). Found in 1986 by Ian McLean. A local species of shaded woodlands.

Sciomyzidae – snail-killing flies

Pherbellia annulipes (Zetterstedt) – **NS**. Trapped in 1985 by D.J. Gibbs. A southern species of ancient woodlands, especially beech, usually on calcareous soils. Only known in Gloucestershire from the Wye Gorge woodlands.

Opomyzidae

Geomyza angustipennis Zetterstedt – **RDB (Rare)**. A great rarity in Britain, known elsewhere only from Kent and possibly Speybridge, this was trapped by D.J. Gibbs in 1989. Ground living.

Anthomyiidae – cleptoparasites (cuckoos) in burrows of sphecid wasps.

Eustalomyia vittipes (Zetterstedt) – **NS**. Trapped in 1985 by D.J. Gibbs.

Muscidae – usually develop in accumulations of nutrient-rich debris, such as carrion, birds nests, etc.

Hydrotaea borussica Stein. **NS**. Trapped in 1985 by D.J. Gibbs.

Note: **CR** = county rare; **CS** = county scarce; **NS** = nationally scarce; **RDB** = included in the Red Data Book list of vulnerable British species.

associates of primary woodland. Silver-washed fritillaries were seen in the managed compartment.

In July and August 1985, D.J. Gibbs spent 3 weeks collecting and trapping invertebrates. The weather was windy and lacked sun, but he recorded 173 species, mostly the Diptera that fell into traps placed at regular intervals along transects II, V and X. The traps provided a quantitative comparison between different types of stand, but no differences were detected between the managed compartment and the reserve, nor between old-growth and young-growth stands within the reserve. A few species of local significance were found.

Fig. 13.11. Nuthatch, *Sitta europaea*, one of the hole-nesting species that are frequently seen in Lady Park Wood. This individual unwisely flew into a conservatory.

Gibbs's comments were more revealing. Facing north and with few sunny clearings, rides and flowers, the reserve was unpromising ground for invertebrates. Indeed, the sunny top ride and the partially shaded old railway line below proved to be the most productive locations. Within the wood, 'a very sunny clearing with abundant flowers' in the managed compartment was most productive. Since then, deer have increased, brambles have decreased and herbs scarcely flower, so one would expect the already limited invertebrate diversity to have declined.

R.A.D. Cameron and B.M. Pokryszko sampled molluscs at three locations within the reserve. At each, they searched a 400 m^2 plot for an hour and took 19 litres of soil and litter for sieving, finding 32 snail species and nine slugs. Their composition and diversity was much the same as the mollusc assemblages in other ancient woods on limestone in the Wye Valley, Herefordshire and the Cotswolds, and possibly similar to 'the faunas we think existed before there was significant human impact on the environment'.

The diversity of butterflies has declined steeply throughout the lower Wye Valley since the 1950s, not just in and around Lady Park Wood. Cooke (1913) recorded that there were 36 species in the gorge early in the 20th century and remembers 'fluttering peacock butterflies, brown speckled fritillaries, and a dozen more species along the river bank'. These may have been augmented by collectors, for a crown employee living in the Biblins Lodge kept 'jars in which are caterpillars waiting metamorphosis' in his sitting room. Parts of the gorge were warm enough to support a population of the introduced map butterfly, until it was exterminated by a patriotic collector during the First World War. However, most of the habitat specialists died out: the purple emperor, for example, hanging on until 1976. We spotted a wood white in 2004, a good year for this species, but neither it nor the fritillary has been seen since. Today, one hardly ever sees a butterfly within the fenced reserve, not even in the gaps.

13.12 Review

Lady Park is a microcosm of British woodlands. Despite the patchy information available, several general themes emerge that apply in some form to most woods:

1. Loss of species associated with open spaces. This is due broadly to lack of management, which has reduced the rate at which open spaces are created. Moreover, the ground condition of those natural open spaces that form instead is quite different from that of gaps formed by silvicultural operations. In many instances, the species survive close by in open spaces outside the reserve.

2. Possible gain of species associated with large trees and dead wood. This might be expected with the increase in the size of trees and the volume and diversity of dead wood, but the evidence is at best circumstantial. In any case, the potential gain in dead-wood associates may be limited by their poor colonising ability and, in the case of epiphytic lichens, impoverishment caused by past pollution.

3. Importance of specialised habitats. In Lady Park, the cliffs and the few seepages enlarge the biodiversity, notably of bryophytes and vascular plants. Elsewhere, watercourses, small mires and small patches of a locally rare soil type enlarge woodland diversity disproportionately to their area.

4. Wild species function on a landscape scale. This is obvious in the case of the larger and more mobile fauna, but it is also true of plants, whose populations within a wood interact with those outside. If the habitats around a particular wood are related, many species come and go, both as individuals and as populations. Any wood is part of a larger-scale patchwork of habitats and processes and is influenced by them.

5. Survival of the original fauna and flora. One often suspects that the species assemblage in an ancient wood is not just a lineal descendent of the assemblage in pre-Neolithic woodland but is still similar. In Lady Park, this has been most confidently voiced for the molluscs, trees and shrubs.

6. Grazing has a substantial impact on the flora and fauna. It reduces flowering and sources of nectar, and thereby changes the balance between herbaceous species directly and fauna indirectly. Grazing must have influenced what is now the reserve in earlier historical periods and possibly also in prehistory. Residual effects of past grazing are suspected, but difficult to prove.

7. People must have had a prolonged impact, most obviously on the larger fauna, eliminating some species and introducing others. This in turn has had, and is having, a substantial impact on stand dynamics. People have also introduced tree diseases and have changed soils. These impacts have accelerated in recent years.

Fig. 13.12. The face of the main cliff has been exposed by a massive fall of Yews, Sessile oak and both lime species, which has provided an opportunity for one of the Hawkweed microspecies to flower.

(a)
(b)
Wood / May '10 / Drought damaged / L. Cavin / F. sylvatica

14 Long-term Ecological Studies

> Many aspects of ecology, especially in woodland, can be understood only through observation continued over decades or centuries. The practice of making observation dependent on research grants and three-year Ph.D. courses prevents these aspects being studied. The philosophy, now fashionable, of designing observations only in order to confirm or disprove some preconceived hypothesis makes it difficult to discover the unexpected. At Madingley Wood, along with Wytham Woods and Lady Park Wood, the long duration of existing studies makes it possible to break out of these limitations.
>
> (Rackham and Coombe, 1996)

Forests operate on timescales of centuries, during which some changes are rapid and profound but most are slow and almost imperceptible. We try to understand them, not only for their intrinsic interest but also for highly practical reasons. What changes are happening now? Is the change unprecedented? To what extent are people responsible? Do we need to do anything to stop, reverse or facilitate change? If we have explicit objectives, are we achieving our aims, and how quickly? What are the consequences for wild species?

Long-term observation of permanent plots is one of many approaches to studying such changes. We can examine pollen cores, lake deposits, historical records, tree rings and other evidence from the past. We can look at chronosequences in contemporary ecosystems – places that are believed to differ only in time and thus show different stages in one sequence of development. We can observe short-term changes and project them forward using models to vary the conditions and estimate the uncertainty of any predictions. We can even reconstruct how stands have developed from evidence in the trees and soils (Henry and Swan, 1974). All these approaches have their pros and cons. In Britain, inferences from historical sources have given us a detailed long-term perspective of how present-day woodlands have developed.

Compared with these, direct observation of long-term change in 'real time' is, on the face of it, a blunt and rather naive approach, and is accompanied by a spectacular list of problems. To name a few: patience is required; originators may need to be altruistic; land tenure must be secure; records must be archived; the theoretical background, methods of recording and analysis will all have moved on by the time the study becomes long term; and the practical implications may emerge too late to be useful.

Nevertheless, many eminent ecologists have concluded that long-term permanent plot studies are a necessary complement to, and test of, other ways of understanding forest dynamics. For example, Hill and Radford:

> The classic, and often the only, way to study vegetation change is to mark out permanent plots on the ground and to make regular observations. There are many unresolved questions, even in (Britain's) much botanised woods (but) to answer (them) we must be both patient and observant. Historical research can provide some of the answers; only time and careful observation can provide the others.
>
> (Hill and Radford, 1986)

In the Pacific Northwest, Acker *et al.* (1998) stated that: 'Long-term observation of permanent sample plots is the only approach that provides real data on patterns and rates of change in forest ecosystems. Information collected on permanent plots provides the ultimate test of the validity of the other approaches.' Specifically, 'long-term observations provide a basis for monitoring effects of global climatic change as well as other changes. (They) are the only means of obtaining data on tree regeneration, growth and death; changes in species composition

Fig. 14.1.a and b. Alistair Jump, who supervised the dendrochronological study of beech and oak growth in Lady Park, holding samples of Liam Cavin's cores (a). Close-up of one of Liam Cavin's cores (b).

and diversity; and decay of dead wood. Other approaches are inferential.' A recent example is a study of carbon storage based on the Lyford plots at Harvard Forest (Eisen and Barker Plotkin, 2015). There, 40 years of records provided an opportunity to show how demographic and ecological changes resulting from succession, natural disturbances and land-use change influenced carbon sequestration. They could measure exchanges between the atmosphere and the forest in the short term, but carbon storage may change in the long term as stands age and change in composition and structure. The Lady Park records have also been used in studies of carbon sequestration (Fig. 14.1: see p. 230).

Permanent plots are not the only method for long-term observation of the environment. One has only to think of records of rainfall, temperature and other aspects of climate, which have been directly observed in recent centuries and extended back with historical records, ice cores and much else. Widespread change has been appreciated from phenological observations, the Common Bird Census, the Nest Record Scheme, butterfly transects and Rothamsted's network of insect traps. Long-term permanent plot studies have been undertaken in other habitats. The prime examples are the various experimental treatments at Rothamsted, including the famous fertiliser treatments of meadow grassland in the Park Grass plots, the continuous cropping of the Broadbalk fields and the prolonged succession to woodland in the Broadbalk and Geescroft Wildernesses. Our understanding of plant demography has been increased by C.O. Tamm's observations on populations of herbs and Terry Wells's repeated recording of orchid populations. Our understanding of bird populations has been increased by long-running observations of pied flycatchers in the Dean and of kittiwakes in the north-west. There are many other examples.

14.1 Permanent Plots in Woodland

The oldest permanent plot in temperate woodlands may be Joseph John's 1847 plot in Boubinsky Prales (formerly Bohemia, but now the Czech Republic) (Řehák, 1968). Dalby Söderskog (Sweden) has been recorded since 1909 (Brunet *et al.*, 2014). The earliest in British semi-natural woodland, which records the southern portion of Wistman's Wood, Devon, started in 1921. The pre-war initiative from which the Lady Park plots emerged appears to have borne little fruit elsewhere, but the plots in the Black Wood of Rannoch (Perthshire), which were started in 1948, form another example of a pre-Nature Conservancy study started by the Forestry Commission. Ecologists at Southampton University started the Denny (New Forest, Hampshire) transect in 1954. Transects on Clairinsh, a wooded island in Loch Lomond (Stirling), were started by the Nature Conservancy in 1961. Plots in Wytham Wood (Oxfordshire) were started by Colyear Dawkins in 1974. Transects in Monks Wood (Cambridgeshire, formerly Huntingdonshire) formed one of several studies started with baselines laid down in the 1980s. Oliver Rackham made meticulous site notes from the outset of his career, which enabled him to plot changes nearly half a century later (Rackham, 2003).

These are examples of studies that have succeeded in the sense that they have survived to become long term and have generated publications. Many more had the potential but have fallen by the wayside, including studies that were started with the clear intention of sustaining them indefinitely. Can we learn from experience, improve the survival rate and mitigate some of the difficulties mentioned earlier? The review by a former director of Rothamsted Experimental Station (Taylor, 1989) is particularly helpful in this regard, and GFP has reviewed experience with woodlands (Peterken and Backmeroff, 1988). One important point to emerge is that long-term permanent plot studies have a human dimension, the understanding of which is critical to their survival.

First, the need for patience: recognised by Mark Hill and Geoff Radford. This is inevitable if we start with a new baseline, but long-term observations become immediate if we can use information recorded in the past as a baseline. Such inherited records may be deliberate (intending to repeat the observations) or opportunist (a survey that was not intended to be repeated is used as a baseline). While the former will usually be quantitative and better designed, the latter may be just as useful for identifying, describing and quantifying change. Informal baselines include photographs, land-use information on maps (Harley, 1979), land surveys, historical descriptions and species lists, all sources where the ecological information may be incidental and qualitative. Old photographs are particularly useful for demonstrating change, partly because they are hardly edited to the particular interests of the recorders. Informal records include medieval documents that gain in longevity what they lose in detail. The practical point is that there is always a case for keeping old records. Many cautionary tales of violation can be told. Thus, the Air Ministry apparently threw out most pre-war air photos, and a set of the 1935 Land Utilisation Survey maps was found in a skip outside the Nature Conservancy Council's HQ, both having been assessed as 'out of date'.

14.2 When do Observations become Long-term?

Change can be measured at the second recording, and variation in the rate of change can first be appreciated at the third recording, which in Lady Park would have been at 10 and 20 years, respectively, if the original plan had been maintained, but which was, in fact, 10 and 32 years. These were certainly enough to demonstrate the pattern of tree growth and mortality in an unmanaged stand, but competition between trees plays out over a tree-generation or more, and that is measured in centuries. Further, some of the factors that influence woodland development happen irregularly, unpredictably and at long intervals. Thus, the 1987 storm in south-east England was the most severe since 1703. In short, observations of woodlands spanning a few years may be illuminating, for example the gap enlargement observed by Runkle and Yetter (1987) over 7 years, but the longer one observes, the more one learns.

Grasslands, in contrast, often change drastically in a year or two, so one might expect observations to become 'long term' sooner. For example, 9 years was seen as long term by Jerrentrup *et al.* (2014). However, many more years may be needed to separate subtle trends from annual fluctuations, and the oldest and most impressive long-term study of them all, the Park Grass plots at Rothamsted, shows that very long-term observations also have value. Started in 1856 to demonstrate the effects of different combinations of mineral fertiliser on meadow growth and composition, their initial aims were achieved within a decade, but they were kept going and by the mid-20th century had shown interesting features that were quite unexpected at the outset, and would probably have been of no interest if they had been anticipated, such as long-term fluctuations in the abundance of individual species and fine-scale genetic differences developing within previously homogeneous populations (Dodd *et al.*, 1995).

Many ecological studies have now become long-term, but how? And do such studies have lessons for those who want to initiate new long-term studies? Equally important, why have many potential and intended long-term studies lapsed?

Logically, one might expect institutions to establish and maintain studies over the long term, but Taylor (1989), who reviewed the Park Grass plots and several other long-term population studies, found that studies became long term more as a result of individuals than institutions. Even the Park Grass plots, which have been maintained by an institution, depended on the interests of individuals within Rothamsted and were threatened by the different interests of others. Today, with the continuing success of, for example, the Common Bird Census, the Butterfly Monitoring scheme and repeats of distribution mapping, combined with the rise of organised citizen science projects such as the great garden birdwatch and phenological records of the first cuckoo in spring and the like, institutions are undoubtedly essential, but individuals still matter. Indeed, the phenological observations provide an object lesson in the complementarity and partnership of individuals and institutions. Widely dismissed as whimsical by professionals, these individual and uncoordinated observations suddenly became important when Tim Sparks realised that they would yield insights into the impacts of climate change, and now they have become institutionalised (Sparks and Carey, 1995; Sparks *et al.*, 2016).

Christa Backmeroff and GFP also found that individuals had been decisive in enabling some permanent plot studies of semi-natural woodlands to become long term. For example, at Wistman's Wood (Dartmoor, Devon), the plot recorded by R. Hansford Worth in 1921 was relocated and recorded by Malcolm and Molly Spooner in 1965 and 1979, and then by us in 1987 and 1997, and at no point were those involved working under instructions from an institution (Mountford *et al.*, 2001). At Denny Wood (Mountford *et al.*, 1999), the transect initiated by John Manners and Joyce Lambert was revived by Peter Edwards and rescued from oblivion by ourselves after Peter moved to Switzerland, all on the initiative of the individuals whose job gave them freedom to develop their topics. These studies, like those at Lady Park, have had to outlast the initiators, who tend to develop other interests, move to new jobs, gain promotion, take on other responsibilities and must one day meet the Grim Reaper. They have also hung by a thread. To take one example, all the records of the Denny study were thrown out by Southampton

University after Peter Edwards moved abroad, and the study survived only because the Nature Conservancy Council had insisted on having copies when it paid for a new recording.

14.3 Experience from Lady Park's Permanent Plots

Our experience of permanent plot and other long-term studies at Lady Park and elsewhere (Box 14.1) bears out the initial point that long-term observations have special characteristics and present multifarious problems, even though there are possibilities for mitigation. In particular, point 16 implies that short-term and long-term ecological research are quite different. The former can follow the accepted pattern of testing a hypothesis with a designed and finite programme of observation and analysis, then stop with a conclusion. Such research is fixed term: it has a beginning and an end. The latter is more like historical research. One accumulates records and new observations and then interprets what one has or observes, making whatever allowance is necessary for incomplete and imprecise information, and reading more into a positive observation than a negative.

14.4 Value of Long-term Permanent Plot Studies

We can summarise our experience under six headings:

1. Testing the assumptions and results of other approaches to understanding change. Long-term observations provide an opportunity to ground-truth the results from other kinds of research by observing what actually happens. Other approaches are quick by comparison, but chronosequences make assumptions about the origins and trajectory of stands of different ages; models of stand development extrapolate from limited (short-term) observations; pollen-based studies show broad patterns of change from which mechanisms can be inferred, etc.; and all should be subject to a reality check. The general point was made in quotations earlier.

2. Provide opportunities for studying the unexpected. In Lady Park, the record enabled us to examine the impact of elm disease and the great drought in the light of the previous performance of the stand and individual trees, and it will provide the same opportunity in relation to ash disease. How we would have welcomed a similar opportunity in south-east England after the 1987 storm! At a time when pollution was being blamed for widespread beech mortality in southern Britain, a long historical perspective at least enabled us to remain open to other interpretation (Cramer, 1984). The rise in concern about climate change was not anticipated by Eustace Jones or Walter Lyford, as far as is known, but their permanent plots added to our knowledge of carbon sequestration by forests.

3. Monitoring environmental change. We need to be aware of what is happening in the environment, and in particular how much change is caused by people. Conservation organisations have to monitor the biodiversity of individual nature reserves and other important sites, as well as large-scale changes in biodiversity. Lady Park, together with a network of other long-term permanent plot studies in woods, has contributed to the understanding of continent-wide environmental change (Verheyen *et al.*, 2017). It could also be used to test the necessarily quick-and-dirty methods for monitoring conditions in Sites of Special Scientific Interest.

4. Illuminate the individuality of woods. We easily lose sight of the individuality of woods. Research seeks generalisations, nature conservation has placed much emphasis on classification in order to express larger patterns, and perforce much of the advice from policy makers and practitioners comes in generalisations. Yet, in-depth understanding of one or a few woods over a long period tends to show generalisations have their limits. Just like people, each wood has its own story to tell.

5. Personal understanding. Much the same thing, perhaps, but our experience of Lady Park has been that the slog of recording gives knowledge in depth, which complements the superficiality of most fieldwork and the breadth of survey. It was this that first got us interested. After a decade of briefly visiting numerous woods throughout Britain, the opportunity to stand in the wood holding charts and records made in earlier years was a revelation. The same could be said for any research, no doubt – it is the researcher who learns most. Even if one knows only one wood in depth, that unrepresentative sample-of-one shows what can be known about any site and alerts the ecologist to the limits of superficial acquaintance. In fact, much of the knowledge gained from one site in depth is general in the sense that one can apply it elsewhere (Billick and Price, 2010; Holmes and Likens, 2016). One good example also enables general points to be explained clearly (Ghazoul, 2015).

6. Demonstration and teaching. Lady Park has received many visitors, including forestry and nature

Box 14.1 Our experience of long-term permanent plot studies

1. They are usually initiated by individuals, who have the interest and vision, not by managers implementing policies. There may be no intention at the outset to continue indefinitely, but researchers often become curious later about how places and populations have changed. As the Williams plots in Lady Park show, it may be others who see the long-term value of an original record.
2. Patience is required, especially in forest types whose growth cycles take centuries.
3. A career-long interest must be maintained, or, if a project is started knowing that the results will be appreciated only by your successors, a degree of altruism.
4. Institutions often support studies in various ways, not least by allowing individual researchers some freedom to develop their programme, and sometimes formally adopt them. The Butterfly Monitoring Scheme originated with Ernie Pollard's interest and vision, combined with the freedom staff at Monks Wood Experimental Station were allowed to try something new.
5. The land on which long-term observations are made, and any management regime involved, must also be committed for the long term. It is unrealistic to embark on a long-term study without secure land tenure.
6. Funding will always be fickle. The 'long, thin' funding required does not fit modern budgeting, nor 3-year postgraduate research grants. No public institution will commit meaningfully to open-ended funding.
7. Long-term studies are usually low priority for institutions, which need to demonstrate more immediate benefits. They may also be unrewarding for individual researchers, whose tenure usually requires quick results.
8. They must eventually pass to successor individuals if they are to continue. The baton must be passed on, and many potential long-term studies have lapsed when 'the baton has been dropped'.
9. The record must survive, which means archiving. To quote *Hamlet*, this is more honoured in the breach than the observance. The Denny experience demonstrates the risks.
10. The knowledge that the record exists and where it lies must survive, which implies an institutional record or visibility through a publication. At Clairinsh (Backmeroff and Peterken, 1989), the site, the record and the need to monitor change on the reserve all survived, but the initiator had become a timber merchant and the site manager of the time had become a director elsewhere.
11. The record must be intelligible without recourse to previous recorders. The methodology and any codes should be explained as part of the record. Direct contact between successive recorders is desirable, if possible, but recorders may not recall all the details many years later.
12. The plots must survive and be able to be found. This requires not just durable plot markers but also liaison with land managers. Our experience is that transects are relocated more readily than plots after periods of apathy. Plots relocated by means of compass bearings will 'move' with magnetic north.
13. The data have to be usable. Systematic data can be stored electronically, but questions remain about how accessible such records will be decades hence. Who now can access data in LOCOSCRIPT? Even without this reservation, there is a case for paper records as independent duplication.
14. The quality of records may not meet modern standards. Even if it is complete, the original sampling design may not conform to current statistical procedures. At Lady Park, different recorders have used different methods to measure trees, which is manageable, but hardly ideal. Methods of design, recording and analysis will change as well, so early observations in a long-term study may be deemed deficient by the time the study becomes long term.
15. Long-term observations can be deemed unscientific, insofar as they fall outside normal, hypothesis-driven research methods. When we offered a paper on the impact of disease on the wych elm population in Lady Park, in which we analysed 25 years of responses to the disease in relation to the growth of the individual elms in the 25 years before the disease arrived, we were told we lacked a hypothesis. But this was inevitable: in 1945, when the hypothesis would have had to be stated, one of the authors had not been born and the other had still not started at primary school. To invent a hypothesis would have been dishonest, and to write up as if there had been one is just an awkward form of words.
16. The theoretical background and research preoccupations of ecology will change by the time observations became meaningfully long term. Thus, long-term studies may be directed at ideas that are no longer of interest: when Lady Park started, most ecologists accepted the idea of a predictable succession to a stable climatic climax, yet now we see things differently. Put another way, we do not know which aspects of long-term observations will be of interest to future ecologists. Starting a long-term woodland study is an act of faith, though one would be wise not to justify it that way when seeking funds.

Continued

Box 14.1 Continued.

One needs clear objectives, even if they turn out to have been a pretext.

17. Explicit objectives may be necessary to start a long-term study, but will be modified or expanded by the time a study becomes long term. The only statement we have of Eustace Jones's initial objectives at Lady Park came in a letter of 16 August 1944 to W.H. Guillebaud: the 'main objectives' were:

> (1) The preservation of an example of limestone woodland with a specially rich and interesting flora and (2) succession of coppice regrowth in old woodland. The latter stages in this will be of special interest – the ultimate fate of coppice shoots; the role of beech and the various calcicole species such as elm and lime.

The 'opportunity to study succession in absence of rabbits was most important. There are plenty of rabbit-infested woods elsewhere.'

18. We cannot predict how a study site will be impacted by natural events and thus cannot predict what will be studied.

19. Any land-management action that waits for long-term observations to be completed may be too late. A long-term study should not be an excuse for prevarication: land managers should take timely action based on the best available information.

conservation groups (Fig 16.2), Master's degree courses in forestry and many individual groups of ecologists. In recent years, we have received foresters from Sweden and ecologists from the USA, Belgium and Switzerland. Even some of our non-ecologist neighbours want to look round! To each, according to his or her interests, we can demonstrate the individual CV of each tree; transitory events that left their mark, such as the 1981 snowfall; conditions leading to an event, such as a tree fall, and what went before; conditions and events that are no longer obvious, such as old gaps, how they formed and how they filled; what tree was once precisely *there*, but is no longer visible; the origin of individual pieces of dead wood; how long an oak stump takes to decay; show old photographs at exactly the position they were taken and thus demonstrate almost subliminally many aspects of change; place a tape round a trunk and be able to say what the girth was 10, 20, … 70 years ago; and much else. These are mostly items that cannot be demonstrated in a wood without long-term observations to back them up, or which can be demonstrated only by making assumptions. Of course, similar opportunities are available in any wood graced with a long-serving forest manager or an elderly, enthusiastic owner, but in Lady Park we can be quantitative; and we can demonstrate with the record points that would have had to be taken on trust from an owner or manager without a detailed record to hand.

Fig. 14.2. Liam Cavin is interviewed live for the breakfast programme on BBC1 as part of a feature related to climate change. On this occasion, Lady Park also featured on the *Today* programme on BBC Radio 4.

Fig. 15.1. Cove forest in Joyce Kilmer reserve, near the Great Smoky Mountains National Park, USA. Dominated by giant tulip trees, *Liriodendron tulipifera*, this stand corresponds to the vision of natural woodland as an intimate mixture of many sizes and species of tree.

15 Natural Woodland in Theory and Practice

The target in Lady Park Wood has been to recreate natural woodland, which, like most ecologists, we take to be woodland whose characteristics have not been influenced by people and have thus been determined by natural factors alone. However, natural woodland so defined has probably not existed in pure form in north-west Europe since the ice retreated and early hunters decimated the megafauna; it has certainly not existed in Britain through historical times, during which woods have been cleared, grazed and felled; and it is not recoverable, for we cannot exclude the effects of past use, nor the indirect effects of people even from strictly protected reserves.

What, then, is the point of a target that is unattainable? Several reasons were listed in the introduction, but fundamentally it is a scientific necessity. We need 'control' woodland, as free as possible from human influence, so that we can assess the impact of people on woodland processes and characteristics. We do not deny that people have long been part of nature – indeed, many revel in the details of how people and woodlands have interacted down the millennia – but we need a clear framework for understanding our impacts.

Most ecologists have a nuanced view of naturalness. We accept that it is not a black and white matter but a continuum from very artificial (such as a Sitka spruce plantation on a Sutherland peat bog) to almost completely natural. We also see this continuum as multidimensional, in the sense that, in any given wood, some features can be more natural than others. Thus, for example, an ancient coppice will have an artificial structure, but a relatively natural composition that varies in concert with soil variation. We also recognise different types of naturalness, past (original woodland before people had a significant impact), present (features we inherit today in woods with a direct link to original natural woodlands) and future (woodland that would develop if we just 'shut the gate' and left it alone), and that they are different. Thus, in Lady Park, the pre-Neolithic 'past natural' composition was probably dominated by the two lime species, but the 'future natural' composition could become a mixture of sycamore, western hemlock and cherry laurel.

Past and present ideas about naturalness have been coloured by F.E. Clements's (1928) understanding of vegetation, which was based around the notion of predictable, natural succession to a stable climax, and this framework underpinned Tansley's interpretation of British woodlands. Tansley was well aware of the long history of planting and coppicing, but clearly wished to see Britain's native woodland as natural. Despite recognising that 'of the great forests which at one time covered a large part of the British Islands, only fragments remain, and most of these have been very much modified by human activity' and that 'much of the existing coppice-with-standards has been planted, and this usually differs from natural oakwood by the relative scarcity of trees other than oak and of shrubs other than hazel', he thought nevertheless that 'in spite of such deviations, this type of wood does not differ essentially in constitution and structure from a natural oakwood, and may be fairly regarded as representing the natural climax forest, though in modified form' (Tansley, 1939). Thus, he acknowledges that the woods we encounter are, at best, modified climax woodland, and very usefully coined the term 'semi-natural', which has been a familiar concept in ecology and nature conservation ever since.

Even though we cannot see natural 'climax' woodland in Britain, we can see it elsewhere in the form of virgin forests in mainland Europe and old-growth stands in North America, or so some suppose. Continental forest ecologists initially regarded virgin forests in pure, almost idealistic, terms. Thus, according to Mauve (1931, translations by Christa Backmeroff), the palaeobotanist Johann

Göppert (1868) gave the first definition of virgin forest as 'forest which has never been exploited, where vegetation has not changed for thousands of years and nature alone formed and destroyed the gigantic trees'. Similar views were still held by some in Mauve's time and Tansley thought 'virgin forest is clearly natural', but Mauve himself accepted Tschermak's (1910) definition as 'any woodland which developed completely without human interference, as a result of ... natural factors for the distribution of plants', which opens the door to 'secondary' virgin forest.

Today, we recognise that the English translation of the German 'Urwald', and similar terms in other languages, as 'virgin' forest is misleading, for many Urwälder are known to be neither original nor untouched by people. Thus, Mionší Prales in southern Slovakia was a wood meadow, Bumbálka Prales (Czeck Republic) is so tiny and hemmed in by meadows and plantations that it could not possibly have escaped modification, and the Hasbrucher and Neuenberger Urwälder (Germany) are former wood pastures, just like the New Forest. GFP has even been shown a 'prales' in southern Bohemia, which turned out to be a 20-year-old, naturally sown alder wood growing in a former oxbow of a managed river. Even the famous Białowieza Forest (Poland) shows many signs of fluctuating past management as wood pasture and wood meadows (Latałowa *et al.*, 2015). Nevertheless, pollen profiles have shown that some have changed little (e.g. Rothwald, Austria; Kral and Mayer, 1968), while the Stužica Prales in the hills where Slovakia, Poland and Byelorussia meet is so remote that one can readily imagine that it has escaped direct modification by people, save for hunting, which will have changed the balance between herbivores and carnivores.

Much the same might be said about the old-growth stands in North America and northern Europe. The American stands may not have been touched by Europeans, but they previously coexisted with pre-Columbian people, who managed the forests by fire and herds of buffalo that restricted their development (Russell, 1983). Many burn easily, whether the fire is set by lightning or by people, which enabled even pre-Columbian people to influence forests on a landscape scale: people and forests co-evolved (Pyne, 1982). In any case, many are now heavily grazed by deer. In Finland, forests were long used by Sami reindeer herders, so that now old-growth forests are understood in terms of traditional uses and modern fire control, as well as natural factors (Keto-Tokoi and Kuuluvainen, 2014). Some old-growth stands, like the central Appalachian Cove Forests, conform to an idealised image of natural woodland as towering mixtures of large trees of several species, living to great ages, and growing among a thicket of saplings and smaller trees that would eventually replace them (Fig. 15.1: see p. 238), but most are rather ordinary. Old natural pine stands look little different from old plantations, and many of the Boreal coniferous forests are almost even-aged as a consequence of past fires. Stand structures found in beech-dominated virgin forests can be found in beech-dominated high forest and coppice in Britain – even natural beech coppice, which can be seen at the treeline in Stuzica Prales, Slovakia, where it is smashed repeatedly by snow.

Tansley completed *The British Islands and their Vegetation* in 1939, a time when most British ecologists seemed to have been unaware of the characteristics of European virgin forests and American old-growth. The forests of mainland Europe were inaccessible even in peacetime, and publications about them were published in German, Czech, Swedish and other Continental languages. The exception was the initiator of the Lady Park study, Eustace Jones, who, like Tansley, had studied and taught at both Oxford and Cambridge. In common with Alex Watt in the Botany School at Cambridge, Mark Anderson in Edinburgh and several Americans, Jones was both forester and ecologist, which detached him from Clementsian ideas of succession and climax vegetation and brought him into contact with Continental forests and down-to-earth foresters.

Jones reviewed the available descriptions of Urwälder in preparation for lectures to forestry students at Oxford. His findings were published only because the editors of *New Phytologist* were short of material at the end of the Second World War, or so he said. Casual though this seems, the result was a review paper (Jones, 1945) that has become a classic (Bradshaw *et al.*, 2011) and was the inspiration for *Natural Woodland* (Peterken, 1996).

Jones found that, far from being stable, virgin forests were subject to a variety of disturbances, some of which were severe enough to destroy stands completely. Thus, natural woodland could take many forms, not just mature stands but also young ones recovering from disturbances. Moreover, even mature stands that escaped destruction were constantly changing in structure and composition under the influence of lesser disturbances. Disturbances as a whole varied in scale from the fall of single trees to immense fires that consumed the forests of entire regions, and, with few exceptions, they happened unpredictably. Natural woodlands were, in fact, unstable at a small and medium scale, but still might be stable at a large scale, where the changes in one place were counterbalanced by changes elsewhere. Against this background, Jones understandably rejected the idea of a predictable succession to a stable, mature, climax woodland. Rather, natural forests were not just dynamic and varied in structure but also they were similar in many ways to managed forests and could be described in terms of the silvicultural systems already well understood by foresters. Natural woodland exhibited a great variety of structures and patterns, not just the small-group dynamics that selection forests ('plenterwald') tried to mimic.

Jones's review, like Watt's pattern and process paper (Watt, 1947) has been widely and enduringly cited. Subsequent research, particularly in North America and Scandinavia, has confirmed that natural forests are almost as diverse in their structures and dynamics as in their composition. Some follow a 'snakes-and-ladders' process in which stands develop according to a predictable sequence (the 'ladder') until the next stand-destroying fire, storm or river-channel movement (the 'snake') sets them back to almost zero. The return time of such disturbances may be decades or centuries, and their incidence will be unpredictable in space and time, though not entirely random, for the age of the stand or its particular location may determine the probability that a stand will be destroyed. Between stand-destroying disturbances, natural forests remain vulnerable to small-scale disturbances that punch gaps in the canopy at an average rate of about 1% per annum, equivalent to an average canopy residence time of 100 years for the trees that get that far. Each forest and each type of forest can be characterised by its disturbance regime, made up of the disturbances to which it is subject, their frequencies and patterns, and the various ways in which one disturbance leads to another.

These ideas must have been fresh in Eustace Jones's mind when he started the recording of Lady Park Wood in the very year that his review of natural structure and dynamics of virgin forest was published. Inevitably, one assumes that he hoped that the project would eventually improve our understanding of the disturbance regimes shaping British woods, though he never said so in so many words. He would have foreseen mainly small-scale, wind-related disturbances generating a fairly steady supply of small gaps, which, in the long run, would produce a small-group structure in which beech and perhaps other shade bearers would dominate. He knew that a catastrophic disturbance might change everything, but would have been quite unable to predict when one would happen.

In the event, a nearly catastrophic disturbance happened after 32 years. The 1976 drought caused major changes immediately, and its effects have endured as drought-damaged beech fail and fall. In addition, as the stands have got older, the variety of disturbance has increased, but thus far wind has not been the most prominent. Disturbances in the form of discrete events (e.g. a drought) have combined with fluctuations in chronic conditions (e.g. grazing by deer) and semi-natural disturbances (e.g. elm disease), while the influence of the slopes has increased as trees have become larger. Each disturbance has exhibited its particular pattern, having more impact on some tree species and some site types than others. This has generated different disturbance regimes in different parts of the wood, determined partly by stand age and site type. Regeneration patterns were largely obscured by fallow deer until 2007, but by 2016 vegetative regrowth and groups of saplings were starting to accelerate in many gaps. A small-group structure was starting to develop, but the long-term, shade-bearing dominant was as likely to be lime as beech.

Lady Park is only a tiny patch in a much larger pattern that has also become clearer in the past 70 years. Other long-term observations of permanent plots confirm that the 1976 drought was felt throughout the beech population of southern Britain, for example in Denny Wood (New Forest, Hampshire) (Mountford *et al.*, 1999), and have hinted that heavy snowfalls may be disruptive at the treeline (Mountford *et al.*, 2001), but elsewhere (for example Monks Wood, Clairinsh island and Black Wood of Rannoch; Peterken and Stace, 1987; Backmeroff and Peterken, 1989; Mountford, 2004) the canopy has remained largely undisturbed. Generally, however, wind has proved to be the main agent of disturbance. This was amply demonstrated in October 1987 by the great storm in south-east England (Peterken, 1996; Rackham, 2003), which levelled 15% of all trees/woodland east of a line from the New Forest to central Norfolk. It shocked not just legions of commuters and gardeners in the greater London region, but also ecologists and conservationists, for no similar storm had passed over in the region since 1703, two centuries before ecology was invented. Wind speeds varied within the storm and were modified by topography, so the impact was notably patchy. While some woods were almost levelled, including the ancient wood pasture at Toys Hill in Kent (Mountford, 2004), others nearby escaped. Within individual woods, such as The Mens (West Sussex), some patches were blown flat, while the rest of the wood stood firm. On scarps facing the storm, large beech were blown uphill and gullies that funnelled the wind concentrated the destruction, while trees on the lee slopes, which seemed to be protected, were damaged by the downdraughts from eddies. The Downs were pockmarked in white where windthrown beech exposed underlying chalk. Alder woods on wet soils were easily blown over, while oaks on the clays mostly remained upright, albeit with their crown branches twisted off, leaving a bottlebrush of wrenched splinters at the top of the trunk. Many other notable storms have passed across Britain; for example, central Scotland in 1968, and blowdowns undoubtedly happen somewhere every year, but 1987 has coloured the thinking of a generation. Interestingly, and for various reasons, its impact is now hard to detect on the ground.

Storms are not the only natural disturbance faced by British woodland. Fire may well have been important naturally in Caledonian pinewoods, but, as Oliver Rackham often said, one cannot set fire to British deciduous woodland, even with a flame-thrower. Drought, as Lady Park has demonstrated, can be influential in beech and birch woods, although its impact has proved to be age dependent. Floods must once have influenced floodplain woodland when river channels were still mobile, and waterlogging can kill mature stands, e.g. the ancient beeches at Denny Wait in the New Forest. Disease increasingly impacts native trees, elms over centuries, oak, alder, beech and ash only recently. And the advent of deer, grey squirrels and now wild boar introduces, or reintroduces, other forms of disturbance.

Despite the variety of disturbances, there is little doubt that, under British conditions, most trees and shrubs die quietly. They are weakened in competition with their neighbours and ultimately excluded by a puff of wind, a mild infection or simply by respiring more than they photosynthesise.

Lady Park is not alone in showing that the arrival or exclusion of deer is a major event in the history of a wood and that any large fluctuation in deer populations is akin to a disturbance. The impact of large herbivores became a matter of fierce debate when Frans Vera, a Dutch ecologist, published his view that, far from being wall-to-wall trees, natural temperate woodland included substantial open areas, not treeless, but open enough to support herbaceous vegetation containing light-demanding plants (Vera, 2000). Large herbivores, in his view, controlled a dynamic mosaic of scrub, closed woodland, parkland and open ground, each state capable of changing into any one of the others. Thus, natural woodland would look more like the unenclosed wood pastures of the New Forest than the close-canopy Urwälder of mainland Europe, the old-growth stands in the eastern USA, or indeed Lady Park Wood.

That early forested environments were partly open was deduced earlier from the relict distribution of some grassland species (Pigott and Walters, 1954) and the superior growth of epiphytic lichens on trees by the edges of glades (Rose and James, 1974). It makes sense to many invertebrate specialists, for, in their adult stage, saproxylic (wood-using) invertebrates need nectar sources that develop only in the open (Siitonen and Ranius, 2015): thus, Vera's hypothesis explains how they thrived in pre-Neolithic forests. Many meadow herbs, particularly the species of wet meadows, were not only present before the Neolithic but also during previous interglacials. Some will grow in shade, but many will not, so the original natural landscape presumably included wet, meadow-like grasslands (Peterken, 2013) that would not have been closely grazed in spring, because herbivore populations were not shielded by supplementary feed from high mortality rates in hard winters. Any inhibition of tree growth and regeneration on wet ground would have been reinforced by a feedback: large herbivores congregate on open areas where the herbage is best, thus keeping them open.

Vera gave no indication of the likely balance between scrub, groves and other states, nor did he differentiate between different locations, but, making reasonable assumptions about the endurance of each state and transitions between them, Kirby (2004) concluded that most of the ground would still be covered by trees. Svenning (2002), after reviewing numerous pollen profiles, concluded that woodland on wet low-lying ground and dry, thin-soiled uplands was probably more open than woodland on other site types. Whitehouse and Smith (2010), who reviewed subfossil remains of beetles in British sites, found indications that early Mesolithic forest was fairly open, but that late Mesolithic forests became closed at a time when elms and limes became prominent. Open spaces nevertheless persisted, most likely as marshes and meadows in valleys. In pre-Neolithic forest on mesic ground, where trees grow and regenerate well and beech and other casters of dense shade tend to dominate, herbivores would have found less to eat and the forest was largely closed.

Possibly the most enduring elements of Vera's cycle would have been the 'groves', which would presumably have been closed and almost-closed stands. Vera says little about how they might function, but they would surely thin like the Lady Park stands and be increasingly vulnerable to a variety of disturbances as they age. The age at which they break up is fairly predictable, but whether they change to parkland or regenerate to a small-group structure would depend on the size of the herbivore population.

Vera's ideas were analysed by Mitchell (2005), Rackham (2006) and contributors to the *British Wildlife* special supplement (June 2009) among others, which, combined, left the impression that natural woodland was more open than most ecologists had previously acknowledged, though not as open as Vera seemed to imply. There were even suggestions from Continental Europe that pre-Neolithic woodland was never completely natural, but was kept open by people in earliest postglacial times (Hédl *et al.*, 2010). Insofar as our experience at Lady Park has anything to add, we note that, in the face of *uncontrolled* deer populations: (i) spiny shrubs and woody debris have thus far afforded negligible protection for saplings; and (ii) lime and hazel, the key elements of pre-Neolithic forests, have been browsed preferentially.

What kind of naturalness are we developing in Lady Park Wood? A Mesolithic hunter operating from King Arthur's Cave on the other side of the Wye would obviously have been disconcerted by the fence and the surrounding land, but within the reserve he would recognise the tree species, for these are much the same as they were, according to pollen analysis from the Levels (Brown *et al.*, 2005), and he would be accustomed to large trees and the decaying litter of fallen timber. He would have relaxed when he realised that wolves and bears had gone, taken the presence of wild pigs for granted, but he would be stalking fallow deer, not red deer, roe deer and aurochs. The Seven Sisters rock towers still present open vantage points, easily accessible from the cave and the Madawg rock shelter, but the flood plain is now grass, not swampy alder woodland. The wood has changed less than its context, and original natural woodland seems just within reach.

Even so, complete restoration of pre-Neolithic woodland is out of the question. Physically, the climate has changed and soils have had 5000 more years to mature. Biologically, the context has changed completely, not only has the fauna changed but also new diseases and a plethora of archaeophytes and neophytes has arrived. The nearest we can approach is what we call present natural woodland (based on native species inherited from pre-Neolithic woodland), and then only by vigorously excluding non-native plants. The larch, Douglas fir and chestnut still present in the reserve are tolerated because they are not spreading, but we would resist cherry laurel, Buddleja, Western red cedar and any others that saw fit to colonise and spread. The present natural woodland of Lady Park may well resemble original natural woodland – the tree species are much the same and the evidence from the remains of prehistoric faunas is that the woodland canopy in the gorge was mostly closed-canopy woodland – but we cannot be sure.

Present natural woodland seems a reasonable target in a reserve where: (i) the woodland seems close to the original composition; (ii) is large enough to include all aspects of the growth cycle; (iii) is surrounded mostly by woodland; and (iv) the terms of tenure offer a reasonable chance that the necessarily long-term objective can be sustained. It is well justified in the few special cases like Lady Park, which provide controls in the scientific sense that enable us to understand managed woodland and advise on more natural forms of forest management. Elsewhere, one might just as well chose other options, such as: (i) manipulating a semi-natural mixture to provide the optimum range of habitats and wildlife, which is what happens in most woodland nature reserves; (ii) permitting a genuine free rein to whatever is there or happens to colonise, i.e. allowing future natural woodland to develop, and accepting that one cannot predict what we will get; and (iii) managing non-native trees to provide a new form of future natural woodland in the form of semi-natural analogues of the forests of the Pacific Northwest, which is effectively what the retention stands and continuous cover policies in upland plantation forest are designed to produce.

16 Near-to-Nature Forestry

Forestry is a broad church now, far removed from the narrow objectives of the 1970s. Then, the aim was to grow utilisable timber as economically as possible, and other considerations were deemed to be a restraint on, or a cost against, timber production. Net discounted revenue calculations were the order of the day, even when they led to poisoning well-established 35-year-old oak plantations and replacing them with Corsican pine. The target was to cover as much of Britain as possible in timber plantations, almost all of which were even-aged and coniferous. In doing so, foresters accepted no limits and barely acknowledged that there might be places where trees were unwelcome. Some saw this as a moral issue – to follow a 'biblical' injunction and make several million trees grow where none grew before – and it was reinforced by the liberal use of herbicides to control 'weeds' and 'scrub', as well as efficient drainage, especially of upland peats, to improve growth rates. As one forester memorably said to GFP, 'If the site ain't right, we'll make it right' – but in his case, it was ironical and said with a smile.

Today, foresters see their profession as multipurpose, much more nuanced, serving a much wider range of society's requirements. Timber and wood production is still the main purpose, and foresters have to work within budget, but public recreation, good landscape, wildlife conservation, natural features, environmental education, archaeological features, even artistic endeavours, are also important, and so are beneficial impacts on the environment, such as clean water, flood mitigation and carbon sequestration, the last achieved by maintaining a healthy stock of timber and minimising peat loss. The Forestry Commission in particular promotes and facilitates public access to its land, and manages its forests so that visitors have an attractive, safe and rewarding experience.

This transformation was matched by changes in the attitudes of conservationists. Instead of merely vilifying the even-aged monoculture cul-de-sac that forestry had entered, many saw the potential benefits for nature conservation of a broader interpretation of forestry, and not just because farmland in contrast had become so hostile to wildlife. Silvicultural interventions diversified stand structure, generated the wildlife-rich phase of stand initiation, generated open spaces that remained insulated from farmland and could protect habitats from excessive grazing. On woodland nature reserves, it might even be possible to generate income from timber sales for conservation management.

Out of this has evolved a renewed interest in managing *with* the ecosystem rather than *against* it. This was formally embodied in the Broadleaves Policy of 1985 and its subsequent refinements and manifest in the expanding range of interests, activities and expertise of foresters. In many ways, this was tantamount to a return to elements of traditional forestry, which worked with site-native trees and valued wildlife, i.e. it was a renewal, not an innovation. Out of this has emerged not just a willingness to work with site-native tree species but also to promote nature-based forestry (Larsen, 2012; Pro Silva Europe Principles, 2012; Brang *et al.*, 2014; Bürgi, 2015; O'Hara, 2016).

Near-to-nature forestry is not new. It was well developed as selection forestry in mainland Europe and even applied in Britain to some Chiltern beechwoods. Part of a spectrum of approaches to forest management (Duncker *et al.*, 2012), it occupies a middle ground between even-aged

Fig. 16.1. Douglas fir in Coed y Brenin Forest, north Wales. Planted in 1928 by the Forestry Commission and latterly thinned, this has developed into an impressive example of continuous cover forestry, where beech and Douglas fir regenerate prolifically and other native broadleaves and introduced conifers are present in small numbers.

plantation forestry with exotic species and woodland reserves from which little or no timber is removed. Mimicking natural woodlands, it fells in groups and single trees; regenerates naturally; preserves old trees; employs species mixtures; favours native species; develops uneven-aged stands; limits soil modification; and avoids using pesticides. But none of these attributes is absolute. Thus, it makes some use of exotic species; allows some planting and monoculture; harvests most trees close to the economic optimum; and permits some wood salvage and some soil drainage. Most forests are thus 'continuous cover', as it is known in Britain, but the permitted measures enable timber production to be sustained, while maintaining soil fertility and biodiversity. If it works, the middle ground is an optimum, not a compromise.

Switching to nature-based forestry requires British foresters to relearn and revise some forestry skills, not just because old skills may have been forgotten but also to find ways of working in the new circumstances, where deer, grey squirrels and wild boar are seemingly uncontrollable and new diseases are imported without let or hindrance. There is also a need to understand what form 'natural' takes in Britain and what aspects can be incorporated within managed woodland.

This is where Lady Park Wood makes, or could make, a contribution. First, at the risk of sounding less than serious, Lady Park demonstrates what happens when foresters neglect woods. Within the fenced reserve, much of the young growth has poor form and lacks dominants (what foresters used to label 'scrub'); there has been little regeneration in gaps; the wood is full of dead and decaying timber; clusters of large trees have fallen together and block tracks; and the place is unsafe in a stiff breeze. On the other hand, Lady Park also demonstrates that foresters have much to learn about managing native broadleaved mixtures. The foresters chose to retain beech disproportionately when thinning, but grey squirrels responded by stripping bark, killing leaders and introducing rot into the trunks. Foresters sought natural regeneration, but they could not control deer at population levels that allowed saplings and coppice shoots to survive, and the thinning operations actually destroyed most of the established regeneration. Even if deer and squirrels had been controlled, the structure the foresters have created is nowhere near to nature.

16.1 Silvicultural Systems

For many years, R.S. Troup's *Silvicultural Systems* (1928) was a standard source for foresters. It described in a relentlessly systematic fashion the several systems that foresters and woodmen had evolved to manage the growth and harvesting of broadleaved woodlands. Some were already obsolete and others were niche systems, suitable for unusual circumstances, but most were widely practised, if not in Britain, then in France, Germany and other European countries. These were the systems to which Eustace Jones (1945) compared the structures generated in natural woodlands.

The various silvicultural systems depend on two main variables (Table 16.1): (i) the amount and pattern of initial and later fellings; and (ii) the origin of the new growth. The latter is a simple continuum, from 100% seedlings and/or plantings to 100% vegetative regrowth, with everything in between being possible, but the felling variable is more complicated. At one end of the scale, the mature stand is clear-felled, leaving nothing. At the other, it is thinned perpetually by removing selected single trees. In between, we have partial initial fellings, which would remove anything from, say, 10% to 90% of the growing stock, with the aim of leaving seed trees, from which the group would be restocked, and conditions that would enable the new regeneration to

Table 16.1. The characteristics of silvicultural systems.

		Partial initial fellings		
	Clear felling	No eventual retentions	Some retentions through next rotation	Perpetual thinning
Seedling	Clear-cut, even-aged high forest	Group, uniform and irregular shelterwood	Two-storied high forest; high forest with standards	Selection forest
Vegetative	Simple coppice		Coppice with standards	Selection coppice

become established. These retained trees might be removed completely once regeneration had been established, resulting in an almost even-aged stand; or they might be retained through the next rotation, resulting in a two-aged stand. Irrespective of the total amount of felling, the pattern might be uniform or irregular. Irregular partial fellings come close to small-group selection fellings, which remove small groups of trees rather than single trees.

The resulting silvilcultural systems, each with traditional names, have several variants, depending on the shape of the initial fellings and the scale of the 'subcompartment'. Thus, selection forest may be either single-tree selection or group selection. The scale to which these systems refer is the compartment or subcompartment. If we dealt on a much smaller scale, for example, group selection would be perceived as a scatter of tiny clear-fellings.

Taking the systems in the four corners of Table 16.1:

- Top, left. Generates uniform, even-aged stands that may be thinned in mid-rotation. Planting is usually necessary, combined with cleaning to reduce self-sown (but unwanted) regeneration. Favours light demanders, but acceptable for shade-tolerant trees. Suited to regimes where a complete change of species is desired and purely commercial objectives prevail.
- Top, right. Generates mixed-age stands while never exposing the soil to full light. Favours shade species, but light demanders may retain a foothold in group selection systems. Usually regenerated from seedlings, but planting is possible. Favours the existing species; unsuited to change of main species. Requires skilled selection of trees for felling and skilful operators. Rotations are not fixed: rather, the volume extracted is balanced against growth rate to maintain a near-constant volume.
- Bottom left. Retains existing species and individuals. Seedling regeneration is usually mixed with coppice growth, but may be cleaned out in mid-rotation. Operations are simple and regeneration is reliable. Changes in species were achieved by: (i) cleaning out unwanted species or (ii) planting wanted species.
- Bottom right. Selection coppice was used only on beech (which coppices weakly), but could be used with other shade species such as limes. Coppice with standards generates a varied structure. It can be operated with only vegetative regeneration – indeed, traditionally, oak standards were generated from stool growth as well as seedling regeneration.

In practice, where vegetative regeneration predominated, simple coppice or coppice with standards were the norms. Standards were always retained through the next coppice rotation. Continuous cover forestry is not strictly a silvicultural system but a mix of the systems towards the top, right-hand corner of the table, combined with small-scale 'subcompartments'.

Each system has, or would have, its consequences for the balance between species in native mixtures. Table 16.2 indicates how the mixture at Lady Park might have responded before the various 20th century afflictions arrived. Now, with fallow deer in excess eliminating most seedlings, disease impairing wych elm, oak and, soon, ash, grey squirrels deforming beech and, to a lesser degree, oak, only limes, birch and aspen are left to function normally. The only 'good news' is the higher frequency with which limes produce fertile seed.

Table 16.2. Responses of the principal trees in Lady Park Wood to various silvicultural systems in the absence of imported diseases and naturalised mammals.

	Mainly seedling (or sucker) regeneration				Mainly vegetative regeneration	
	Clear-cut	Shelterwood	Two-storied high forest	Selection forest	Simple coppice	Coppice with standards
Oak	+*	+++	+++	+	+++	++++
Ash	++*	++++	++++	++(s)	++++	+++
Beech	+	+++	++++	++++(s)	++	++
Limes	****	+	+	++++(v)	++++	++++
Wych elm	+*	+	+	+++	+++	+++
Birches	++++	+++	+++	+	++++	+++
Aspen	^^^^	++	+++		++++	+++

Note: In the clear-cutting column, *shows the power of vegetative growth from the stumps of felled mature trees or ^suckers from root systems; (s) = mainly seedling regeneration; (v) = mainly vegetative regeneration.

16.2 Using Natural Regeneration to Restock

The most obvious way in which foresters incorporate natural elements into managed broadleaved woodland is by accepting natural regeneration when restocking after a clear-fell or accepting natural regeneration within continuous cover stands, but at Lady Park we have received a demonstration of the hazards and limitations of regeneration from stools, seed and suckers.

A stand originating from coppice can, after felling, generate a well-stocked stand of timber trees if: beech, oak, ash and lime stools outnumber hazel; and no more than 30 years have elapsed since the last coppicing. A high density of hazel and other small trees will occupy space that could have been occupied by timber trees, and a long-outgrown coppice will have less vigorous stools at lower density than in-rotation coppice. Any 'blanks' in the coppice will be filled with suckers and seedling regeneration if seedlings can establish themselves quickly or – better – have already established themselves under the old stand. Even then, they must outpace bramble and other low growth. The composition of a new stand is predictable insofar as it originates from stools and suckers, but where stools are sparse and weak it depends on seed sources, colonising distances and fickle mast years.

By 1942/4, over 70 years had passed since the last coppicing, which left the young-growth stands with a reduced density and vigour of stools. Some beech, lime and oak stools responded strongly, but new growth was severely browsed by fallow deer. Birch, ash and sallow of seedling origin with hazel of coppice origin thus dominated the new growth, while the scatter of beech, oak and lime seedlings struggled in the thicket. As the thicket developed, deer turned their attention to saplings, especially oak. The potential timber trees were strongly clustered, due to the pre-1942 clustering of lime, beech and oak stools and the habit of ash to regenerate in groups. At 32 years growth, the 1976 drought killed large numbers of birch, which should have released potential timber trees, but by then many had been weakened by prolonged shade and grey squirrels. The small number of young beech saplings, established under the thicket, might also have prospered, but most were killed by voles.

Nevertheless, despite these and later hazards, the combined density of timber trees *sensu lato* was 475 trees/ha in 2013/15, seven decades after the stand was naturally regenerated (Table 16.3). Oak, however, was poorly represented; the form of many beech and oak was poor, due to early browsing and then repeated bouts of bark stripping by grey squirrels; and parts of the stand were occupied mainly by hazel.

If the stand had been earmarked for timber, it would have been thinned to favour timber trees and facilitate advanced natural regeneration, and in 1983 exactly this regime was implemented in the young growth in the managed compartment. The outcome has been even more damage to beech, marginal increases in the growth rates of timber trees, and almost complete failure of natural regeneration. Even so, experience between 1983 and 1992 showed that regeneration was possible, albeit principally as stool sprouts from felled and singled trees. And, experience in the fenced reserve shows that natural seedling regeneration of beech, ash and occasionally limes is possible if deer are excluded, though bramble inhibits it in many gaps.

This outcome would be judged a moderate success if the aim had been to regenerate a mixed stand of locally native tree species. Despite deer and the initial proliferation of bramble, a diverse stand grew up in which all tree and shrub species played a part and few non-natives colonised from nearby plantations. But, if the aim had been a crop of timber, the outcome is poor, even though it has cost nothing directly. The main problems, though, have been self-inflicted wounds in the form of DED and now ash disease, and a failure to control deer and grey squirrels, without whom the stock and form of timber trees would have been much better.

In the circumstances represented by Lady Park, planting would increase the proportion of timber trees and may be the only way to maintain oak

Table 16.3. Potential timber trees in young-growth stands (2.0496 ha) in 2013/15 according to their origins and canopy position.

Canopy position in 2002	Canopy		Subcanopy		
Origin	Maiden	Coppice	Maiden	Coppice	Total
Ash	172	18	82	4	276
Beech	3	26	29	86	144
Birch	173	27	47	3	250
Limes	8	146	7	107	268
Oak	4	19	1	12	36

as a significant part of the mixture. Wall-to-wall planting would be unnecessary, just a scatter of planted saplings, leaving enough space for natural regeneration to fill the gaps. The planted stock would ideally be site-native species grown from local stock. The actual species and the balance between them would be selected according to management objectives – timber or something else. If mixtures are planted, a degree of clustering of individual species would reflect patterns in the natural woodland.

16.3 Allowing Stands to Thin Naturally

Unmanaged stands thin rapidly after canopy closure and continue to thin steadily throughout their growth. As they get older, they approach 30–35 m^2/ha basal area and thereafter fluctuate; that is, in the absence of massive disturbance, mortality is balanced by the growth of surviving trees, much of the stand remains closed and small gaps fill mainly by crown expansion. The outcome at any given point is either a stand dominated by long-lived forest dominants (oak, beech, lime, ash) that may survive intact for over 200 years or an even-aged stand of short-lived pioneers (birch, aspen, sallow, gean), which will collapse in 50–70 years, or something in between.

As a method of generating a crop of native trees, natural thinning has several downsides: (i) most trees have narrow crowns and none may become dominant, so elite trees are rare and large dimensions are slow to develop; (ii) thinnings are not utilised, though they contribute to dead wood; (iii) a stand of tall, slender trees is especially vulnerable to wind; (iv) many saplings of forest dominant species are shaded out by the rapid early growth of pioneers; and (v) there is no selection for desirable species.

Thinning should therefore be beneficial if the best-looking stems are retained and selection favours the species that should yield the most valuable timber, but the experience at Lady Park has proved the opposite. The best of the retained lime, oak, ash, birch and beech have grown only a little faster than they would have grown without thinning, so the benefits of thinning may not justify the cost. Moreover, beech, the favoured species, has been ruined as timber by grey squirrels, and might have been less damaged if the stand had not been thinned. In the event, ash, lime and oak should have been favoured.

16.4 Using Natural Gap Creation Patterns as a Guide to Felling Patterns

Small gaps form all the time when canopy trees are excluded by competition and when large branches fall from the largest trees, but these gaps close by crown expansion, i.e. they are the natural counterpart of forestry thinning. Large gaps form irregularly, usually by wind, accentuated by the skittle effect of falling trees knocking over neighbours, and these gaps may enlarge later if trees on the margins, now more exposed and possibly damaged in the original fall, come down too. These gaps will not close by crown expansion, but are large enough for established saplings and other underwood individuals, as well as new regeneration, to grow into the canopy.

The gaps generated after the 1976 drought, combined with some earlier falls below the cliff, formed a highly irregular pattern, with a tendency for gaps on the steeper slopes to be elongated downslope (see Fig. 5.11 and Chapter 12, this volume). The areas of four gaps formed since 2000 in the old-growth were 290–597 m^2 (mean *c.*410 m^2); and the areas of three gaps in the young growth were 761–1063 m^2 (average *c.*901 m^2). Such gaps are relatively small-scale, tantamount to group selection fellings, but other natural disturbances can produce far larger gaps. Elsewhere, the 1987 storm laid waste some mature woods, leaving no more than 5% of the original cover standing, but overall its impact was patchy, irregular and full of unexpected variations in intensity, caused by variation in land form and eddies and other irregularities in the wind flow (Peterken, 1996; Rackham, 2003).

Most gaps in natural woodland are small-scale, but they are enough to allow advance regeneration to become established and well-placed saplings to reach for the canopy. Reflecting this, near-to-nature forestry would thin to encourage advance regeneration and fell small coupes to release advance regeneration and established saplings. This, of course, was the essence of traditional group selection systems.

16.5 Building up Dead-wood Volumes

Perhaps the most striking feature of Lady Park's old-growth stands is the volume of dead wood, in all sizes and in all stages of decay (see Chapter 12, this volume). There was a substantial input after 1976, but estimates were made only in 1995, when the volumes were found to be 50–130 m³/ha in the old-growth and 28–45 m³/ha in the young growth. Since then, the old-growth volume has probably remained much the same, but the young-growth volume must have increased. In the young growth, most of this volume is on the ground, but in the old-growth the larger trees still carry a good deal of dead branch wood and many canopy trees have partially rotted trunks.

While these volumes indicate that dead-wood volumes can build up to natural levels in little more than a century, this is only possible if no timber is extracted. In managed stands, foresters cannot hope to match the higher volumes found in natural stands, except very locally, but they can at least measure how close they have got. In the lower Wye Valley, managed stands have little more than 10 m³/ha, and even where efforts are made to retain dead wood, volumes rarely exceed 30 m³/ha, and then only in the smaller size classes. A closer approach to natural conditions requires that a few larger trees are retained until they die, then are left to rot.

16.6 Biosecurity: Countering Pests and Diseases

Grey squirrels have vastly reduced the prospects for many native timber trees, especially beech and oak, but in unthinned natural stands they have largely ignored ash and lime, and have allowed slow-growing beech poles in the underwood and subcanopy to grow relatively unmolested. In the similar but thinned adjacent stand, they have destroyed the beech as timber trees.

Fallow deer also have the potential to destroy woodland almost completely, as the recent failure of regeneration in the managed compartment shows. Muntjac, too, have just arrived, so we wait with trepidation to find out whether they will repeat their performance in another National Nature Reserve, Monks Wood, where natural regeneration was destroyed and the ground vegetation was converted from broadleaved herbs to grasses (Cooke, 2006). A version of wild boar have also arrived recently, but in this case their disturbance of the soil surface and the pits they excavate would enhance regeneration if deer were not also numerous. All these species have been introduced, though deer and pigs were once part of the natural fauna, albeit with large carnivores to keep them under control.

Fungal blights have increased rapidly as scourges of native woodland (Rackham, 2015), from the silver-leaf fungus on oak, through elm disease, the blight of alders and the forthcoming ash disease. All have affected the prospects of their target trees adversely without eliminating them. Thus, oaks still show a limited capacity to regenerate from seed; not all the alders have been killed; and wych elms have enjoyed limited protection from DED by growing in the underwood, and still set a good deal of seed. Like beech, if one's aim is to grow a fairly large elm, this can still be done, if slowly.

The common feature of all these pests and diseases is that they are not native. Even if one stretches a point to accept fallow deer and hybrid pigs as native, their populations have been allowed to reach damaging levels. With ash disease in the offing, it is easy to become fatalistic about our capacity to wreck native woodland. Our experience at Lady Park reinforces Oliver Rackham's view that deer and our lax biosecurity are the greatest threats to the woods.

16.7 Varying Stand Structure to Suit Site Type

As the old-growth stands grew out of the uniform structure imposed on Lady Park by traditional coppicing, different structures developed in different topographical positions. Under the influences of the great drought and instability on the steep slopes, as well as the residual effects of past management, the upslope old-growth developed gaps; the old-growth above the cliff remained closed; and the stand below the cliff became open with a dense underwood. These differences confirm Eustace Jones's finding that the structure and dynamics of natural woodland is not the same everywhere; and that some variety should be incorporated into the silvicultural regimes of any wood that is designed to be 'near to nature'.

If we consider the original natural conditions and agree with Vera (see Chapter 15, this volume) that large herbivores had a controlling influence of woodland structure and dynamics, the silvicultural regime would incorporate a good deal of open space and scrub. However, adopting Kirby's (2004) assumptions, much of the ground would still be occupied by 'groves', which we take to be closed high forest. If we consider contemporary virgin forests and old-growth, the prevailing form would be small-group selection forest with an average gap creation rate of 1% per annum – implying that the trees that reached the canopy would remain there for 100 years. Larger-scale disturbances would interrupt gap creation at irregular and unpredictable intervals, generating even-aged and two-storied high forest. Locally, there would be major differences, notably on flood plains and steep slopes: the former would be subject to floods, channel movement and beavers, resulting in more openings and extensive even-aged stands, the latter would be subject to frequent tree falls, shortening the natural rotation.

Of course, much of this is speculative – discovering more is one reason for the Lady Park recording – but some broad-brush guidance can be offered to managers of native woodland:

- Irregular, small-group fellings would be generally appropriate, with occasional larger fellings, especially on flood plains, wet ground and steep slopes.
- Rotations would be generally longer on flat ground and deeper soils, but shorter on steep slopes.
- Uniform thinning on a large scale is inappropriate: the resulting structures do not happen in nature.
- Rides with scrub margins would provide the habitats indicated by the Vera cycle, as well as open spaces centred on wet ground and dry sites.

16.8 Matching Species to Site

This is meat and drink to foresters, but it is worth mentioning. In Lady Park, it is clear that ash can grow anywhere, but grows best on the deeper soils. Oak and service are best suited to the deeper, acid soils above the cliff. Large-leaved lime is best on limestone outcrops, while small-leaved lime grows well on any site. Wych elm and gean grow best on the deeper, more alkaline soils; aspen on the more acid soils. Beech will grow anywhere, but seems to be best on the drier soils above the cliff, despite its vulnerability to drought. The birches will grow anywhere, but the two species are best left to decide for themselves where they will grow. Alder, of course, is the opposite, thriving only on the flood plain. Experience beyond Lady Park has enabled ecologists and foresters to build a set of preferences for each site type in each region. Even without this, the ancient semi-natural woods are a guidebook in themselves, for the species are already matched to site.

Such guidance has, however, become largely academic. Unless new controls can be found, pests and diseases rule out most tree species in any wood where the primary objective is to grow useable timber. In amenity woods and nature reserves, where poor-quality growth will be easier to tolerate, one has a wider choice whenever planting is deemed to be necessary. Often, however, natural regeneration can be left to make the choice for us.

16.9 Constant Re-evaluation of the Preferred Timber Species

In Lady Park, oak was the preferred species under traditional management; beech became the favourite after 1897 and was still the favourite in 1992, even though it was then being torn apart by squirrels. Once grey squirrel control had proved to be impractical and too expensive, and beech had proved vulnerable to drought after 1976, ash should have been favoured, since it has not been touched by drought or squirrels. Now that ash disease is approaching, lime seems the least vulnerable tree. Birch, wych elm and field maple are ruled out, being worthless timber or subject to drought, squirrels or disease, but gean may be a reasonable bet. These changes in the 'best bet' have happened within little more than a broadleaved rotation.

Changes in perception like this have been characteristic of British forestry, not just in native woodlands. How many times have we heard that mature oak stands were planted for the navy, which, by the time they matured, had long since moved on from the *Fighting Temeraire* and the like? How many ancient woods were planted with conifers in the 1960s and 1970s, only to be reverted – thank goodness – to native broadleaves in the 1990s and 21st century? Even if we consider just native broadleaves, preferences keep changing – an enthusiasm for gean here, an edict against beech there.

If there are lessons here, they are: (i) to maintain a diversity of species in the canopy; and (ii) take biosecurity seriously.

16.10 Continuous Cover Forestry

Continuous cover forestry (CCF) is a new theme in forestry; or, to be more accurate, a modern revival of an old theme. In essence, it is the antithesis of clear-fell and replant even-aged forestry, a collection of various forms of perpetual thinning, which retain a mature stand and thereby does not change the landscape or the woodland habitats abruptly when timber is felled. Now that the Welsh Assembly Government is committed to implementing CCF in 50% of Welsh forests, it has become important (Fig. 16.1: see p. 244).

Like its early 20th century precursor, Dauerwald, CCF is not strictly a silvicultural system but more an approach or attitude to forestry. The scale and pattern of the felling depends on the species involved and the local circumstances. With shade bearers such as beech and Douglas fir, the felling can be fine scale, but with light demanders such as oak, larger groups must be felled, but not quite as many as the seeding felling of an oak shelterwood. It requires particular skills and incurs extra costs, but this is the price of environmental gains.

Lady Park was established as a research reserve long before CCF became a theme of forestry, but it hints that adopting CCF is far from straightforward. Imagining the fenced reserve as a group selection stand quickly shows that felling would have to be careful not to damage retained trees, and timber extraction would need to take particular care of established regeneration and retained trees. When the mature stands were opened by the 1976 drought, they accelerated the rate of subsequent windthrow. The thinned parts of the managed compartment show that without fencing there is no regeneration, and the fenced reserve shows that, even with fencing, regeneration is slow and not guaranteed in gaps. While the wood remains attractive and the managed compartment was cosy enough to induce a traveller community to live there for a month or two in 2013, there are also environmental downsides, notably the reduction of open-space species due to shade, grazing and litter accumulation.

So, there is a case for limiting CCF to sheltered sites and places where clear-felling has a large downside or few benefits. In upland forests, one can get the optimum benefits by combining CCF and long rotations in sheltered sites in valleys, with short rotations of clear-fells on the uplands (Peterken *et al.*, 1992). In lowland woods, CCF would reduce diversity if it were applied everywhere, because woods now act as the main havens for the semi-natural grasslands that have been largely eliminated from farmland. Here, we need woods where rides are kept open, the ride margins are coppiced and limited clear-felling is the norm.

16.11 Habitat Networks

Another new theme in forestry is habitat networks. Since the 1990s, the Forestry Commission in England, Wales and Scotland has sought to reverse the fragmentation of woodland that was the norm until the early 20th century. Many benefits should result, including more resilient populations of woodland wildlife species.

Lady Park is about as far as one can get in Britain from being a small, isolated wood. It is part of a much larger wood, which is itself part of an extensive tract of forest-dominated landscape in the Wye Gorge, Forest of Dean and Trellech plateau, stretching 30 km from north to south and 35 km from east to west. This has enabled many, perhaps most, of the species evicted from Lady Park to survive in the immediate vicinity, from which they could recolonise when and if conditions within the reserve again suited them. This colonisation is relatively easy because the colonising distances are short and entirely within woodland. Further, by being part of a large and diverse woodland, species that were not within the wood 70 years ago have had an opportunity to establish themselves in the wood as habitat conditions changed.

Suppose Lady Park had been surrounded by arable, leys or improved grassland, like so many lowland woods. It would have been large by ancient woodland standards (Spencer and Kirby, 1992), but few of the species evicted in the past 70 years would have survived nearby. They might survive somewhere in the district, but colonising distances would have been larger and the route would have been across hostile territory.

The general point is that wildlife populations in large woods are resilient, especially if they are also diverse, whereas those in small, isolated woods are vulnerable to irreversible local extinctions brought about by, for example, droughts or forestry operations. Expressed another way, wildlife populations in a small part of a large wood are likely to be part of larger populations, whereas populations in a small, isolated wood of the same size are likely to be functionally isolated. Inherently mobile species can overcome this isolation, but many woodland species are poor colonists. Habitat fragmentation affects woodland species more than the species of most other habitats.

Habitat networks attempt to reverse fragmentation by enlarging existing woods and restoring and creating habitat links to other woods. This

Fig. 16.2. The Institute of Chartered Foresters visits Lady Park on 12 September 2000 as part of its study tour in the Lower Wye Valley, one of many visits by groups of foresters and ecologists.

will take decades and centuries, and cannot be an unmitigated success. It cannot resolve wide-scale problems, such as the biological homogenisation of temperate woods through management change, nitrate pollution and the globalisation of species and diseases.

Even Lady Park demonstrates the problems, from change in the ground vegetation associated with increased nitrogen through to the ease with which unwanted aliens have colonised – sycamore, larch, Indian balsam, butterfly bush, grey squirrels, muntjac, wild pigs and fallow deer – and this is part of a larger dilemma. If we achieve good habitat connectedness on a landscape scale, we facilitate the movement of both the 'good' and the 'bad'. Indeed, the unwanted species will inevitably be good colonists, whereas many native species will be such poor colonists that they may make only limited use of new habitat links. So, we should be careful what we wish for. While functioning habitat networks must be a net benefit, isolation has helped some populations to survive (e.g. Scilly island elms; red squirrels on the Isle of Wight and Brownsea Island in Poole Harbour, Dorset) and some ancient woods to remain sycamore free. And, in the very long term, isolation enables new species to evolve.

17 Rewilding, Remoteness and Wilderness

Eustace Jones would not have said this in 1944, but Lady Park Wood has, in effect, become a test of woodland rewilding. Of course, many other woods have been allowed to run wild, but this one has been recorded as it has grown wilder and thus provides both an informed example of what happens and a basis for assessing whether rewilding is worthwhile and wise.

Rewilding is a modern idea based on old themes, which has come to prominence now because we have realised increasingly that our separation from nature is harming both nature and us. According to the *Rewilding Britain* website (www.rewildingbritain.org.uk/), rewilding 'is ecological restoration and a little bit more', which allows nature to look after itself, thereby 'securing the good things that nature provides' – clean air and water, carbon storage, flood control and 'amazing experiences'. Where land rewilds, rewilders see habitats expanding, wildlife multiplying, people reconnecting with nature and 'communities flourishing with new opportunities'. Most of the projects they mention are in the sparsely populated uplands of the north and west, where woodland regeneration, tree planting and reduction of sheep and deer populations are the order of the day, but their interest extends to the agricultural and densely populated lowlands, where restoration of the river Wandle (Surrey) and free-range grazing in the Knepp Estate (Sussex) demonstrate that rewilding can encompass a variety of enterprises. At its purest, rewilding would reintroduce the fauna that we have driven to extinction, including wolves and other fierce carnivores, and let nature develop as it will. At its more nuanced, it merely seeks *wilder* environments than we currently allow.

This is manifestly a constructive, forward-looking, large-scale and ambitious approach to environmental conservation. It contrasts with long-established nature conservation measures, which it regards as unduly bureaucratic, backward looking and intent mainly on maintaining fragments of traditional agriculture and the habitats it bequeathed to us, serving up carefully managed habitats – 'gardening' – rather than the wilderness experience that many people crave. Rewilding also embodies social and economic criticism of wasteful subsidies for upland sheep husbandry and the elitist monopoly of large tracts of hill land by sporting interests. There is also an ethical imperative, for rewilding seeks to restore land to an earlier, undamaged condition, supposedly fertile, tree clad and biodiverse.

Can Lady Park tell us anything about rewilding? It is part of a very large ancient woodland set in rugged countryside, which would make a reasonable candidate for rewilding, but the reserve itself is too small to be seen as wilderness by itself. Nevertheless, after 70 years of rewilding, it can illuminate some facets, especially if experience at Lady Park is linked to experience elsewhere. Does wildlife really benefit? Will visitors like what develops? Can rewilded woodlands ever be truly natural? Can they offer remoteness, one of the key features of a wilderness experience, and what does this mean anyway? The answers to these and other questions reinforce our view that rewilding should take many different forms and degrees according to context.

17.1 Does Rewilding Benefit Wildlife?

The theory is sound enough. Remove excessive grazing by deer or sheep, restore naturalistic grazing patterns, regenerate woodland to a more natural structure, reintroduce wolves, beavers and other key fauna, and the prospects for wildlife should improve, yet experience at Lady Park – admittedly without wolves – has been mixed in this regard.

Fig. 17.1. Glen Nevis. Remnants of native pine woodland and ravine ash woodland have been incorporated in even-aged, self-sown birch woodland. Among spectacular mountains, the trail runs along a narrow ravine ledge above a rushing torrent up to wide, open alluvial flats and an enormous cataract, the whole a popular wild experience.

Small-scale diversity of the stands themselves has declined with the attenuation of populations of shrubs and light-demanding trees. The ground flora is less diverse, mainly because most species associated with open spaces have been ousted, but also because many shade-bearing herbs have declined, and in some cases vanished. Butterflies, too, have largely been excluded. On the other hand, bats – especially barbastelles – have seemingly prospered, some birds have done well (tree creeper, woodpeckers), fungi seem richer inside the reserve than outside, and the accumulation of dead wood should eventually favour saproxylic invertebrates. Meanwhile, molluscs and perhaps bryophytes have evidently remained rich. The paucity of earlier records and very incomplete recording of invertebrates limits what we can say, but the general picture seems clear enough: rewilding has not been an unmitigated blessing for wildlife.

On the scale of the entire Highmeadow and Doward Woods, together with the associated open ground, this profit-and-loss account would be different. While many of the plants evicted from Lady Park survive elsewhere within these woods, the whole area is impoverished floristically, in the sense that some species have become extinct and many are much more restricted than they once were. Enclaves of grassland on limestone have become scrub; rock faces have become more shaded by the growth of trees at their foot; and even the colourful natural rockery gardens survive only in small patches on a few outcrops. Butterflies once prospered in the constant supply of artificial gaps and early succession stands provided by coppicing, but have declined in the conditions provided by modern forestry, even though forest roads remain permanently open. The rich quarry flora found by the first field meeting of the British Ecological Society in 1910 has now been replaced by just another patch of naturally regenerated woodland.

Obviously, the Highmeadow and Doward Woods have not been rewilded, but they have changed enough to show what would happen if management *sensu lato* ceased completely. Deer and wild pigs would proliferate, so some ground would remain open, but the open spaces on which perhaps half of the flora and fauna depends would be reduced to a minimum, and those that remain would be grazed tight. Dead wood and the associated species would abound, but the Wye would remain a constant source of non-native species brought down from its extensive catchment. In terms of biodiversity, rewilding of these habitats on this scale would still be a mixed blessing at best.

17.2 Remoteness and Wilderness

One of the reasons why Aldo Leopold (1949) wanted to preserve wilderness was for recreation as 'a means of perpetuating, in sport form, the more virile and primitive skills in pioneering travel and subsistence', such as hunting, fishing and travel on foot, and this made sense in America, where the landscapes are large and the frontier is still embedded in national identity. In Britain, likewise, many people head for the hills and the wild coast, seeking our muted form of wilderness and the benefits associated with tranquillity and remoteness. These are the 'amazing experiences' that modern British rewilders want to enhance.

The history of Lady Park Wood and its surroundings shows that remoteness and wilderness change. In the 18th century, the visitors taking the Wye Tour would set out from Ross in the morning, drift under the Yat Rock and pass Lady Park and the Seven Sisters rocks in the afternoon, before gliding calmly into Monmouth, where they would spend the night. Imbued with picturesque ideas, they were seeking sublime experiences on a river that was, and still is, one of the wildest in southern Britain. The gorge and the surrounding hills have always seemed remote, even though the Wye Valley was a place of industry at the time and the Wye remained an important transport artery. As late as the 11th century, the boundaries between Herefordshire, Gloucestershire and Monmouthshire remained incoherent, and well into the 20th century we were not sure whether Monmouthshire counted as part of Wales. Moreover, the district has always been unusually well wooded for Britain, and Lady Park forms part of one of the largest individual woods.

Today, the lower Wye Valley, the southernmost portion of the Welsh borderland, is hardly remote in an absolute sense. Like almost all of lowland Britain, it is pervaded by a dense network of public roads, and on sunny summer weekends it rings with the shouts of canoeists on the Wye and campers paddling round the shoals, while cyclists force one to step aside on the track below the wood. But the upper gorge is largely free of public roads and, as it happens, Lady Park Wood is further from a public road than anywhere else in the district. For much of the time, the wood is peaceful and tranquil, out of earshot of motor cars, a place where the bark of a rutting buck in autumn can put the lone walker on high alert and where on misty, late November afternoons, the wood is so gloomy, still and quiet that one jumps when a blackbird scatters some leaves. As Arthur Cooke, a resident of Chepstow, wrote just before the First World War:

> Pleasant it is to wander at random in these lovely (Highmeadow) woods, passing long hours alone and undisturbed; for the ordinary Forest visitor comes not here, and even in the immediate neighbourhood of Symonds Yat (now a tourist honeypot with a tea house and visitor centre) it is only on some fine bank holiday that we shall find the tripper in their lonely depths.
>
> (Cooke, 1913)

Lady Park had not always felt this remote. In 1770, when William Gilpin passed in a boat, he noticed the smoke of charcoal burners in or near Lady Park. (The colliers continued nearby until the late 19th century (see Fig. 2.1)). Small-time miners and their families dug for iron ore in the wood. Just upstream, a mill forged iron and mined away some of the limestone rock towers. The woods were full of woodmen, one of whom lived in the Biblins Lodge opposite. The grassland on the flood plain was mown as a meadow. The main track through the reserve was once just an ordinary part of the local road network. And, the river was a thoroughfare up which boats were towed and down which tourists floated, sometimes alighting, like Joseph Farrington, to talk to the cottager (Newby, 1998).

In 1873, Lady Park became even less remote – a branch railway line was built through the gorge. The nearest station was upstream at Symonds Yat, and no doubt this increased the frequency of visitors. A siding was constructed on the flood plain just below the reserve to handle wagons taking away the timber felled on the woods belonging to the Office of Woods and their successors, the Forestry Commission. What's more, the foresters constructed a timber sling across the Wye to carry logs from the Doward to the trains, otherwise they could not have been extracted. Some tourists were intrepid enough to leave their boats and scale the cliff to visit Lady Park cave (Box 17.1).

Today, Lady Park Wood and the adjacent Wye Gorge have again become (relatively) remote, but, ironically, their very remoteness attracts crowds. The narrow field on the flood plain opposite is a campsite reserved for youths and young adults; the woods are permeated by a network of walking tracks; and in about 2006, a cycle track was built along the old railway line, so at weekends the neighbourhood resounds with excitement. On the other hand, the sense of remoteness remains, and not just for the lone, out-of-season visitor. Several groups of forestry students have been impressed with the long drive from the forest gate before they reach the reserve. Teenagers in the Biblins camp cannot use their mobile phones: evidently, there have been serious cases of sensory deprivation and disorientation as a result – a truly wilderness experience! When the local fallow deer felt unsafe, they repaired to the reserve. This last had serious consequences, for the reserve became

Box 17.1 John Bellows's account of visiting Lady Park Cave in 1880

> The mouth of the cave is 129 feet above the river in the face of a steep cliff of the Mountain Limestone, and about 20 feet below its crest. The approach to it was until lately, from above; a break-neck track, hardly safe for a mountain goat, can still be seen winding down amongst the roots of the overhanging thicket. We mount by five successive flights of ladders: as we near the top, a startled wood-pigeon dashes out from the tree-top with a great fluster, all but knocking our hats off. We are not sorry to have reached the solid rock: there has evidently been a land-slip or a local slip of some of the blocks and earth of the upper portion of the cliff, ... There is a 'mine' not far off: the yew trees along the cliff might have told us that. We notice brush at the entrance of the cavern, and we pick out some nuggets of pure iron ore from its sides. ... we are satisfied that it has been worked for iron. The percolation of lime-water, dripping, perhaps for hundreds of years, has encrusted the once jagged surfaces with stalagmite, obliterating the tool-marks ... No remains of extinct animals have been found here.
>
> (Bellows, 2013)

absurdly overgrazed, leading to the intense discussions about fencing and how this would affect the objectives of the recording in the reserve.

So the idea and actuality of remoteness and wilderness are relative and fluctuating. Mountains and parts of the coast are wildest, difficult to reach, manifestly rugged, risky and obviously shaped by nature. Their antithesis is farmland, but even here Robert McFarlane found a kind of wilderness in a beech plantation near Cambridge (Macfarlane, 2007). Semi-natural woods have a special place. Neither remote nor wild, they are places where, walking alone, one can feel wilderness, even in a city. Years ago, a visitor to Overhall Grove in the arable plains of Cambridgeshire told GFP that visiting the wood made her expand. This is the wilderness effect: experience the wild and one has a complement, contrast, antidote to the crowded, rectilinear, unnatural environments in which most of us live.

17.3 What Kind of Wildness Can We Achieve?

Rewilding in its purest form seeks to re-establish a natural ecosystem, but it faces several limitations. It cannot happen at once: in fact, any wood will take decades to become half-wild. It cannot erase quarries, roads and other physical modifications of the landform. However hands-off we are, we cannot exclude the current effects of people. And, the natural environment towards which the system will develop cannot be the original, pre-Neolithic condition, and may be entirely novel.

In Lady Park, we have long recognised these limitations (Peterken and Mountford, 2005). Edge effects are significant; many species operate on a scale larger than the reserve; widespread pollution affects what happens within the reserve; and the impacts of fallow deer within the reserve are determined by the degree of control exercised outside. Most of these factors would be minimised in a very large rewilding scheme with a self-contained catchment, but the wide-scale impacts of people would remain.

We have also been forced to decide what quality of naturalness we want. The choice lies between future naturalness (the condition that would develop if we adopted unqualified non-intervention policies for the reserve) and original naturalness (the condition of pre-Neolithic woodland before people became significant) (Peterken, 1996). The former would minimise the direct control by people, but the outcome might be a mixture of sycamore, western hemlock and cherry laurel. We chose the latter because we wanted to study how woods worked with the mixture of species that would have occupied the ground if people had never influenced the wood – a partially attainable target in this instance, because the dominant trees of pre-Neolithic woodland were still present and very few non-native trees and shrubs had colonised, though limited because we could not control disease, grey squirrels, pollution and climate change. This choice, however, required us to intervene directly to remove (the mercifully few) invading sycamores and other non-natives.

Fallow deer also presented another dilemma, for they were clearly capable of preventing natural regeneration and devastating the ground flora. Leaving aside the point that they were originally introduced, we accepted that deer of some kind were legitimate components of a near-natural reserve, but should we accept them at the high population density tolerated by Forest Enterprise or exclude them with a fence? Too many deer would be just as unnatural as none at all, but at least we could claim to be testing Frans Vera's concept of how natural woodlands functioned in the presence of large herbivores. We chose the alternative when we had the chance, partly because deer would get in anyway when trees fell across the fence – they can get out over deer-leaps – and intermittent incursions seemed closer to the moderate deer population that would allow at least some regeneration.

Equivalent dilemmas can arise in any rewilding scheme. They may matter only to ecologists, but managers will need to decide whether to eliminate exotics and reintroduce supposedly lost tree species. They must also decide the overall composition when rewilding commences with tree planting, as it has at Carrifran Wildwood in the Southern Uplands of Scotland (Ashmole and Ashmole, 2009).

17.4 Will People Like What They Get?

Rewilding is a beguiling idea, and many will enjoy the process, but one wonders whether most people will be happy with the result. Uplands and other less intensively farmed landscapes are already natural enough for most people and they enjoy the long views and freedom to roam. We are fascinated by the shape of the land and the detail of landforms, much as William Gilpin (1782) admired the intimate detail of rocks and the configuration of the ground and watercourses as part of the picturesque experience.

The outcome of rewilding is usually more woodland, which blocks out views and obscures the detail of landforms. Large, dense forests were a byword for disorientation, and remain so even with sat navs and trails. The trails themselves restrict the freedom to roam. Open spaces are occluded by scrub, and eventually woodland and the lack of openings can become oppressive: even Americans find the Appalachian Trail claustrophobic (Bryson, 1998). Indeed, the British, who have become culturally bound to open countryside, may be particularly sensitive: during a year living in a rewilded landscape in Massachusetts, USA, where semi-natural woodland had largely replaced farmland, GFP found that country walks always led towards the few remaining open spaces.

The early stages will be attractive. Thus, for example, Glen Nevis (Fig. 17.1: see p. 254) has filled naturally with a beautiful sea of birch, diversified with patches of old pine woodland, alders and ash. But it will grow older: in a few decades, birches will be toppling over paths, dead wood will accumulate and there will be an air of decay and decrepitude. Walkers traversing Lady Park before 1976 would have passed through healthy mature and burgeoning mixed woodland, a delight to the eye, but now they have to circumnavigate great tree falls, accept the large numbers of dead trees and ignore the danger presented by large, often decrepit trees on slopes. This danger is small, but real. In April 2014, a large branch falling from an old beech missed a group of staff from Natural Resources Wales by about 2 m – ironically, they were discussing dead wood at the time. In December 1999, a man walking home at night along the old railway track at the base of the wood was crushed by an ash falling out from the reserve, which had been brought down by snow accumulating on ivy that had grown into the crown, and which (as we later discovered) had been weakened by the death of a principal root. We do not enter the reserve when the wind blows, and even this would not have preserved us on 17 October 2013 if we had been standing by the healthy beech that fell on a calm sunny afternoon (see Chapter 7, Section 7.6, Case study 7.4, this volume).

That said, visitor sentiment may change, and a proportion of visitors today find Lady Park attractive and welcome the frisson of danger. Let's see!

17.5 Rewilding in Practice

Lady Park highlights some reasons to be cautious about rewilding, and there are others. James Fenton (2015) maintains that much of the uplands might naturally be open, not the tree-covered landscape that is assumed by those rewilders who plant trees and regenerate woodland. George Monbiot's (2013) forcefully expressed view that sheep pasturage in upland Wales is unsustainable and should be replaced by rewilded hills met – as he himself describes – resistance from some of the present occupants, who were understandably unwilling to leave the land and lose the cultural links and associations derived from a long rural farming tradition. Further, many visitors value the open landscape, long views and the chance to see the structure of the land. Simon Fairlie (2013) has argued that all farming can be regarded as unsustainable. Even lowland agriculture, on which we really do depend for food, depends on imported fertiliser and energy, and generates outputs that contribute to global warming. Viewed in that light, rewilding depends on food imports from countries that may be wrecking their own environment to produce it; contributes to global warming; and increases the threat to our food sovereignty.

Nevertheless, rewilding is positive, adventurous and provides a desirable new growth point for conservation around which people of many backgrounds can be freshly enthusiastic. Recognising the constraints, most proposals amount to partial, rather than pure, rewilding – wilder, not wild – and the form chosen depends on the land available and its context.

Rewilding embraces a continuum of possibilities. The following reference points go beyond what is usually included within rewilding.

1. Pure rewilding. That is, allowing free rein to natural processes after restoring large carnivores and herbivores and other lost keystone species. There would be no target habitats. Whatever happens would be accepted, including developments that elsewhere would elicit interventions. Thus, massive blowdowns would not be regarded as damaging; diseases would not be controlled; and the fate of individual animals would be 'left to nature'.

Pure rewilding can only be attempted on a scale large enough to support populations of large carnivores. Public access would have to be limited. It can only be envisaged where human populations are small in parts of the Highland and Western Oceanic regions, the necessary scale is available and we inherit natural landforms and a wide range of semi-natural habitats, but even there it meets resistance and will probably need to be 'diluted'. Beaver reintroductions would go some way towards rewilding rivers.

2. Extensive pasturage with 'domestic' stock. This takes its inspiration from the famous Oostvaardersplassen (the Netherlands), where red deer and near-wild breeds of cattle and horses have been allowed to use a polder without restriction (Vera, 2009). People control populations, i.e. take the role of large carnivores, but habitats have been allowed to develop naturally, i.e. without interventions and with no targets. The outcome so far has been extremely rich bird life.

This form of rewilding must be undertaken on at least a moderate scale in order to allow herbivores to develop naturalistic roaming lifestyles; to cover a wide range of topography; and to have enough scale for habitat fluctuations to take place without losing niches in the process. Any existing semi-natural habitat, including woodland, can be treated in this way, and it is equally suited to habitat restoration and expansion schemes. Excellent examples are under way in fenland at Wicken (Cambridgeshire) and in various combinations of woodland, plantation, farmland and moorland in Ennerdale (Cumbria) and Knepp Castle Estate (West Sussex) (Boxes 17.2 and 17.3).

Ancient wood pastures managed with minimal silvicultural interventions count as another form. Indeed, many conservationists see such places as demonstrations of Vera's concept of natural woodland in action. The prime example is the New Forest (Hampshire), where continuing common pasturage has maintained the glades, heaths and other open areas. The Forest also demonstrates that periodic reductions in the intensity of pasturage have been necessary for natural regeneration to prosper (Peterken and Tubbs, 1965), but if grazing lapses completely, the open spaces will eventually fill with trees, a point reinforced by Epping Forest (Essex and Greater London). In the Highlands, rewilding in, for example, Glenfeshie and Creag Meagaidh, has required some reduction in grazing intensity.

3. Inconspicuous management of woodland as mature stands designed to increase actual or perceived wildness. This can take several forms, all of which require discrete and unobtrusive management.

One form, traditionally known as 'selection forestry' and now subsumed under near-to-nature or continuous cover forestry (CCF), appears to be light-touch woodland management, but is actually carefully controlled mimicry of the dynamics of some natural woodlands (see Chapter 16, this volume). It is particularly suited to conspicuous rural woods, where extensive clear-felling would change the landscape abruptly, and urban woods, which must also maintain a permanent presence while being used by the public, and it can even be applied in small woods. Woods dominated by

Box 17.2 Ennerdale

Ennerdale, a remote, uninhabited valley in western Cumbria, has been rewilded since 2003 by a partnership between the National Trust, Forestry Commission, United Utilities and Natural England (Browning, 2015). Its 4300 ha includes extensive tracts of upland sheep pasture and mountain rockscape running up to Great Gable, Red Pike and Pillar, but the core is the river Liza and its tributaries running through extensive conifer plantations. Wildlife includes red squirrels, pied flycatchers, redstarts, salmon, Arctic charr and the reintroduced marsh fritillary.

The partnership will not abandon the valley completely to nature, but manage it at low intensity, working with natural processes, all the time working with and for public participation. Fences have been removed to allow extensive grazing and to enhance the sense of wildness. Plantations, which date from 1925 onwards, are being felled in stages and partially restocked with juniper and native broadleaves. Cattle have been introduced to help diversify former sheep pasture and generate a 'dynamic mosaic' of mixed species, open glades, dense stands, big trees and dead wood. Semi-natural mixed birch woodland, which survives on steep slopes above Ennerdale Water, is expanding naturally.

This is a very long-term, open-ended project, which has generated only limited change so far. Sycamore and other introduced broadleaves are prominent in second-growth stands on richer soils within the plantations. Sitka spruce and other conifers are regenerating well. The river Liza remains the most natural low-ground feature, dynamically modifying channels during spates. With large conifer logs beached on shoals and more trees falling into the stream as they are undermined, parts of the valley remind one of the Hoh and Queets valleys in Washington's Olympic National Park.

Box 17.3 Knepp

The 1400 ha Knepp Castle Estate in West Sussex was formerly farmed conventionally, but it lies on Weald Clay, whose heavy soils make arable cropping barely profitable, and the forestry operation suffered from the collapse of timber prices in the 1990s. Responding to this, the owners, Charlie Burrell and Isabella Tree, embarked in 2001 on an 'open-ended experiment' (Burrell and Tree, 2015). Ploughing and fertiliser applications ceased; some of the land was sown with wildflower mixtures; and the whole estate was stocked with free-ranging combinations of Old English Longhorn cattle, fallow deer, roe deer, red deer, Exmoor ponies and Tamworth pigs. After 15 years, the southern section, which reverted directly from arable, is now a mosaic of grassland, marsh, tall-herb 'meadows', scrub, open grown trees, wood pasture and woodland, all reminiscent of both urban edgelands and the New Forest valleys, while the land sown to grass remains as grassland with scattered bushes. In response, wildlife diversity has increased spectacularly: the farmland flora has recovered and the estate now supports important breeding populations of purple emperors, turtle doves, nightingales and much else (Marren, 2016).

This is essentially a radical change from conventional farming to organic meat production and ancillary businesses, with attendant benefits for recreation, biodiversity, ecosystem services, health (of both people and livestock) and estate finances. The 'rewilding' label gives it a novel and progressive identity, but it benefits further from the imaginative and enthusiastic approach of the owners, the presence of numerous source habitats within and around the estate, and collaboration with numerous experts and agencies.

The estate woodland is not yet wholly entered into the rewilding project. While some is open to livestock and has developed a clear browse line, much is still behind fences. Some coppice has been restored; conifer plantations have been thinned; rides have been maintained; vistas have been cut; rhododendron has been removed; and dead wood is left standing. The native mix is pedunculate oak, ash, field maple, hazel and hornbeam, with alder containing some oaks of 200 years or more. Regeneration within woods is limited, but the developing scrub on old fields protects numerous oak saplings that should eventually dominate.

beech and other shade bearers are particularly suited. CCF has been adopted in many state forests in Wales, including parts of Coed-y-Brenin forest where Douglas fir plantations made in 1928 have been retained, thinned and regenerated naturally into monumental attractions (see Fig. 16.1). Many woodland nature reserves are managed with minimal silvicultural interventions, often explicitly to develop a more natural structure out of formerly managed stands, but sometimes because vigorous interventions would be damaging or expensive.

4. Nature reserve management. Even though today's rewilders sometimes castigate reserves managed to retain traditional habitats as backward-looking 'gardening', traditional nature reserves are a valid part of the rewilding spectrum. After all, even carefully managed nature reserves remain relatively wild when compared with their intensively used surroundings, and the 'gardening' has been necessary to protect wild populations that are too small and isolated to survive without deliberate aid. Such reserves may become less necessary if habitat networks are restored and work as we hope – a cause embraced by rewilders – but that is a long-term project indeed.

Individual ancient woods have been listed, and since 1985 have benefited from special consideration under national forestry policies. Most have ceased to be managed in the traditional manner, and many have been left to rewild by default. As a result, the meadow-like grasslands of rides and other open spaces have been reduced or lost, and the species associated with them likewise, most obviously the butterflies (Asher *et al.*, 2001) but also up to half the flora (Peterken and Francis, 1999). This is particularly serious for biodiversity generally, because semi-natural grassland has been largely extinguished from lowland farmland, and the road verges have become eutrophicated and thereby impoverished. Moreover, once the species associated with rides have been ousted by wall-to-wall trees, there will be nowhere for them to return from. This isolation will also have prevented the species of mature timber habitats from taking advantage of de facto rewilding of small ancient woods in farmland. Rewilding of small woods in much of the British lowlands is thus likely to diminish biodiversity if it is not part of a much larger rewilding enterprise.

5. Urban habitat creation. The notion of relative rewilding can be extended to urban and peri-urban woodlands. They will not actually be wild, but to an urban general public they afford the necessary contrast to the built environment and increased opportunities for contact with nature right where people live. Fragments of the rural countryside that have become isolated within a matrix of urban development have been safeguarded (so far) from repeated attempts to develop (i.e. destroy) them, and new habitats have been created on unused, derelict land. Natural links have been reinforced along canals, roads and railways to the countryside beyond the urban, which can connect with wider habitat networks. As David Goode (2014) amply demonstrated, all this has been accompanied by community involvement and a much-needed positive conservation message – what we can do to improve, rather than what we must not do to avoid damage.

Taken together, these reference points show that rewilding should not be a purist, all-or-nothing enterprise but an adaptable and subtle attitude, responding to context and offering benefits in all circumstances from the remotest Highlands to city centres. It recognises that naturalness, or wildness, can take several forms, from original natural to future natural, and that rewilding can be applied to both.

Exponents of rewilding recognise the importance of restoring a network of forest habitats, and thus biological communications between woodlands, e.g. linking Abernethy, Glenfeshie and Mar Lodge pinewoods across the Cairngorms. Forests were once the matrix habitat, and, even accepting that forests included a good deal of open ground, trees formed the matrix environment. Unlike the species of naturally isolated and ephemeral habitats, woodland species could survive without being good colonisers, spreading by small moves within generally benign habitat. Now the opposite is the case. Woods are mostly small and isolated from each other, and as a result populations of woodland species that cannot thrive outside woodland have become fragmented, and some, the so-called ancient woodland indicators, are almost confined to remnants of original woodland.

In this context, flood plains and riparian land are prime candidates for some rewilding, especially where there is still a remnant of the natural, braided and mobile channels, such as the Lower Spey and short reaches of the Usk and Severn, for these are the natural, long-distance links in the landscape. Rewilded valleys should not be wholly wooded, though woodland would be a large component, dead wood would accumulate in back channels and rewilded reaches would be linked by partially wooded buffer strips. Apart from the general benefits of rewilding, water quality would be improved, though practical problems with up- and downstream land uses and the management of debris would have to be solved.

On flood plains, and indeed anywhere else, alleviating isolation does not need a network of closed canopy woodland but a full range of forest habitats, including open ground, distributed so that species are able to move over long distances. This must be done efficiently – maximum connectedness with minimum extra habitat – and here corridors of mixed habitat with a substantial forest element along watercourses, steep slopes and coastal fringes would be particularly appropriate.

Tailpiece

At Lady Park, we have been studying the messy reality of how a mixed-species wood has grown when allowed to develop naturally. The result is not as smooth as the formulations that emerge from models of stand growth, nor as neat and tidy as the performance of a relatively simple stand, such as that studied by Coomes and Allen (2007). Indeed, one is tempted to regard Lady Park as 'real life' and models as wishful thinking, until one recognises that the story of the past 70 years is just one of an infinite number of possible narratives that might have emerged from the wood, each of which would have given us a slightly different understanding of how this wood works. Alternative real-life narratives can be read in other woods kept under close, long-term study, but they do not cover all possible combinations of circumstances, events and responses.

All approaches contribute to our understanding. Models explore the underlying pattern; simple woods help to validate the models; and complex woods show more of the uncertainty, randomness and individuality involved. Just like life, in fact: one person's story is their reality, but any individual's life could have been very different had the cookie crumbled differently, while novels, poetry and painted images point to a deeper reality.

References

Acker, S.A., McKee, W.A., Harmon, M.E. and Franklin, J.F. (1998) Long-term research on forest dynamics in the Pacific Northwest: a network of permanent forest plots. In: Dalmeier, F. and Comiskey, J.A. (eds) *Forest Biodiversity in North, Central and South America, and the Caribbean*. Man and the Biosphere series, 21. Parthenon Publishing Group, Paris, pp. 93–106.

Ainsworth, A.M. (2004) *Developing tools for assessing fungal interest in habitats. I. Beech woodland saprotrophs*. English Nature Research Reports No 597, Peterborough, UK.

Allen, K.A., Lehsten, V., Hale, K. and Bradshaw, R. (2016) Past and future drivers of an unmanaged carbon sink in European temperate forest. *Ecosystems* 19(3), 545–554.

ApSimon, A.M., Smart, P.L., Macphail, R., Scott, K. and Taylor, H. (1992) King Artur's Cave, Whitchurch, Herefordshire: reassessment of a Middle and Upper Paleolithic, Mesolithic and Beaker site. *Proceedings of University of Bristol Spelaeological Society* 19, 183–249.

Asher, J., Warren, M., Fox, R., Harding, P., Jeffcoate, G. and Jeffcoate, S. (2001) *The Millennium Atlas of Butterflies in Britain and Ireland*. Oxford, Oxford University Press, Oxford, UK.

Ashmole, M. and Ashmole, P. (2009) *The Carrifran Wildwood Story*. Borders Forest Trust, Jedburgh, Scotland, UK.

Backmeroff, C.E. and Peterken, G.F. (1989) Long-term changes in the woodlands of Clairinsh, Lock Lomond. *Transactions of the Botanical Society of Edinburgh* 45, 253–297.

Baeten, L., Vangansbeke, P., Hermy, M., Peterken, G., Vanhuyse, K. and Verheyen, K. (2012) Distinguishing between turnover and nestedness in the quantification of biological homogenisation. *Biodiversity and Conservation* 21, 1399–1409.

Baker, C.A., Moxey, P.A. and Oxford, P.M. (1978) Woodland continuity and change in Epping Forest. *Field Studies* 4, 645–669.

Bartelink, H., van Dijk, G. and van Hiele, R. (1987) *Lady Park Wood. Analysis of spontaneous developments in a mixed broadleaved woodland on limestone*. Report No 87-25. Wageningen Agricultural University, Wageningen, the Netherlands.

Bellows, J. (2013) *A Week's Holiday in the Forest of Dean*. Reprinted with a preface by Ian Standing. Holborn House, Blakeney, UK.

Bernhardt-Römermann, M., Baeten, L., Craven, D., De Frenne, P., Hédl, R., *et al.* (2015) Drivers of temporal changes in temperate forest plant diversity vary across spatial scales. *Global Change Biology* 21, 3726–3737.

Billick, I. and Price, M.V. (eds) (2010) *The Ecology of Place*. University of Chicago, Chicago and London.

Birks, H.J.B., Deacon, J. and Peglar, S. (1975) Pollen maps for the British Isles 5,000 years ago. *Proceedings of the Royal Society of London*, B189, 87–105.

Bradley, A.G. (1910) *The Wye*. Adam and Charles Black, London.

Bradshaw, R.H.W. (1981) Quantitative reconstruction of local woodland vegetation using pollen analysis from a small basin in Norfolk, England. *Journal of Ecology* 69, 941–955.

Bradshaw, R.H.W., Josefsson, T., Clear, J.L. and Peterken, G.F. (2011) The structure and reproduction of the virgin forest: a review of Eustace Jones (1945). *Scandinavian Journal of Forest Research* 26, 45–53.

Brang, P., Spathelf, P., Larsen, J.B., Bauhus, J., Boncina, A., *et al.* (2014) Suitability of close-to-nature silviculture for adapting temperate European forests to climate change. *Forestry*, 87, 492–503.

British Wildlife (2009) Naturalistic grazing and re-wilding in Britain. Perspectives from the past and future directions. *British Wildlife* 20 No 5 (Special Supplement), 1–63.

Brown, A., Bell, M., Timpany, S. and Nayling, N. (2005) Mesolithic to Neolithic and medieval coastal environmental change: intertidal survey at Woolaston, Gloucestershire. *Archaeology in the Severn Estuary* 16, 67–83.

Browning, G. (2015) Wild Ennerdale: looking forward to the 'future-natural'. *Conservation Land Management* 13(4), 4–8.

Brunet, J., Bukina, Y., Hedwall, P., Holmström, E. and von Oheimb, G. (2014) Pathogen induced disturbance and succession in temperate forests: evidence from a 100-year data set in southern Sweden. *Basic and Applied Ecology* 15, 114–121.

Brunet, J., Hedwall, P., Holmström, E. and Wahlgren, E. (2016) Disturbance of the herbaceous layer after an invasion of a eutrophic forest by wild boar. *Nordic Journal of Botany* 34, 120–128.

Bryson, B. (1998) *A Walk in the Woods*. Transworld, London.

Bürgi, M. (2015) Close-to-nature forestry. In: Kirby, K.J. and Watkins, C. (eds) *Europe's Changing Woods and Forests, From Wildwood to Managed Landscapes*. CAB International, Wallingford, UK, pp. 107–115.

Burrell, C. and Tree, I. (2015) The Knepp Wildland Project – process-led conservation. *Conservation Land Management* Summer 2015, 9–13.

Cavin, L. and Jump, A. (2017) Highest drought sensitivity and lowest resistance to growth suppression is found in the range core of the widespread temperate tree *Fagus sylvatica* L. not the equatorial range edge. *Global Change Biology* 23, 362–379.

Cavin, L., Mountford, E.P., Peterken, G.F. and Jump, A.S. (2013) Extreme drought alters competitive dominance within and between tree species in a mixed forest stand. *Functional Ecology* 27, 1424–1435.

Christensen, M., Hahn, K., Mountford, E.P., Ódor, P., Standovár, T., *et al.* (2005) Dead wood in European beech (*Fagus sylvatica*) forest reserves. *Forest Ecology and Management* 210, 267–282.

Clements, F.E. (1928) *Plant Succession and Plant Indicators*. Carnegie Institute of Washington, Washington, DC.

Cmd 7122 (1947) *Conservation of Nature in England and Wales*. Report of the Wild Life Conservation Special Committee. HMSO, London.

Cmd 7814 (1949) *Nature Reserves in Scotland.* Final Report. HMSO, London.

Conventz, H. (1909) *The Care of Natural Communities*. Cambridge University Press, Cambridge, UK.

Cooke, A.O. (1913) *The Forest of Dean*. Constable, London.

Cooke, A.S. (2006) *Monitoring Muntjac Deer* Muntjacus reevesii *and Their Impacts in Monks Wood National Nature Reserve*. English Nature Research Report 681, Peterborough, UK.

Coomes, D.A. and Allen, R.B. (2007) Mortality and tree-size distributions in natural mixed-age forests. *Journal of Ecology* 95, 27–40.

Coppins, A.M. and Coppins, B.J. (2002) *Indices of Ecological Continuity for Woodland Epiphytic Lichen Habitats in the British Isles*. British Lichen Society, Wimbledon, UK.

Coppins, S. and Coppins, B.J. (2012) *Atlantic Hazel: Scotland's Special Woodlands*. Atlantic Hazel Action group. ISBN-13: 978-0957203402.

Courtney, P. (1983) The rural landscape of eastern and lower Gwent c.AD 1070–1750. PhD thesis. University of Cardiff, Wales, UK.

Cramer, H.H. (1984) On the predisposition to disorders of Middle European Forests. *Bayer GMBH Leverkusen Pflanzenschutz-Nachrichten* 37, 97–207.

De Frenne, P. and 35 others (2013) Microclimate moderates plant responses to macroclimate warming. *Proceedings of the National Academy of Sciences of the United States of America* 110, 18561–18565.

Dodd, M., Silvertown, J., McConway, K., Potts, J. and Crawley, M. (1995) Community stability: a 60-year record of trends and outbreaks in the occurrence of species in the Park Grass experiment. *Journal of Ecology* 83, 277–285.

Duncker, P.S., Barreiro, S.M., Hengeveld, G.M., Lind, T., Mason, W.L., *et al.* (2012) Classification of forest management approaches: a new conceptual framework and its applicability to European forestry. *Ecology and Society* 17(4), 51. Available at: http://dx.doi.org/10.5751/ES-05262-170451 (accessed 6 February 2017).

Eisen, K. and Barker Plotkin, A. (2015) Forty years of forest measurements support steadily increasing aboveground biomass in a Quercus-dominant northeastern forest. *Journal of the Torrey Botanical Society* 142, 97–112.

Ellenberg, H. (1988) *Vegetation Ecology of Central Europe*, 4th edn. Cambridge University Press, Cambridge, UK.

Elwes, H.J. and Henry, A. (1906–13) *The Trees of Great Britain and Ireland.* Privately printed, Edinburgh. Republished 1969, S.R. Publishers, Wakefield, UK.

Evans, T.G. (2007) *Flora of Monmouthshire.* Chepstow Society, Chepstow, UK.

Fairlie, S. (2013) Rewilding and food security. *The Land* 14, 23–25.

Fenton, J. (2015) Is rewilding destroying the remaining naturalness of the Scottish Highlands? Paper presented to the Wild Thing? Conference, Sheffield, September 2015. Available at: www.james-hc-fenton.eu.

Fraxigen (2005) *Ash Species in Europe: Biological Characteristics and Practical Guidelines for Sustainable Use.* Oxford Forestry Institute, University of Oxford, Oxford, UK.

Ghazoul, J. (2015) *Forestry, A Very Short Introduction.* Oxford University Press, Oxford, UK.

Gilpin, W. (1782) *Observations on the River Wye and Several Parts of South Wales, etc. Relative to Picturesque Beauty: Made in the Summer of the Year 1770.* London (2nd edition, R. Blamire, London, 1789).

Godwin, H. (1975) *The History of the British Flora.* Cambridge University Press, Cambridge, UK.

Goode, D. (2014) *Nature in Towns and Cities.* New Naturalist No 127. Collins, London.

Göppert, J.H.R. (1868) *Skizzen zur Kenntnis der Urwälder Schlesiens and Bohmens.* Nova acta der Leopold-Akademie der Naturforscher, Germany.

Green, P. and Peterken, G.F. (1997) Variation in the amount of dead wood in the woodlands of the Lower Wye Valley, UK, in relation to the intensity of management. *Forest Ecology and Management* 98(3), 229–238.

Greig, J. (1982) Past and present lime woods of Europe. In: Bell, M. and Limbrey, S. (eds) *Archaeological Aspects of Woodland Ecology.* British Archaeological Reports International Series 146, Oxford, UK, pp. 23–55.

Hale, K. (2015) Long-term carbon storage in a semi-natural British woodland. PhD thesis. University of Liverpool, UK.

Harley, J.B. (1979) *Ordnance Survey and land-use mapping: parish books of reference and the county series: 2500 maps, 1855–1918.* Geo Abstracts, Norwich, UK.

Hart, C.E. (1966) *Royal forest. A History of Dean's Woods as Producers of Timber.* Clarendon Press, Oxford, UK.

Hédl, R., Kopecký, M. and Komárek, J. (2010) Half a century of succession in a temperate oakwood: from species-rich community to mesic forest. *Diversity and Distributions* 16, 267–276.

Henry, J.D. and Swan, J.M.A. (1974) Reconstructing forest history from live and dead plant material – an approach to the study of forest succession in southwest New Hampshire. *Ecology* 55, 772–783.

Herms, D.A. and McCullough, D.G. (2014) Emerald ash borer invasion of North America: history, biology, ecology, impacts, and management. *Annual Review of Entomology* 59, 13–30.

Hill, H.C. (1897) *Report on the Forest of Dean, with Suggestions for its Management.* HMSO, London.

Hill, M.O. and Radford, G.L. (1986) *Register of permanent vegetation plots.* Institute of Terrestrial Ecology, Monks Wood Experimental Station, Huntingdon, UK.

Hill, M.O., Mountford, J.O., Roy, D.B. and Bunce R.G.H. (1999) *Ellenberg's Indicator Values for British Plants.* Centre for Ecology and Hydrology, Wallingford, UK.

Holmes, R.T. and Likens, G.E. (2016) *Hubbard Brook. The Story of a Forest Ecosystem.* Yale University Press, New Haven and London.

Iversen, J. (1960) Problems of the early post-glacial forest development in Denmark. *Danmarks Geologiske Undersøgelse.* IV Series 4, No 3, 1–32.

Jerrentrup, J.S., Wragg-Mönning, N., Röver, K-U. and Isselstein, J. (2014) Grazing intensity affects insect diversity via sward structure and heterogeneity in a long-term experiment. *Journal of Applied Ecology* 51, 968–977.

Jones, E.W. (1945) The structure and reproduction of the virgin forest of the north temperate zone. *New Phytologist* 44, 130–148.

Jones, E.W. (1959) Biological flora of the British Isles. *Quercus* L. *Journal of Ecology* 47, 169–222.

Keto-Tokoi, P. and Kuuluvainen, T. (2014) *Primaeval Forests of Finland.* Maahenki, Helsinki.

Kirby, K.J. (2004) A model of a natural wooded landscape in Britain as influenced by large herbivore activity. *Forestry* 77, 405–420.

Kirby, K.J., Reid, C.M., Thomas, R.C. and Goldsmith, F.B. (1998) Preliminary estimates of fallen dead wood and standing dead trees in managed and unmanaged forests in Britain. *Journal of Applied Ecology* 35, 148–155.

Kral, F. and Mayer, H. (1968) Pollenanalytische überprüfung des Urwaldcharacters in den Naturwaldreservation Rothwald and Neuwald (Niederösterreichische Kalkalpen). *Forstwissenschaftliches Centralblatt* 3, 129–192.

Larsen, J.B. (2012) Close-to-nature forest management: the Danish approach to sustainable forestry. In: Garcia, J.M. and Diez Casero, J.J. (eds) *Sustainable Forest Management – Current Research*. InTech, Rijeka, Croatia. ISBN: 978-953-51-0621-0. Available at: https://www.intechopen.com/books/sustainable-forest-management-current-research/sustainable-forestry-through-close-to-nature-management (accessed 26 May 2017).

Latałowa, M., Zimmy, M., Jędrzejewska, B. and Samojlik, T. (2015) Białowieża primaeval forest: a 2000-year interplay of environmental and cultural forces in Europe's best preserved temperate woodland. In: Kirby, K.J. and Watkins, C. (eds) *Europe's Changing Woods and Forests, From Wildwood to Managed Landscapes*. CAB International, Wallingford, UK, pp. 243–264.

Ledzian, M.D. (1991) *Forest Families*. British Association for Cemeteries in South Asia, Putney, UK.

Leopold, A. (1949) *A Sand County Almanac*. Oxford University Press, Oxford, UK.

Macfarlane, R. (2007) *The Wild Places*. Granta Publications, London.

Macfarlane, R. (2015) *Landmarks*. Hamish Hamilton, London.

Machen, E. (1849) First Report of the Select Committee on the Woods, Forests and Land Revenues of the Crown. *House of Commons Journal* 1849, pp. 144–145.

Marren, P. (2016) The great rewilding experiment at Knepp Castle. *British Wildlife* 27, 333–339.

Mauve, K. (1931) Ueber Bestandesaufbau, Zuwachsverhältnisse und Verjüngung im galizischen Karpathen-Urwald. *Mitteilungen aus Forstwirtschaft und Forstwissenschaft* 2, 257–311.

Mitchell, F.G.J. (2005) How open were European primeval forests? Hypothesis testing using paleoecological data. *Journal of Ecology* 93, 168–177.

Monbiot, G. (2013) *Feral. Rewilding the Land, Seas and Human Life*. Allen Lane, London (Penguin, 2014).

Mountford, E.P. (1997) A decade of grey squirrel bark-stripping damage to beech in Lady Park Wood, UK. *Forestry* 70, 17–29.

Mountford, E.P. (2002) Fallen dead wood levels in the near-natural beech forest at La Tilliae reserve, Fontainbleau, France. *Forestry* 75, 203–208.

Mountford, E.P. (2004) Long-term patterns of mortality and regeneration in near-natural woodland. PhD thesis. Harper Adams University College, Open University, UK.

Mountford, E.P. (2006) Long-term patterns and impacts of grey squirrel debarking in Lady Park Wood young-growth stands (UK). *Forest Ecology and Management* 232, 100–113.

Mountford, E.P. and Peterken, G.F. (1999) Effects of stand structure, composition and treatment on bark-stripping of beech by grey squirrels. *Forestry* 72(4), 379–386.

Mountford, E.P., Peterken, G.F., Edwards, P.J. and Manners, J.G. (1999) Long-term change in growth, mortality and regeneration of trees in Denny Wood, and old-growth wood-pasture in the New Forest (UK). *Perspectives in Plant Ecology, Evolution and Systematics* 2, 223–272.

Mountford, E.P., Backmeroff, C.E. and Peterken, G.F. (2001) Long-term patterns of growth, mortality, regeneration and natural disturbance in Wistman's Wood, a high altitude oakwood on Dartmoor. *Report and Transactions of the Devon Association for the Advancement of Science* 133, 227–262.

Newby, E. (1998) *The Diary of Joseph Farrington*. Yale University Press, New Haven and London.

O'Hara, K.L. (2016) What is close-to-nature silviculture in a changing world? *Forestry* 89, 1–6.

Oliver, C.D. and Larsen, B.C. (1996) *Forest Stand Dynamics*. Wiley, New York.

Parker, G.R., Leopold, D.J. and Eichenberger, J.K. (1985) Tree dynamics in an old-growth, deciduous forest. *Forest Ecology and Management* 11, 31–57.

Peterken, G.F. (1981) *Woodland Conservation and Management*. Chapman and Hall, London.

Peterken, G.F. (1996) *Natural Woodland*. Cambridge University Press, Cambridge, UK.

Peterken, G. (2008a) *Wye Valley*. New Naturalist No 105. Collins, London.

Peterken, G. (2013) *Meadows*. British Wildlife Publishing, Gillingham, UK.

Peterken, G.F. (2015) Woodland history in the British Isles – an interaction of environmental and cultural forces. In: Kirby, K.J. and

Watkins, C. (eds) *Europe's Changing Woods and Forests.* CAB International, Wallingford, UK, pp. 265–278.

Peterken, G.F. and Backmeroff, C. (1988) *Long-term Monitoring in Unmanaged Woodland Nature Reserves.* Research and Survey No 9. Nature Conservancy Council, Peterborough, UK.

Peterken, G.F. and Francis, J. (1999) Open spaces as habitats for vascular ground flora species in the woods of central Lincolnshire, U.K. *Biological Conservation* 91, 55–72.

Peterken, G.F. and Game, M. (1984) Historical factors affecting the number and distribution of vascular plant species in the woodlands of central Lincolnshire. *Journal of Ecology* 72, 155–182.

Peterken, G.F. and Jones, E.W. (1987) Forty years of change in Lady Park Wood: the old-growth stands. *Journal of Ecology* 75, 477–512.

Peterken, G.F. and Jones, E.W. (1989) Forty years of change in Lady Park Wood: the young growth stands. *Journal of Ecology* 77, 401–429.

Peterken, G.F. and Mountford, E. (1995) Lady Park Wood reserve – the first half century. *British Wildlife* 6(4), 205–213.

Peterken, G.F. and Mountford, E.P. (1996) Effects of drought on beech in Lady Park Wood, an unmanaged mixed deciduous woodland. *Forestry* 69, 117–128.

Peterken, G.F. and Mountford, E.P. (1998) Long-term change in an unmanaged population of wych elm subjected to Dutch elm disease. *Journal of Ecology* 86, 205–218.

Peterken, G.F and Mountford, E.P. (2005) Natural woodland reserves: 60 years of trying at Lady Park Wood. *British Wildlife* 17, 7–16.

Peterken, G.F. and Stace, H. (1987) Structure and development of the Black Wood of Rannoch. *Scottish Forestry* 41(1), 29–44.

Peterken, G.F. and Tubbs, C.R. (1965) Woodland regeneration in the New Forest, Hampshire, since 1650. *Journal of Applied Ecology* 2, 159–170.

Peterken, G.F., Ausherman, D., Buchenau, M. and Forman, R.T.T. (1992) Old-growth conservation within British upland conifer plantations. *Forestry* 65, 127–144.

Peterken, S. (2008b) *Landscapes of the Wye Tour.* Logaston Press, Little Logaston, UK.

Pigott, C.D. (1981) The status, ecology and conservation of *Tilia platyphyllos* in Britain. In: Synge, H. (ed.) *The Biological Aspects of Rare Plant Conservation.* Wiley, Chichester, UK, pp. 305–318.

Pigott, C.D. (1991) *Tilia cordata* Miller. Biological flora of the British Isles No 174. *Journal of Ecology* 79, 1147–1207.

Pigott, C.D. (2012) *Lime-trees and Basswoods. A Biological Monograph of the Genus* Tilia. Cambridge University Press, Cambridge, UK.

Pigott, C.D. and Walters, S.M. (1954) On the interpretation of the discontinuous distributions shown by certain British species of open habitats. *Journal of Ecology* 42, 95–116.

Power, S.A. (1994) Temporal trends in twig growth of *Fagus sylvatica* L. and their relationships with environmental factors. *Forestry* 67, 13–30.

Price, C.R. (2003) *Late Pleistocene and Early Holocene Small Mammals in South-west Britain: Environmental, and Taphonomic Implications and their Role in Archaeological Research.* BAR British Series 347. Archaeopress, Oxford, UK.

Pro Silva (2012) *Pro Silva principles.* Pro Silva Europe, Truttenhausen, France.

Pyne, S.J. (1982) *Fire in America. A Cultural History of Wildland and Rural Fire.* Princeton University Press, Princeton, New Jersey.

Rackham, O. (1975) *Hayley Wood, its History and Ecology.* Cambridge and Isle of Ely Naturalists' Trust Ltd, Cambridge, UK.

Rackham, O. (1980) *Ancient Woodland, its History, Vegetation and Uses in England.* Edward Arnold, London (new edition, 2003, Castlepoint Press, Dalbeattie, UK)

Rackham, O. (1989) *The Last Forest: The Story of Hatfield Forest.* Dent, London.

Rackham, O. (1992) Mixtures, mosaics and clones: the distribution of trees within European woods and forests. In: Cannell, M.G.R., Malcolm, D.C. and Robertson, P.A. (eds) *The Ecology of Mixed-species Stands of Trees.* Special Publication No11 of the British Ecological Society. Blackwell, Oxford, UK, pp. 1–21.

Rackham, O. (2003) *Ancient Woodland, its History, Vegetation and Uses in England.* New edition, Castlepoint Press, Dalbeattie, UK. (First edition, 1980. Edward Arnold, London.)

Rackham, O. (2006) *Woodlands.* New Naturalist 100. Collins, London.

Rackham, O. (2015) *The Ash Tree.* Little Toller Books, Dorset, UK.

Rackham, O. and Coombe, D.E. (1996) Madingley Wood. *Nature in Cambridgeshire* 38, 27–56.
Řehák, J. (1968) Srovnávaci výzkum v Boubinském pralese. *Lesnická práce* 47, 206–212.
Rich, T.C.G., Houston, L., Robertson, A. and Proctor, M.C.F. (2010) *Whitebeams, Rowans and Service Trees of Britain and Ireland*. Botanical Society of the British Isles (in association with the National Museum Wales), London.
Rodwell, J.S. (ed.) (1991) *British Plant Communities, 1. Woodlands and Scrub*. Cambridge University Press, Cambridge, UK.
Rose, F. and James, P.W. (1974) Regional studies in the British lichen flora I. The corticolous and lignicolous species of the New Forest, Hampshire. *Lichenologist* 6, 1–72.
Runkle, J.R. (1982) Patterns of disturbance in some old-growth mesic forests of eastern North America. *Ecology* 63, 1533–1546.
Runkle, J.R. and Yetter, T.C. (1987) Treefalls revisited: gap dynamics in the southern Appalachians. *Ecology* 68, 417–424.
Russell, E.W.B. (1983) Indian-set fires in the forests of the northeastern United States. *Ecology* 64, 78–88.
Sheail, J. (1976) *Nature in Trust. The History of Nature Conservation in Britain*. Blackie, Glasgow and London.
Sheil, D., Burslem, D. and Alder, D. (1995) The interpretation and misinterpretation of mortality rate measures. *Journal of Ecology* 83, 331–333.
Siitonen, J. and Ranius, T. (2015) The importance of veteran trees for saproxylic insects. In: Kirby, K.J. and Watkins, C. (eds) *Europe's Changing Woods and Forests, From Wildwood to Managed Landscapes*. CAB International, Wallingford, UK, pp. 140–153.
Silvertown, J., Poulton, P., Johnston, A.E., Edwards, G., Heard, M. and Biss, P.M. (2006) The Park Grass Experiment 1956–2006: its contribution to ecology. *Journal of Ecology* 94, 801–814.
Solla, A. and Gil, L. (2002) Xylem vessel diameter as a factor in resistance of *Ulmus minor* to *Ophiostoma novo-ulmi*. *Forest Pathology* 32, 123–134.
Sollars, E.S.A., Harper, A.L., Kelly, L.J., Sambles, C.M., Ramirez-Gonzalez, R.H., *et al.* (2017) Genome sequence and genetic diversity of European ash trees. *Nature* 541, 212–217.
Sparks, T.H. and Carey, P.D. (1995) The responses of species to climate over two centuries: an analysis of the Marsham phenological record, 1736–1947. *Journal of Ecology* 83, 321–329.
Sparks, T., Atkinson, S. and Lewthwaite, K. (2016) The speed of spring 2015: results from nature's calendar. *British Wildlife* 27, 265–268.
Spencer, J.W. and Kirby, K.J. (1992) An inventory of ancient woodland for England and Wales. *Biological Conservation* 62, 77–93.
Standing, I. (1981) *Iron Making in the Forest of Dean by the Rev. H.G. Nicholls 1866. Introduction*. Forest Bookshop, Coleford, UK.
Stephens, G.R. and Ward, J.S (1992) *Sixty years of natural change in unmanaged mixed hardwood forests*. Bulletin 902, The Connecticut Agricultural Experimental Station, New Haven.
Straw, N.A., Williams, D.T., Kulinich, O. and Gninenko, Y.I. (2013) Distribution, impact and rate of spread of emerald ash borer *Agrilus planipennis* (Coleoptera: Buprestidae) in the Moscow region of Russia. *Forestry* 86, 515–522.
Svenning, J.C. (2002) A review of natural vegetation openness in north-western Europe. *Biological Conservation* 104, 133–148.
Symonds, H.H. (1936) *Afforestation in the Lake District*. Dent, London.
Tansley, A.G. (1939) *The British Islands and their Vegetation*. Cambridge University Press, Cambridge, UK.
Taylor, L.R. (1989) Objective and experiment in long-term ecological research. In: G.E. Likens (ed.) *Long-term Studies in Ecology*. Springer, New York, pp. 20–70.
Tegel, W., Seim, A., Hakelberg, D., Hoffmann, S., Panev, M., *et al.* (2014) A recent growth increase of European beech (*Fagus sylvatica* L.) at its Mediterranean distribution limit contradicts drought stress. *European Journal of Forest Research* 133(1), 61–71.
Troup, R.S. (1928) *Silvicultural Systems*. Clarendon Press, Oxford, UK.
Tschermak, L. (1910) Einiges über den Urwald vonwaldbaulichenGesichtespunkten. *Centralbl. f. d. ges. Forstwesen*. 36, 310.
Tubbs, C.R. (1986) *The New Forest*. New Naturalist No 73, Collins, London.
Vanhuyse, K., Vangansbeke, P. and Peterken, G.F. (2012) Changes since 1979 in the ground flora of Lady Park Wood, and what they imply. *Natur Cymru* 43, 15–19.
Vera, F.M.W. (2000) *Grazing Ecology and Forest History*. CAB International, Wallingford, UK.

Vera, F.W.M. (2009) Large-scale nature development – the Oostvaardersplassen. *British Wildlife* 20, Special supplement, 28–36.

Verheyen, K., Baeten, L., De Frenne, P., Bernhardt-Römerman, M., Brunet, J., *et al.* (2012) Driving factors behind the eutrophication signal in understorey plant communities of deciduous temperate forests. *Journal of Ecology* 100, 352–365.

Verheyen, K., De Frenne, P., Baeten, L., Waller, D., Hedl, R., *et al.* (2017) Combining biodiversity resurveys across regions to advance global change research. *BioScience* 67, 73–83.

Ward, J.S. (2013) Influence of disturbance on stand development during forest succession. In: Miller, G.W., Schuler, T.M., Gottschalk, K.W., Brooks, J.R., Grushecky, S.T., *et al.* (eds) *Proceedings, 18th Central Hardwood Forest Conference; 2012 March 26–28; Morgantown, WV; Gen. Tech. Rep. NRS-P-117*. US Department of Agriculture, Forest Service, Northern Research Station, Newtown Square, Pennsylvania, pp. 93–104.

Watt, A.S. (1947) Pattern and process in the plant community. *Journal of Ecology* 35, 1–22.

Westoby, M. (1984) The self-thinning rule. *Advances in Ecological Research* 14, 167–225.

Whitehouse, N.J. and Smith, D. (2010) How fragmented was British Holocene woodland? Perspectives on the 'Vera' grazing debate from the fossil beetle record. *Quaternary Science Reviews* 29, 539–553.

Wilkinson, G. (1978) *Epitaph for the Elm*. Hutchinson, London.

Woods, R.G. (2010) *A Lichen Red Data List for Wales*. Plantlife, Salisbury, UK.

Zimmermann, J., Hauck, M., Dulamshoren, C. and Leuchner, C. (2015) Climate-related growth decline affects *Fagus sylvatica*, but not other broad-leaved tree species in central European mixed forests. *Ecosystems* 18, 560–572.

One of *The Arborealists*, a group of professional tree and woodland artists, working in Lady Park, April 2017. In 2018, they plan to exhibit their artistic responses to the wood beside an ecological interpretation based on this long-term study.

Vascular Plant Species

Trees and shrubs – please see index for page numbers

Herbs, graminoids and ferns

Nomenclature according to: Preston, C.D., Pearman, D.A. and Dines, T.D., editors, (2002). *New atlas of the British and Irish flora*. Oxford University Press, Oxford.

Index

Page numbers in **bold** type refer to figures, tables and boxed text.
'LPW' throughout the index stands for 'Lady Park Wood'.